Introduction to Complex Variables and Application

MARK J. ABLOWITZ

University of Colorado Boulder

ATHANASSIOS S. FOKAS

University of Cambridge and University of Southern California

CAMBRIDGE
UNIVERSITY PRESS

CAMBRIDGE
UNIVERSITY PRESS

University Printing House, Cambridge CB2 8BS, United Kingdom

One Liberty Plaza, 20th Floor, New York, NY 10006, USA

477 Williamstown Road, Port Melbourne, VIC 3207, Australia

314–321, 3rd Floor, Plot 3, Splendor Forum, Jasola District Centre, New Delhi – 110025, India

79 Anson Road, #06–04/06, Singapore 079906

Cambridge University Press is part of the University of Cambridge.

It furthers the University's mission by disseminating knowledge in the pursuit of education, learning, and research at the highest international levels of excellence.

www.cambridge.org
Information on this title: www.cambridge.org/9781108832618
DOI: 10.1017/9781108961806

© Cambridge University Press 2021

First published 2021

Printed in the United Kingdom by TJ Books Limited, Padstow Cornwall

A catalogue record for this publication is available from the British Library.

ISBN 978-1-108-83261-8 Hardback
ISBN 978-1-108-95972-8 Paperback

Contents

Preface

The study of complex variables is both beautiful from a purely mathematical point of view, and provides a powerful tool for solving a wide array of problems arising in applications. It is perhaps surprising that to explain real phenomena, mathematicians, scientists, and engineers often resort to the "complex plane." In fact using complex variables one can solve many problems that are either very difficult or virtually impossible to solve by other means. The text provides a broad treatment of both the fundamentals and the applications of this subject.

This text can be used in an introductory undergraduate course. Alternatively, it can be used in a beginning graduate-level course and as a reference. Indeed, this book provides an introduction to the study of complex variables. It also contains a number of applications which include evaluation of integrals, methods of solution to certain ordinary and partial differential equations, and ideal fluid flow. It also provides a broad discussion of conformal mappings and many of their applications. In fact, applications are discussed throughout the book. Our point of view is that students are motivated and enjoy learning the material when they can relate it to applications.

To aid the instructor we have denoted with an asterisk certain sections which are more advanced. These sections can be read independently or can be skipped. However, in teaching the course we have found that the more advanced sections can be effectively used as a source of valuable material for student projects. Every effort has been made to make this book self-contained. Thus, advanced students using this text will have the basic material at their disposal without dependence on other references.

We realize that many of the topics presented in this book are not usually covered in complex variables texts. This includes the generalized Cauchy integral formula, ODEs in the complex plane, the solution of linear PDEs by integral transforms, conformal mappings of polygons with circular sides, etc. Actually some of these topics, when studied at all, are only included in advanced graduate-level courses.

However, we believe that these topics arise so frequently in applications that early exposure is useful. It is fortunate that it is indeed possible to present this material in such a way that it can be understood with only the foundation presented in the introductory chapters of this book.

We are indebted to our families who have endured all too many hours of our absence. We are thankful to B. Fast and C. Smith for an outstanding job of word processing the manuscript and to B. Fast who has so capably used mathematical software to verify many formulae and produce figures.

Several colleagues helped us with the preparation of this book. B. Herbst made many suggestions and was instrumental in the development of the computational section. C. Schober, L. Luo, and L. Glasser worked with us on many of the exercises.

We are deeply appreciative that (the late) David Benney encouraged us to write this book. We would like to take this opportunity to thank those agencies who have over the years consistently supported our research efforts. Actually this research led us to several of the applications presented in this book. We thank the Air Force Office of Scientific Research, the National Science Foundation, the Engineering and Physical Research Council of the UK, and in particular Arje Nachman, Program Director (Air Force Office of Scientific Research), for his continual support.

1

Complex Numbers and Elementary Functions

This chapter introduces complex numbers, elementary complex functions, and their basic properties. It will be seen that complex numbers have a simple two-dimensional character which submits to a straightforward geometric description. While many results of real variable calculus carry over, some very important novel and useful notions appear in the calculus of complex functions. Applications to differential equations are briefly discussed as well.

1.1 Complex Numbers and Their Properties

In this text we shall use Euler's notation for the imaginary unit number:

$$i^2 = -1. \tag{1.1.1}$$

A complex number is an expression of the form

$$z = x + iy. \tag{1.1.2}$$

Here x is the real part of z, namely Re (z); and y is the imaginary part of z, namely Im (z). So, for example, the complex number $z_0 = 2 + 3i$ has real part Re $(z_0) = 2$ and imaginary part Im $(z_0) = 3$. If $y = 0$, we say that z is real, and if $x = 0$, we say that z is purely imaginary. We often denote z an element of the complex numbers as $z \in \mathbb{C}$, where x, an element of the real numbers is denoted by $x \in \mathbb{R}$. Geometrically, we represent Eq. (1.1.2) in a two-dimensional coordinate system called the **complex plane** (see Figure 1.1).

The real numbers lie on the horizontal axis and pure imaginary numbers on the vertical axis. The analogy with two-dimensional vectors is immediate. A complex number $z = x + iy$ can be interpreted as a two-dimensional vector (x, y).

It is useful to introduce another representation of complex numbers, namely polar coordinates (r, θ):

$$x = r \cos \theta \qquad y = r \sin \theta, \qquad r \geq 0. \tag{1.1.3}$$

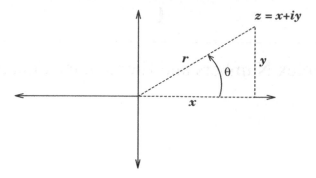

Figure 1.1 The complex plane ("z-plane")

Hence, the complex number z can be written in the alternative polar form:

$$z = x + iy = r(\cos \theta + i \sin \theta). \tag{1.1.4}$$

The radius r is denoted by

$$r = \sqrt{x^2 + y^2} \equiv |z| \tag{1.1.5a}$$

(note: $\equiv$ denotes equivalence) and naturally gives us the notion of the **absolute value** of z, denoted by $|z|$, that is, it is the length of the vector associated with z. The value $|z|$ is often referred to as the **modulus** of z. The angle θ is called the **argument** of z, and is denoted by arg z. We take the standard convention that counterclockwise is the positive direction. When $z \neq 0$, the values of θ can be found from Eq. (1.1.3) via standard trigonometry:

$$\tan \theta = y/x, \tag{1.1.5b}$$

where the quadrant in which x, y lie is understood as given. We note that $\theta \equiv$ arg z is **multivalued** because $\tan \theta$ is a periodic function of θ with period π. Given $z = x+iy$, $z \neq 0$ we identify θ to have one value in the interval $\theta_0 \leq \theta < \theta_0+2\pi$, where θ_0 is an arbitrary number; others differ by integer multiples of 2π. We shall take $\theta_0 = 0$. For example, if $z = -1+i$, then $|z| = r = \sqrt{2}$ and $\theta = \frac{3\pi}{4} + 2n\pi$, $n = 0, \pm1, \pm2, \ldots$. The previous remarks apply equally well if we use the polar representation about a point $z_0 \neq 0$. This just means that we translate the origin from $z = 0$ to $z = z_0$. In general θ lies in the region: $\theta_0 \leq \theta < \theta_0 + 2n\pi$ where n is an integer. The standard case is to take $\theta_0 = 0, n = 0$ which is sometimes called the principal branch; here $0 \leq \theta < 2\pi$. So if we are given the complex number $z_1 = -\frac{1}{2} - i\frac{\sqrt{3}}{2}$, then $\tan \theta_1 = \sqrt{3}$. Since z_1 lies in the third quadrant we have that $\theta_1 = \pi + \pi/3 = 4\pi/3$. On the other hand if we take the branch with $\theta_0 = -\pi, n = 0$ then θ lies in the region: $-\pi \leq \theta < \pi$. In the latter case corresponding to z_1 we now have $\theta_1 = -\pi + \pi/3 = -2\pi/3$.

At this point it is convenient to introduce a special exponential function. The polar exponential is defined by

$$\cos\theta + i\sin\theta = e^{i\theta}. \qquad (1.1.6)$$

Hence, Eq. (1.1.4) implies that z can be written in the form

$$z = re^{i\theta}. \qquad (1.1.4')$$

This exponential function has all of the standard properties we are familiar with in elementary calculus and is a special case of the complex exponential function to be introduced later in this chapter. For example, using well-known trigonometric identities, Eq. (1.1.6) implies

$$e^{2\pi i} = 1, \qquad e^{\pi i} = -1, \qquad e^{\pi i/2} = i, \qquad e^{3\pi i/2} = -i,$$
$$e^{i\theta_1}e^{i\theta_2} = e^{i(\theta_1+\theta_2)}, \qquad (e^{i\theta})^m = e^{im\theta}, \qquad (e^{i\theta})^{1/n} = e^{i\theta/n}.$$

Example 1.1.1 Let $z = \frac{1}{\sqrt{2}} + \frac{i}{\sqrt{2}}$; put this into polar form: $z = re^{i\theta}, 0 \le \theta < 2\pi$. Using $z = x+iy$ with $x = \frac{1}{\sqrt{2}}, y = \frac{1}{\sqrt{2}}$, we find $r^2 = x^2+y^2 = \frac{1}{2}+\frac{1}{2} = 1$ and $\tan\theta = \frac{y}{x} = 1$; hence $\theta = \frac{\pi}{4}$ since the point (x, y) is in the first quadrant. Thus $z = e^{i\pi/4}$. On the other hand if $z = \sqrt{3}e^{-i\pi/6}$ we have that:

$$z = \sqrt{3}(\cos\pi/6 - i\sin\pi/6) = \sqrt{3}\left(\frac{\sqrt{3}}{2} - \frac{i}{2}\right) = \frac{3}{2} - \frac{i\sqrt{3}}{2}.$$

Example 1.1.2 Show that the equation $|z - z_0| = R$ with $z_0 = x_0 + iy_0$ represents a circle with center (x_0, y_0), radius R.

Indeed, $z - z_0 = x - x_0 + i(y - y_0)$; therefore $|z - z_0|^2 = (x - x_0)^2 + (y - y_0)^2 = R^2$ which is a circle with the designated center and radius.

With the properties discussed so far, one can solve an equation of the form

$$z^n = a = |a|e^{i\phi} = |a|(\cos\phi + i\sin\phi), \qquad n = 1, 2, \ldots .$$

Using the periodicity of $\cos\phi$ and $\sin\phi$, we have

$$z^n = a = |a|e^{i(\phi+2\pi m)} \qquad m = 0, 1, \ldots, n - 1$$

and find the n roots

$$z = |a|^{1/n}e^{i(\phi+2\pi m)/n} \qquad m = 0, 1, \ldots, n - 1.$$

For $m \ge n$ the roots repeat.

If $a = 1$, these are called the n roots of unity: $1, \omega, \omega^2, \ldots, \omega^{n-1}$, where $\omega = e^{2\pi i/n}$. So if $n = 2, a = -1$, we see that the solutions of $z^2 = -1 = e^{i\pi}$ are $z = \{e^{i\pi/2}, e^{3i\pi/2}\}$, or $z = \pm i$. In the context of real numbers there are no solutions to $z^2 = -1$, but in the context of complex numbers this equation has two solutions. Later in this book we shall show that an nth-order polynomial equation, $z^n + a_{n-1}z^{n-1} + \cdots + a_0 = 0$, where the coefficients $\{a_j\}_{j=0}^{n-1}$ are complex numbers, has n and only n solutions (roots) in the complex domain, counting multiplicities (for example, we say that $(z - 1)^2 = 0$ has two solutions, and that $z = 1$ is a solution of multiplicity two).

The **complex conjugate** of z is defined as

$$\bar{z} = x - iy = re^{-i\theta}. \tag{1.1.7}$$

Two complex numbers are said to be equal if and only if their real and imaginary parts are respectively equal; namely, writing $z_k = x_k + iy_k$, for $k = 1, 2$, then

$$z_1 = z_2 \implies x_1 + iy_1 = x_2 + iy_2 \implies x_1 = x_2, y_1 = y_2.$$

Thus $z = 0$ implies $x = y = 0$.

Addition, subtraction, multiplication, and division of complex numbers follow from the rules governing real numbers. Thus, noting $i^2 = -1$, we have

$$z_1 \pm z_2 = (x_1 \pm x_2) + i(y_1 \pm y_2), \tag{1.1.8a}$$

and

$$z_1 z_2 = (x_1 + iy_1)(x_2 + iy_2) = (x_1 x_2 - y_1 y_2) + i(x_1 y_2 + x_2 y_1). \tag{1.1.8b}$$

In fact we note that from Eq. (1.1.5a),

$$z\bar{z} = \bar{z}z = (x + iy)(x - iy) = x^2 + y^2 = |z|^2. \tag{1.1.8c}$$

This fact is useful for division of complex numbers:

$$\begin{aligned}
\frac{z_1}{z_2} &= \frac{x_1 + iy_1}{x_2 + iy_2} = \frac{(x_1 + iy_1)(x_2 - iy_2)}{(x_2 + iy_2)(x_2 - iy_2)} \\
&= \frac{(x_1 x_2 + y_1 y_2) + i(x_2 y_1 - x_1 y_2)}{x_2^2 + y_2^2} \\
&= \frac{x_1 x_2 + y_1 y_2}{x_2^2 + y_2^2} + \frac{i(x_2 y_1 - x_1 y_2)}{x_2^2 + y_2^2}.
\end{aligned} \tag{1.1.8d}$$

It is easily shown that the commutative, associative and distributive laws of addition and multiplication hold.

Geometrically speaking, addition of two complex numbers is equivalent to that of the parallelogram law of vectors (see Figure 1.2).

The useful analytical statement

$$||z_1| - |z_2|| \le |z_1 + z_2| \le |z_1| + |z_2| \tag{1.1.9}$$

has the geometrical meaning that no side of a triangle is greater in length than the sum of the other two sides – hence the term for inequality Eq. (1.1.9) is the **triangle inequality**.

Equation (1.1.9) can be proven as follows:

$$|z_1 + z_2|^2 = (z_1 + z_2)(\overline{z_1} + \overline{z_2}) = z_1\overline{z_1} + z_2\overline{z_2} + z_1\overline{z_2} + \overline{z_1}z_2$$
$$= |z_1|^2 + |z_2|^2 + 2\text{Re}(z_1\overline{z_2}).$$

Hence

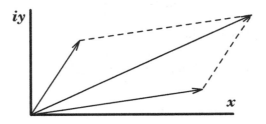

Figure 1.2 Addition of vectors

$$|z_1 + z_2|^2 - (|z_1| + |z_2|)^2 = 2\,(\text{Re}(z_1\overline{z_2}) - |z_1|\,|z_2|) \le 0, \qquad (1.1.10)$$

where the inequality follows from the fact that

$$x = \text{Re}\,z \le |z| = \sqrt{x^2 + y^2}$$

and $|z_1\overline{z_2}| = |z_1||z_2|$.

Equation (1.1.10) implies the right-hand inequality of Eq. (1.1.9) after taking a square root. The left-hand inequality follows by redefining terms. Let

$$W_1 = z_1 + z_2, \qquad W_2 = -z_2.$$

Then the right-hand side of Eq. (1.1.9) (just proven) implies that

$$|W_1| \le |W_1 + W_2| + |-W_2| \quad \text{or}$$
$$|W_1| - |W_2| \le |W_1 + W_2|,$$

which then proves the left-hand side of Eq. (1.1.9) if we assume that $|W_1| \ge |W_2|$; otherwise, we can interchange W_1 and W_2 in the above discussion and obtain

$$||W_1| - |W_2|| = -(|W_1| - |W_2|) \le |W_1 + W_2|.$$

Similarly, note the immediate generalization of Eq. (1.1.9)

$$|z_1 + z_2 + \cdots + z_n| = |\sum_{j=1}^{n} z_j| \le \sum_{j=1}^{n} |z_j| = |z_1| + |z_2| + \cdots + |z_n|.$$

Example 1.1.3 Show that $|\text{Im} z| \le |z|$.

We have that

$$\text{Im} z = y \le |x + iy| = \sqrt{x^2 + y^2} = |z|.$$

Alternatively, this inequality can also be found from

$$\text{Im} z = y = \frac{z - \bar{z}}{2i}.$$

Then, from the triangle inequality

$$|\text{Im} z| \le \frac{|z| + |\bar{z}|}{2} \le |z|,$$

since $|\bar{z}| = |z|$.

1.1.1 Problems for Section 1.1

1. Express each of the following complex numbers in polar exponential form:

$$\text{(a) } 1 \qquad \text{(b) } -i \qquad \text{(c) } 1 + i$$

$$\text{(d) } \frac{1}{2} + \frac{\sqrt{3}}{2}i \qquad \text{(e) } \frac{1}{2} - \frac{\sqrt{3}}{2}i$$

2. Express each of the following in the form $a + bi$, where a and b are real:

$$\text{(a) } e^{2+i\pi/2} \qquad \text{(b) } \frac{1}{1+i} \qquad \text{(c) } (1+i)^3 \qquad \text{(d) } |3 + 4i|$$

   (e) Define $\cos(z) = (e^{iz} + e^{-iz})/2$, and $e^z = e^x e^{iy}$. Evaluate $\cos(i\pi/4 + c)$, where c is real.

3. Solve for the roots of the following equations:

$$\text{(a) } z^3 = 4 \qquad \text{(b) } z^4 = -1$$

$$\text{(c) } (az + b)^3 = c, \text{ where } a, b, c > 0 \qquad \text{(d) } z^4 + 2z^2 + 2 = 0$$

4. Establish the following results:

$$\text{(a) } \overline{z + w} = \bar{z} + \bar{w} \qquad \text{(b) } |z - w| \le |z| + |w| \qquad \text{(c) } z - \bar{z} = 2i \, \text{Im} \, z$$

$$\text{(d) } \text{Re} \, z \le |z| \qquad \text{(e) } |w\bar{z} + \bar{w}z| \le 2|wz| \qquad \text{(f) } |z_1 z_2| = |z_1||z_2|$$

5. There is a partial correspondence between complex numbers and vectors in the plane. Consider a complex number $z = a + bi$ and a vector $\mathbf{v} = a\hat{\mathbf{e}}_1 + b\hat{\mathbf{e}}_2$, where $\hat{\mathbf{e}}_1$ and $\hat{\mathbf{e}}_2$ are unit vectors in the horizontal and vertical directions. Show that the laws of addition $z_1 \pm z_2$ and $\mathbf{v}_1 \pm \mathbf{v}_2$ yield equivalent results as do the magnitudes $|z|^2$, $|\mathbf{v}|^2 = \mathbf{v} \cdot \mathbf{v}$. (Here $\mathbf{v} \cdot \mathbf{v}$ is the usual vector dot product.) Explain why there is no general correspondence for laws of multiplication or division.

6. Discuss the solutions to the equation: $|z - 1| = 2$.
7. Use

$$\frac{1}{z} = \frac{\bar{z}}{|z|^2}$$

to show that Re $(\frac{1}{z})$ and Re z have the same sign for all z.

1.2 Elementary Functions, Stereographic Projections

1.2.1 Elementary Functions

As a prelude to the notion of a function we present some standard definitions and concepts. A circle with center z_0 and radius r is denoted by $|z - z_0| = r$. A **neighborhood** of a point z_0 is the set of points z for which

$$|z - z_0| < \epsilon, \tag{1.2.1}$$

where ϵ is some (small) positive number. Hence a neighborhood of the point z_0 is all the points inside the circle of radius ϵ, not including its boundary. An annulus $r_1 < |z - z_0| < r_2$ has center z_0, with inner radius r_1 and outer radius r_2. A point z_0 of a set of points S is called an **interior point** of S if there is a neighborhood of z_0 entirely contained within S. The set S is said to be an **open set** if all the points of S are interior points. A point z_0 is said to be a **boundary point** of S if every neighborhood of $z = z_0$ contains at least one point in S and at least one point not in S.

A set consisting of all points of an open set and none, some or all of its boundary points is referred to as a **region**. A region is said to be **bounded** if there is a constant $M > 0$ such that all points z of the region satisfy $|z| \leq M$, that is, they lie within this circle. A region is said to be **closed** if it contains all of its boundary points. A region that is both closed and bounded is called **compact**. Thus the region $|z| \leq 1$ is compact because it is both closed and bounded. The region $|z| < 1$ is open and bounded. The half plane Re $z > 0$ (see Figure 1.3) is open and unbounded.

Let $z_1, z_2, \ldots, z_n$ be points in the plane. The $n - 1$ line segments $\overline{z_1 z_2}, \overline{z_2 z_3}, \ldots, \overline{z_{n-1} z_n}$, taken in sequence, form a connected broken line. An open region is said to be **connected** if any two of its points can be joined by a connected broken line that is contained in the region. (There are more detailed definitions of connectedness, but this simple one will suffice for our purposes.) For an example of a connected region see Figure 1.4.

A disconnected region is exemplified by all the points interior to $|z| = 1$ and exterior to $|z| = 2$: $S = \{z : |z| < 1, |z| > 2\}$.

A connected open region is called a **domain**. For example the set (see Figure 1.5)

$$S = \{z = re^{i\theta} : \theta_0 < \arg z < \theta_0 + \alpha\}$$

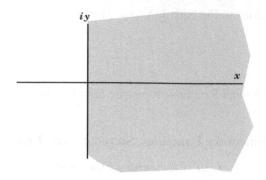

Figure 1.3 Half plane

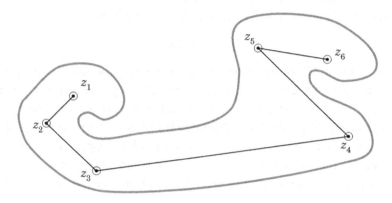

Figure 1.4 Connected region

is a domain which is unbounded.

Because a domain is an open set, we note that no boundary point of the domain can lie in the domain. Notationally, we shall refer to a region as $\mathcal{R}$; the closed region

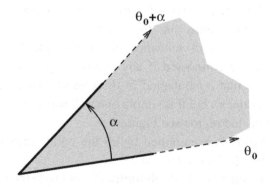

Figure 1.5 Domain – a sector

containing $\mathcal{R}$ and all of its boundary points is sometimes referred to as $\overline{\mathcal{R}}$. If $\mathcal{R}$ is closed, then $\mathcal{R} = \overline{\mathcal{R}}$. The notation $z \in \mathcal{R}$ means z is a point contained in $\mathcal{R}$. Usually we denote a domain by $\mathcal{D}$.

If for each $z \in \mathcal{R}$ there is a unique complex number $w(z)$ then we say $w(z)$ is a **function** of the complex variable z, frequently written as

$$w = f(z) \tag{1.2.2}$$

in order to denote the function f. Often we simply write $w = w(z)$, or just w. The totality of values $f(z)$ corresponding to $z \in \mathcal{R}$ constitutes the **range** of $f(z)$. In this context the set $\mathcal{R}$ is often referred to as the **domain of definition** of the function f. While the domain of definition of a function is frequently a domain, as defined earlier for a set of points, it does not need to be so.

By the above definition of a function we disallow multivaluedness; no more than one value of $f(z)$ may correspond to any point $z \in \mathcal{R}$. In Sections 2.2 and 2.3 we will deal explicitly with the notion of multivaluedness and its ramifications.

The simplest function is the **power** function:

$$f(z) = z^n, \qquad n = 0, 1, 2, \ldots . \tag{1.2.3}$$

Each successive power is obtained by multiplication: $z^{m+1} = z^m z$, for $m = 0, 1, 2, \ldots$. A **polynomial** is defined as a linear combination of powers

$$P_n(z) = \sum_{j=0}^{n} a_j z^j = a_0 + a_1 z + a_2 z^2 + \cdots + a_n z^n, \tag{1.2.4}$$

where the a_j are complex numbers; that is, $a_j \in \mathbb{C}$. Note that the domain of definition of $P_n(z)$ is the entire z-plane simply written as $z \in \mathbb{C}$. A **rational** function is a ratio of two polynomials $P_n(z)$ and $Q_m(z)$, where $Q_m(z) = \sum_{j=0}^{m} b_j z^j$. Thus

$$R(z) = \frac{P_n(z)}{Q_m(z)}, \tag{1.2.5}$$

and the domain of definition of $R(z)$ is the z-plane, *excluding* the points where $Q_m(z) = 0$. For example, the function $w = 1/(1 + z^2)$ is defined in the z-plane excluding $z = \pm i$. This is written as $z \in \mathbb{C} \setminus \{i, -i\}$.

In general, the function $f(z)$ is complex and, when $z = x + iy$, we can write $f(z)$ in the complex form:

$$w = f(z) = u(x, y) + i v(x, y). \tag{1.2.6}$$

The function $f(z)$ is said to have the real part u, with $u = \text{Re } f$, and the imaginary part v, with $v = \text{Im } f$. For example,

$$w = z^2 = (x + iy)^2 = x^2 - y^2 + 2ixy,$$

which implies

$$u(x, y) = x^2 - y^2 \quad \text{and} \quad v = 2xy.$$

Just as with real variables we have the standard operations on functions. Given two functions $f(z)$ and $g(z)$, we can define addition, $f(z) + g(z)$, multiplication $f(z)g(z)$, and composition $f[g(z)]$ of complex functions.

It is convenient to define some of the more common functions of a complex variable – which, as with polynomials and rational functions, will be familiar to the reader.

Motivated by the formula $e^{a+b} = e^a e^b$ for real variables, we define the exponential function

$$e^z = e^{x+iy} = e^x e^{iy}.$$

Noting the polar exponential definition (used already in Section 1.1, Eq. (1.1.6))

$$e^{iy} = \cos y + i \sin y,$$

we see that

$$e^z = e^x(\cos y + i \sin y). \tag{1.2.7}$$

Equation (1.2.7) and standard trigonometric identities yield the properties

$$e^{z_1+z_2} = e^{z_1} e^{z_2} \quad \text{and} \quad (e^z)^n = e^{nz}, \qquad n = 1, 2, \ldots. \tag{1.2.8}$$

We also note that

$$|e^z| = |e^x| \, |\cos y + i \sin y| = e^x \sqrt{\cos^2 y + \sin^2 y} = e^x$$

and

$$\overline{(e^z)} = e^{\bar{z}} = e^{x-iy} = e^x(\cos y - i \sin y).$$

The trigonometric functions $\sin z$ and $\cos z$ are defined as

$$\sin z = \frac{e^{iz} - e^{-iz}}{2i}, \tag{1.2.9}$$

$$\cos z = \frac{e^{iz} + e^{-iz}}{2} \tag{1.2.10}$$

and the usual definitions of the other trigonometric functions are taken:

$$\tan z = \frac{\sin z}{\cos z}, \quad \cot z = \frac{\cos z}{\sin z}, \quad \sec z = \frac{1}{\cos z}, \quad \csc z = \frac{1}{\sin z}. \tag{1.2.11}$$

All of the usual trigonometric properties such as

$$\begin{aligned}
\sin(z_1 + z_2) &= \sin z_1 \cos z_2 + \cos z_1 \sin z_2, \\
\sin^2 z + \cos^2 z &= 1, \ldots
\end{aligned} \tag{1.2.12}$$

follow from the above definitions.

The hyperbolic functions are defined analogously

$$\sinh z = \frac{e^z - e^{-z}}{2}, \tag{1.2.13}$$

$$\cosh z = \frac{e^z + e^{-z}}{2}, \tag{1.2.14}$$

$$\tanh z = \frac{\sinh z}{\cosh z}, \quad \coth z = \frac{\cosh z}{\sinh z}, \quad \operatorname{sech} z = \frac{1}{\cosh z}, \quad \operatorname{csch} z = \frac{1}{\sinh z}.$$

Similarly, we have the usual identities such as

$$\cosh^2 z - \sinh^2 z = 1. \tag{1.2.15}$$

The above identity can be verified by using equations (1.2.13–1.2.14):

$$\left(\frac{e^z + e^{-z}}{2}\right)^2 - \left(\frac{e^z - e^{-z}}{2}\right)^2 = \left(\frac{e^{2z}}{4} + \frac{1}{2} + \frac{e^{-2z}}{4}\right) - \left(\frac{e^{2z}}{4} - \frac{1}{2} + \frac{e^{-2z}}{4}\right)$$

$$= 1.$$

From these definitions we see that, as functions of a complex variable, $\sinh z$ and $\sin z$ (and $\cosh z$ and $\cos z$) are simply related:

$$\begin{aligned}
\sinh iz &= i \sin z, & \sin iz &= i \sinh z, \\
\cosh iz &= \cos z, & \cos iz &= \cosh z.
\end{aligned} \tag{1.2.16}$$

By now it is abundantly clear that the elementary functions defined in this section are natural generalizations of the conventional ones familiar from real variables. Indeed, the analogy is so close that it provides an alternative and systematic way of defining functions, which is entirely consistent with the above and allows the definition of a much wider class of functions. This involves introducing the concept of **power series**. In Chapter 3 we shall look more carefully at series and sequences. However, because power series of real variables are already familiar to the reader, it is useful to introduce the notion here.

A power series of $f(z)$ about the point $z = z_0$ is defined as

$$f(z) = \lim_{n \to \infty} \sum_{j=0}^{n} a_j(z - z_0)^j = \sum_{j=0}^{\infty} a_j(z - z_0)^j, \tag{1.2.17}$$

where a_j, z_0 are constants.

Convergence is of course crucial. For simplicity we shall state (motivated by real variables but without proof at this juncture) that Eq. (1.2.17) converges, via the ratio test, whenever

$$\lim_{n \to \infty} \left| \frac{a_{n+1}}{a_n} \right| |z - z_0| < 1. \qquad (1.2.18)$$

That is, it converges inside the circle $|z - z_0| = R$, where

$$R = \lim_{n \to \infty} \left| \frac{a_n}{a_{n+1}} \right|,$$

when this limit exists (see also Section 3.2). If $R = \infty$, we say the series converges for all finite z; if $R = 0$, we say the series converges only for $z = z_0$. We refer to R as the **radius of convergence**.

The elementary functions discussed above have the following power series representations:

$$e^z = \sum_{j=0}^{\infty} \frac{z^j}{j!}, \qquad \sin z = \sum_{j=0}^{\infty} \frac{(-1)^j z^{2j+1}}{(2j+1)!}, \qquad \cos z = \sum_{j=0}^{\infty} \frac{(-1)^j z^{2j}}{(2j)!},$$

$$\qquad (1.2.19)$$

$$\sinh z = \sum_{j=0}^{\infty} \frac{z^{2j+1}}{(2j+1)!}, \qquad \cosh z = \sum_{j=0}^{\infty} \frac{z^{2j}}{(2j)!},$$

where $j! = j(j-1)(j-2) \cdots 3 \cdot 2 \cdot 1$ for $j \geq 1$, and $0! \equiv 1$. The ratio test shows that these series converge for all finite z. We remark that the series (1.2.19) agree with what we know in real variables. This follows from deep theorems on analytic continuation to be discussed in Chapter 3.

Complex functions arise frequently in applications. Later, in Section 1.4 we discuss how exponential functions are used to solve differential equations. And in Chapter 4 we analyse partial differential equations via transform methods that employ complex variables. In differential equations the notion of stability is central. For example, in the investigation of stability of physical systems we derive equations for small deviations from rest or equilibrium states. The solutions of the perturbed equation often have the form e^{zt}, where t is real (e.g. time) and z is a complex number satisfying an algebraic equation (or a more complicated transcendental system). We say that the system is unstable if there are any solutions with Re $z > 0$, because $|e^{zt}| \to \infty$ as $t \to \infty$. We say the system is **marginally stable** if there are no values of z with Re $z > 0$, but some with Re $z = 0$. (The corresponding exponential solution is bounded for all t.) The system is said to be **stable** and **damped** if all values of z satisfy Re $z < 0$ because $|e^{zt}| \to 0$ as $t \to \infty$.

A function $w = f(z)$ can be regarded as a **mapping** or transformation of the points in the z-plane ($z = x + iy$) to the points of the w-plane ($w = u + iv$).

In real variables in one dimension this notion amounts to understanding the graph $y = f(x)$; that is, the mapping of the points x to $y = f(x)$. In complex variables the situation is more difficult owing to the fact that we really have four dimensions – hence a graphical depiction such as in the real one-dimensional case is not feasible. Rather, one considers the two complex planes, z and w, separately and asks how the region in the z-plane transforms or maps to a corresponding region or **image** in the w-plane. Some examples follow.

Example 1.2.1 The function $w = z^2$ maps the upper half z-plane including the real axis, Im $z \geq 0$, to the entire w-plane (see Figure 1.6). This is particularly clear when we use the polar representation $z = re^{i\theta}$. In the z-plane, θ lies inside $0 \leq \theta < \pi$, whereas in the w-plane, $w = r^2 e^{2i\theta} = Re^{i\phi}$, $R = r^2$, $\phi = 2\theta$ and ϕ lies in $0 \leq \phi < 2\pi$.

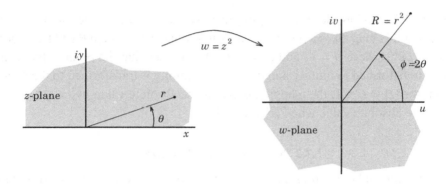

Figure 1.6 Map of $z \to w - z^2$

Example 1.2.2 The function $w = \bar{z}$ maps the upper half z-plane Im $z > 0$ into the lower half w-plane (see Figure 1.7). Namely, $z = x + iy$ and $y > 0$ imply that $w = \bar{z} = x - iy$. Thus $w = u + iv \Rightarrow u = x, v = -y$.

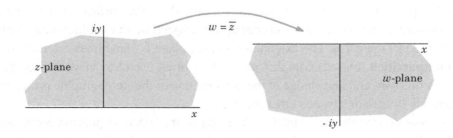

Figure 1.7 Conjugate mapping

The study and understanding of complex mappings is very important, and we will see that there are many applications. In subsequent sections and chapters we shall more carefully investigate the concept of mappings; we shall not go into any more detail or complication at this juncture.

It is often useful to add the **point at infinity** (usually denoted by ∞ or z_∞) to our, so far open, complex plane. As opposed to a finite point where the neighborhood of z_0, say, is defined by Eq. (1.2.1), here the neighborhood of z_∞ is defined by those points satisfying $|z| > 1/\epsilon$ for all (sufficiently small) $\epsilon > 0$. A convenient way of defining the notion of infinity in an unbounded region is to let $z = 1/t$ and then to say that the neighborhood of $t = 0$ corresponds to the neighborhood of infinity. An unbounded region R can be thought of as containing the point z_∞. We say a function has values at infinity if it is defined in a neighborhood of z_∞. We will see below that there is a mapping which allows us to make the complex plane compact. The complex plane with the point z_∞ included is referred to as the **extended complex plane**.

Example 1.2.3   Consider the mapping $w = \frac{1}{z^2}$. Let $z = re^{i\theta}$. Then $w = \frac{1}{r^2}e^{-2i\theta}$. We see that this map takes a point with absolute value r, argument θ and transforms it to $w = \rho e^{i\phi}$ where $\rho = \frac{1}{r^2}$ and $\phi = -2\theta$. We also see that the point at the origin $z = 0$ is transformed to infinity and the point at infinity is transformed to the origin. Further, for $|z| = r < 1$, the mapping expands the absolute value $|w| = \frac{1}{r^2} > 1$ and shrinks the absolute value when $|z| = r > 1$.

1.2.2 Stereographic Projection

Consider a unit sphere sitting on top of the complex plane with the south pole of the sphere located at the origin of the z-plane (see Figure 1.8). In this subsection we show how the extended complex plane can be mapped onto the surface of a sphere whose south pole corresponds to the origin and whose north pole is mapped to the point z_∞. All other points of the complex plane can be mapped in a one-to-one fashion to points on the surface of the sphere by using the following construction. Connect the point z in the plane with the north pole using a straight line. This line intersects the sphere at the point P. In this way each point $z = x + iy$ on the complex plane corresponds uniquely to a point P on the surface of the sphere. This construction is called the **stereographic projection** and is diagrammatically illustrated in Figure 1.8. The extended complex plane is sometimes referred to as the **compactified** (closed) complex plane. It is often useful to view the complex plane in this way, and knowledge of the construction of the stereographic projection is valuable in certain advanced treatments.

So, more concretely, the point $P : (X, Y, Z)$ on the sphere is put into correspondence with the point $z = x + iy$ in the complex plane by finding on the surface of the sphere, (X, Y, Z), the point of intersection of the line from the north pole of the

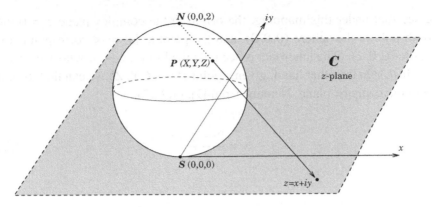

Figure 1.8 Stereographic projection

sphere, $N : (0, 0, 2)$, to the point $z = x + iy$ on the plane. The construction is as follows. We consider three points in the three-dimensional setup:

$$N = (0, 0, 2) : \qquad \text{north pole}$$
$$P = (X, Y, Z) : \qquad \text{point on the sphere}$$
$$C = (x, y, 0) : \qquad \text{point in the complex plane.}$$

These points must lie on a straight line, hence the difference of the points $P - N$ must be a real scalar multiple of the difference $C - N$; that is,

$$(X, Y, Z - 2) = (sx, sy, -2s), \tag{1.2.20}$$

where $s \neq 0$ is a real number. The equation of the sphere is given by

$$X^2 + Y^2 + (Z - 1)^2 = 1. \tag{1.2.21}$$

Equation (1.2.20) implies

$$X = sx, \qquad Y = sy, \qquad Z = 2 - 2s. \tag{1.2.22}$$

Inserting Eq. (1.2.22) into Eq. (1.2.21) yields, after manipulations,

$$s^2(x^2 + y^2 + 4) - 4s = 0. \tag{1.2.23}$$

This equation has as its only non-vanishing solution

$$s = \frac{4}{|z|^2 + 4}, \tag{1.2.24}$$

where $|z|^2 = x^2 + y^2$. Thus, given a point $z = x + iy$ in the plane, we have on the sphere the unique correspondence:

$$X = \frac{4x}{|z|^2 + 4}, \qquad Y = \frac{4y}{|z|^2 + 4}, \qquad Z = \frac{2|z|^2}{|z|^2 + 4}. \tag{1.2.25}$$

We see that under this mapping, the origin in the complex plane $z = 0$ yields the south pole of the sphere $(0, 0, 0)$, and the point at $|z| = \infty$ corresponds to the north pole $(0, 0, 2)$. (The latter fact is seen via the limit $|z| \to \infty$ with $x = |z| \cos \theta$, $y = |z| \sin \theta$.) On the other hand, given a point $P = (X, Y, Z)$ we can find its unique image in the complex plane. Namely, from Eq. (1.2.22),

$$s = \frac{2 - Z}{2} \tag{1.2.26}$$

and

$$x = \frac{2X}{2 - Z}, \qquad y = \frac{2Y}{2 - Z}. \tag{1.2.27}$$

The stereographic projection maps any locus of points in the complex plane onto a corresponding locus of points on the sphere and vice versa. For example, the image of an arbitrary circle in the plane, is a circle on the sphere that does not pass through the north pole. Similarly, a straight line corresponds to a circle passing through the north pole (see Figure 1.9). Here a circle on the sphere corresponds to the locus of points denoting the intersection of the sphere with some plane: $AX + BY + CZ = D$, A, B, C, D constant. Hence on the sphere the images of straight lines and of circles are not really geometrically different from one another. Moreover, the images on the sphere of two nonparallel straight lines in the plane intersect at *two* points on the sphere – one of which is the point at infinity. In this framework, *parallel* lines are circles that touch one another at the point at infinity (north pole). We lose Euclidean geometry on a sphere.

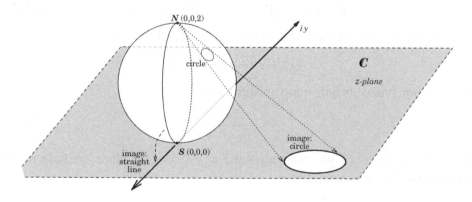

Figure 1.9 Circles and lines in stereographic projection

1.2.3 Problems for Section 1.2

1. Sketch the regions associated with the following inequalities. Determine if the region is open, closed, bounded, or compact.

   (a) $|z| \leq 1$ (b) $|2z + 1 + i| < 4$ (c) $\operatorname{Re} z \geq 4$

   (d) $|z| \leq |z + 1|$ (e) $0 < |2z - 1| \leq 2$.

2. Sketch the following regions. Determine if they are connected, and what the closure of the region is if they are not closed.

   (a) $0 < \arg z \leq \pi$.   (b) $0 \leq \arg z < 2\pi$.

   (c) $\operatorname{Re} z > 0$ and $\operatorname{Im} z > 0$.

   (d) $\operatorname{Re}(z - z_0) > 0$ and $\operatorname{Re}(z - z_1) < 0$ for two complex numbers z_0, z_1.

   (e) $|z| < \dfrac{1}{2}$ and $|2z - 4| \leq 2$.

3. Use Euler's formula for the exponential and the well-known series expansions of the real functions e^x, $\sin y$, and $\cos y$, to show that

$$e^z = \sum_{j=0}^{\infty} \frac{z^j}{j!}.$$

Hint: Use

$$(x + iy)^k = \sum_{j=0}^{k} \frac{k!}{j!(k - j)!} x^j (iy)^{k-j}.$$

4. Use the series representation

$$e^z = \sum_{j=0}^{\infty} \frac{z^j}{j!}, \quad |z| < \infty,$$

to determine series representations for the following functions:

   (a) $\sin z$;   (b) $\cosh z$.

Use these results and properties of the exponential to deduce where the power series for $\sin^2 z$ and $\operatorname{sech} z$ can be expected to converge.

5. Use any method to determine series expansions for the following functions:

   (a) $\dfrac{\sin z}{z}$;   (b) $\dfrac{\cosh z - 1}{z^2}$;   (c) $\dfrac{e^z - 1 - z}{z}$.

6. Let $z_1 = x_1$ and $z_2 = x_2$, with x_1, x_2 real, and the relationship

$$e^{i(x_1 + x_2)} = e^{ix_1} e^{ix_2},$$

to deduce the known trigonometric formulae

$$\sin(x_1 + x_2) = \sin x_1 \cos x_2 + \cos x_1 \sin x_2,$$
$$\cos(x_1 + x_2) = \cos x_1 \cos x_2 - \sin x_1 \sin x_2.$$

Therefore show

$$\sin 2x = 2 \sin x \cos x,$$
$$\cos 2x = \cos^2 x - \sin^2 x.$$

7. Discuss the following transformations (mappings) from the z-plane to the w-plane; here z is the entire finite complex plane.

(a) $w = z^3$.   (b) $w = 1/z$.   (c) $w = 1/z^3$.

8. Consider the transformation

$$w = z + 1/z, \qquad z = x + iy, \qquad w = u + iv.$$

Show that the image of the points in the upper half z-plane ($y > 0$) that are exterior to the circle $|z| = 1$ corresponds to the entire upper half plane $v > 0$.

9. Consider the following transformation

$$w = \frac{az + b}{cz + d}, \qquad \Delta = ad - bc \neq 0.$$

(a) Show that the map can be inverted to find a unique (single-valued) z as a function of w everywhere.

(b) Verify that the mapping can be considered as the result of three successive maps:

$$z' = cz + d, \qquad z'' = 1/z', \qquad w = -\frac{\Delta}{c}z'' + \frac{a}{c},$$

where $c \neq 0$ and is of the form

$$w = \frac{a}{d}z + \frac{b}{d},$$

when $c = 0$.

The following problems relate to the subsection on stereographic projection.

10. To what curves on the sphere do the lines $\text{Re} z = x = 0$ and $\text{Im} z = y = 0$ correspond?

11. Describe the curves on the sphere to which any straight lines on the z-plane correspond.

12. Show that a circle in the z-plane corresponds to a circle on the sphere. (Note the remark following the reference to Figure 1.9 in Section 1.2.2.)

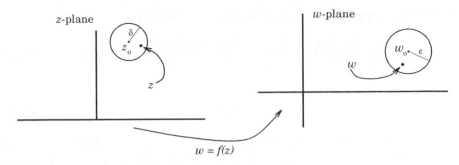

Figure 1.10 Mapping of a neighborhood

1.3 Limits, Continuity, and Complex Differentiation

The concepts of limits and continuity are similar to that of real variables. In this sense our discussion can serve as a brief review of many previously understood notions. Consider a function $w = f(z)$ defined at all points in some neighborhood of $z - z_0$, except possibly for z_0 itself. We say $f(z)$ has the limit w_0 if as z approaches z_0, $f(z)$ approaches w_0 (z_0, w_0 finite). Mathematically we say

$$\lim_{z \to z_0} f(z) = w_0, \tag{1.3.1}$$

if for every (sufficiently small) $\epsilon > 0$ there is a $\delta > 0$ such that

$$|f(z) - w_0| < \epsilon \quad \text{whenever} \quad 0 < |z - z_0| < \delta, \tag{1.3.2}$$

where the absolute value is defined in Section 1.1 (see, for example, Eqs. (1.1.4) and (1.1.5a)).

This definition is clear when z_0 is an interior point of a region $\mathcal{R}$ in which $f(z)$ is defined. If z_0 is a boundary point of $\mathcal{R}$, then we require Eq. (1.3.2) to hold only for those $z \in \mathcal{R}$.

Figure 1.10 illustrates these ideas. Under the mapping $w = f(z)$, all points interior to the circle $|z - z_0| = \delta$ with z_0 deleted are mapped to points interior to the circle $|w - w_0| = \epsilon$. The limit will exist only in the case when z approaches z_0 (that is, $z \to z_0$) in an *arbitrary direction* from within the region $\mathcal{R}$, then this implies that $w \to w_0$.

This limit definition is standard. Indeed this limit procedure is the same as that of two-dimensional real-variable calculus. As an example, let us consider the following.

Example 1.3.1 Show that

$$\lim_{z \to i} 2 \left(\frac{z^2 + iz + 2}{z - i} \right) = 6i. \tag{1.3.3}$$

We must show that given $\epsilon > 0$, there is a $\delta > 0$ such that

$$\left|2\left(\frac{z^2 + iz + 2}{z - i}\right) - 6i\right| = \left|2\left(\frac{(z-i)(z+2i)}{(z-i)}\right) - 6i\right| < \epsilon \qquad (1.3.4)$$

whenever

$$0 < |z - i| < \delta. \qquad (1.3.5)$$

Since $z \neq i$, inequality (1.3.4) implies that $2|z - i| < \epsilon$. Thus if $\delta = \epsilon/2$, Eq. (1.3.5) ensures that Eq. (1.3.4) is satisfied. Therefore Eq. (1.3.3) is demonstrated.

Example 1.3.2 Show that

$$\lim_{z \to i} z^2 = -1.$$

We must find a $\delta > 0$ where $0 < |z - i| < \delta$ which implies $|z^2 + 1| < \epsilon$. Noting that $|z^2 + 1| = |(z + i)(z - i)| = |z - i + 2i||z - i|$, we have, by the triangle inequality,

$$|z^2 + 1| \leq (|z - i| + 2)|z - i| < (\delta + 2)\delta.$$

Thus, if we take $(\delta + 2)\delta \leq \epsilon$ we insure that $|z^2 + 1| < \epsilon$ whenever $0 < |z - i| < \delta$.

Finally we note that for sufficiently small ϵ, δ, we have $\delta^2 < \delta$ and therefore $(\delta + 2)\delta < \delta + 2\delta = \epsilon$. Thus $\delta = \frac{\epsilon}{3}$ is sufficient to establish the required limit.

This limit definition can also be applied to the point $z = \infty$. We say that

$$\lim_{z \to \infty} f(z) = w_0 |z| > \frac{1}{\epsilon^{1/3}} \qquad (1.3.6)$$

for finite w_0, if for every (sufficiently small) $\epsilon > 0$ there is a $\delta > 0$ such that

$$|f(z) - w_0| < \epsilon \quad \text{whenever} \quad |z| > \frac{1}{\delta}. \qquad (1.3.7)$$

Example 1.3.3 Show that $\lim_{z \to \infty}(\frac{1}{z^3} + 1) = 1$.

In this example $f(z) = \frac{1}{z^3} + 1$, $w_0 = 1$. So we must find $\delta > 0$ such that

$$\left|\frac{1}{z^3} + 1 - 1\right| = \left|\frac{1}{z^3}\right| < \epsilon,$$

whenever $|z| > \frac{1}{\delta}$. It follows that from $|1/z^3| < \epsilon$ that $|z| > |z| > 1/\epsilon^{1/3}$. So by taking $\delta = \epsilon^{1/3}$ we have demonstrated the limit.

We assert that the following properties are true. (Their proof is an exercise of the limit definition and follows that of real variables.) If for $z \in \mathcal{R}$ we have two functions $w = f(z)$ and $s = g(z)$ such that

$$\lim_{z \to z_0} f(z) = w_0, \qquad \lim_{z \to z_0} g(z) = s_0,$$

then

$$\lim_{z \to z_0} (f(z) + g(z)) = w_0 + s_0,$$

$$\lim_{z \to z_0} (f(z)g(z)) = w_0 s_0 \text{ and } \qquad \lim_{z \to z_0} \frac{f(z)}{g(z)} = \frac{w_0}{s_0}, \qquad s_0 \neq 0.$$

Similar conclusions hold for sums and products of a finite number of functions. As mentioned in Section 1.2, the point $z = z_\infty = \infty$ is often dealt with via the transformation

$$t = \frac{1}{z}.$$

The neighborhood of $z = z_\infty$ corresponds to the neighborhood of $t = 0$. So the function $f(z) = 1/z^2$ near $z = z_\infty$ behaves like $f(1/t) = t^2$ near zero; that is, $t^2 \to 0$ as $t \to 0$, or $1/z^2 \to 0$ as $z \to \infty$.

In analogy to real analysis, a function $f(z)$ is said to be **continuous** at $z = z_0$ if

$$\lim_{z \to z_0} f(z) = f(z_0), \tag{1.3.8}$$

where z_0, $f(z_0)$ are finite. Equation (1.3.8) implies that $f(z)$ exists in a neighborhood of $z = z_0$ and that the limit, as z approaches z_0, of $f(z)$ is $f(z_0)$ itself. In terms of ϵ, δ notation, given $\epsilon > 0$, there is a $\delta > 0$ such that $|f(z) - f(z_0)| < \epsilon$ whenever $|z - z_0| < \delta$. The notion of continuity at infinity can be ascertained in a similar fashion. Namely, if $\lim_{z \to \infty} f(z) = w_\infty$, and $f(\infty) = w_\infty$, then the definition for continuity at infinity, $\lim_{z \to \infty} f(z) = f(\infty)$, is the following: Given $\epsilon > 0$ there is a $\delta > 0$ such that $|f(z) - w_\infty| < \epsilon$ whenever $|z| > 1/\delta$.

The theorems on limits of sums and products of functions can be used to establish that sums and products of continuous functions are continuous. It can also be shown that compositions of continuous functions are continuous; i.e., (we frequently use the abbreviations: i.e.: that is and e.g.: for example) $\lim_{z \to z_0} f(g(z)) = f(g(z_0))$ where f is continuous in the neighborhood of $g(z_0)$. It should also be pointed out that since $|f(z) - f(z_0)| = |\overline{f}(z) - \overline{f}(z_0)|$, the continuity of $f(z)$ at z_0 implies the continuity of the complex conjugate $\overline{f}(z)$ at $z = z_0$. (Recall the definition of the complex conjugate, Eq. (1.1.7).) Thus if $f(z)$ is continuous at $z = z_0$, then

$$\text{Re } f(z) = \left(f(z) + \overline{f}(z)\right)/2,$$
$$\text{Im } f(z) = \left(f(z) - \overline{f}(z)\right)/2i,$$
$$\text{and } |f(z)|^2 = \left(f(z)\overline{f}(z)\right)$$

are all continuous at $z = z_0$.

We shall say a function $f(z)$ is **continuous in a region** if it is continuous at every point of the region. Usually we simply say that $f(z)$ is continuous when the associated region is understood. Considering continuity in a region $\mathcal{R}$ generally requires that $\delta = \delta(\epsilon, z_0)$; that is, δ depends on both ϵ and the point $z_0 \in \mathcal{R}$. The function $f(z)$ is said to be **uniformly continuous** in a region $\mathcal{R}$ if $\delta = \delta(\epsilon)$; that is, δ is independent of the point $z = z_0$.

As in real analysis, a function which is continuous in a compact (closed and bounded) region $\mathcal{R}$ is uniformly continuous and bounded; that is, there is a $C > 0$ such that $|f(z)| < C$. (The proofs of these statements follow from the analogous statements of real analysis.) Moreover, in a compact region, the modulus $|f(z)|$ actually attains both its maximum and minimum values on $\mathcal{R}$; this follows from the continuity of the real function $|f(z)|$.

Example 1.3.4 Show that the continuity of the real and imaginary parts of a complex function $f(z)$ implies that $f(z)$ is continuous.

$$f(z) = u(x, y) + iv(x, y).$$

We know that

$$
\begin{aligned}
\lim_{z \to z_0} f(z) &= \lim_{\substack{x \to x_0 \\ y \to y_0}} (u(x, y) + iv(x, y)) \\
&= u(x_0, y_0) + iv(x_0, y_0) = f(z_0),
\end{aligned}
$$

which completes the proof. It also illustrates that we can appeal to real analysis for many of the results in this section.

Conversely, we have

$$|u(x, y) - u(x_0, y_0)| \le |f(z) - f(z_0)|,$$
$$|v(x, y) - v(x_0, y_0)| \le |f(z) - f(z_0)|,$$

because $|f|^2 = |u|^2 + |v|^2$, in which case continuity of $f(z)$ implies continuity of the real and imaginary parts of $f(z)$. Explicitly, this follows from the fact that, given $\epsilon > 0$, there is a $\delta > 0$ such that $|f(z) - f(z_0)| < \epsilon$ whenever $|z - z_0| < \delta$ (and noting that $|x - x_0| < |z - z_0| < \delta$, $|y - y_0| < |z - z_0| < \delta$).

For example, the function

$$f(z) = z^2 = f(x, y) = x^2 - y^2 + 2ixy$$

is continuous for all x, y, since, with $f(x, y) = u(x, y) + iv(x, y)$, we have that the real and imaginary parts of $f : u(x, y) = x^2 - y^2$, $v(x, y) = 2xy$ are polynomials and are continuous for all x, y.

Let $f(z)$ be defined in some region $\mathcal{R}$ containing the neighborhood of a point z_0. The **derivative** of $f(z)$ at $z = z_0$, denoted by $f'(z_0)$ or $\dfrac{df}{dz}(z_0)$, is defined by

$$f'(z_0) = \lim_{\Delta z \to 0} \left(\frac{f(z_0 + \Delta z) - f(z_0)}{\Delta z} \right), \qquad (1.3.9)$$

provided this limit exists. We sometimes say that f is **differentiable** at z_0.

Another standard definition for the derivative is:

$$f'(z_0) = \lim_{z \to z_0} \left(\frac{f(z) - f(z_0)}{z - z_0} \right). \qquad (1.3.10)$$

Eq. (1.3.10) follows from Eq. (1.3.9) by taking $\Delta z = z - z_0$. If $f'(z_0)$ exists for all points $z_0 \in \mathcal{R}$, then we say $f(z)$ is differentiable in $\mathcal{R}$ – or just differentiable, $\mathcal{R}$ being understood. If $f'(z_0)$ exists, then $f(z)$ is continuous at $z = z_0$. This follows from

$$\lim_{z \to z_0} (f(z) - f(z_0)) = \lim_{z \to z_0} \left(\frac{f(z) - f(z_0)}{z - z_0} \right) \lim_{z \to z_0} (z - z_0)$$
$$= f'(z_0) \lim_{z \to z_0} (z - z_0) = 0.$$

A continuous function is not necessarily differentiable. Indeed it turns out that differentiable functions possess many special properties.

On the other hand, because we are dealing now with complex functions, which have a two-dimensional character, there can be new kinds of complications not found in functions of one real variable. A prototypical example follows.

Consider the function

$$f(z) = \bar{z}. \qquad (1.3.11)$$

Even though this function is continuous, as discussed earlier, we now show that it does not possess a derivative. Consider the difference quotient:

$$\lim_{\Delta z \to 0} \frac{\overline{(z_0 + \Delta z)} - \overline{z_0}}{\Delta z} = \lim_{\Delta z \to 0} \frac{\overline{\Delta z}}{\Delta z} \equiv q_0. \qquad (1.3.12)$$

This limit does not exist because a unique value of q_0 cannot be found; indeed it depends on how Δz approaches zero. Writing $\Delta z = re^{i\theta}$, $q_0 = \lim_{\Delta z \to 0} e^{-2i\theta}$. So if $\Delta z \to 0$ along the positive real axis ($\theta = 0$), then $q_0 = 1$. If $\Delta z \to 0$ along the positive imaginary axis then $q_0 = -1$ (because $\theta = \pi/2$, $e^{-2i\theta} = -1$), etc. Thus we find the surprising result that the function $f(z) = \bar{z}$ is not differentiable anywhere (i.e., for any $z = z_0$) even though it is continuous everywhere! In fact this situation will be seen to be the case for general complex functions unless the real and imaginary parts of our complex function satisfy certain compatibility conditions (see Section 2.1). Differentiable complex functions, often called **analytic functions**, are special and important.

On the other hand the notation $f(z) = \bar{z}$ can be misleading. Indeed, an alternative notation is $f(z, \bar{z}) = \bar{z}$. A general concept to be discussed later is: in order for the function f to be differentiable with respect to z it is necessary that f be a function only of z as opposed to z and $\bar{z}$.

Despite the fact that the formula for a derivative is identical in form to that of the derivative of a real-valued function, $f(z)$, a significant point to note is that $f'(z)$ follows from a two-dimensional limit ($z = x + iy$ or $z = re^{i\theta}$). Thus for $f'(z)$ to exist, the relevant limit must exist independent of the direction from which z approaches the limit point z_0. For a function of one real variable we only have two directions: $x < x_0$ and $x > x_0$.

If f and g have derivatives, then if follows by similar proofs to those of real variables that

$$(f + g)' = f' + g',$$
$$(fg)' = f'g + fg',$$
$$\left(\frac{f}{g}\right)' = (f'g - fg')/g^2, \qquad g \neq 0,$$

and if $f'(g(z))$ and $g'(z)$ exist, then

$$[f(g(z))]' = f'(g(z))g'(z).$$

In order to differentiate polynomials, we need the derivative of the elementary function $f(z) = z^n$, n a positive integer,

$$\frac{d}{dz}(z^n) = nz^{n-1}. \qquad (1.3.13)$$

This follows from

$$\frac{(z + \Delta z)^n - z^n}{\Delta z} = nz^{n-1} + a_1 z^{n-2}\Delta z + a_2 z^{n-3}\Delta z^2 + \cdots + \Delta z^{n-1} \rightarrow nz^{n-1}$$

as $\Delta z \rightarrow 0$, where $a_1, a_2, \ldots$ are the appropriate binomial coefficients of $(a + b)^n$.

Thus we have as corollaries to this result,

$$\frac{d}{dz}(c) = 0, \qquad c = \text{constant} \qquad (1.3.14a)$$

$$\frac{d}{dz}\left(a_0 + a_1 z + a_2 z^2 + \cdots + a_m z^m\right) = a_1 + 2a_2 z + 3a_3 z^2 + \cdots + ma_m z^{m-1}. \qquad (1.3.14b)$$

Moreover, with regard to the (purely formal at this point) power series expansions discussed earlier, in Chapter 3 we will find that

$$\frac{d}{dz}\left(\sum_{n=0}^{\infty} a_n z^n\right) = \sum_{n=0}^{\infty} na_n z^{n-1} \qquad (1.3.15)$$

inside the radius of convergence of the series.

We also note that the derivatives of the usual elementary functions behave in the same way as in real variables. Namely

$$\frac{d}{dz}e^z = e^z, \qquad \frac{d}{dz}, \sin z = \cos z, \qquad \frac{d}{dz}\cos z = -\sin z,$$

$$\frac{d}{dz}\sinh z = \cosh z, \qquad \frac{d}{dz}\cosh z = \sinh z,$$

(1.3.16)

etc. The proofs can be obtained from the fundamental definitions. For example,

$$\frac{d}{dz}e^z = \lim_{\Delta z \to 0}\frac{e^{z+\Delta z} - e^z}{\Delta z} = e^z \lim_{\Delta z \to 0}\left(\frac{e^{\Delta z} - 1}{\Delta z}\right)$$

$$= e^z,$$

(1.3.17)

where we note that

$$\lim_{\Delta z \to 0}\frac{e^{\Delta z} - 1}{\Delta z} = \lim_{\substack{\Delta x \to 0 \\ \Delta y \to 0}}\left(\frac{(e^{\Delta x}\cos \Delta y - 1) + ie^{\Delta x}\sin \Delta y}{(\Delta x + i\Delta y)}\right) = 1.$$

(1.3.18)

One can put Eq. (1.3.18) into real/imaginary form and use polar coordinates for Δx, Δy. This calculation is also discussed in the problems given for this section. Later we shall establish the validity of the power series formulae for e^z (see Eq. (1.2.19)), from which Eq. (1.3.18) follows immediately (since $e^z = 1 + z + z^2/2 + \cdots$) without need for the double limit. The other formulae in Eq. (1.3.16) can also be deduced using the relationships (1.2.9), (1.2.10), (1.2.13), and (1.2.14).

1.3.1 Problems for Section 1.3

1. Evaluate the following limits:

   (a) $\lim_{z \to i}(z + 1/z)$ (b) $\lim_{z \to z_0} 1/z^m$, m integer

   (c) $\lim_{z \to i} \sinh z$ (d) $\lim_{z \to 0}\frac{\sin z}{z}$ (e) $\lim_{z \to \infty}\frac{\sin z}{z}$

   (f) $\lim_{z \to \infty}\frac{z^2}{(3z + 1)^2}$ (g) $\lim_{z \to \infty}\frac{z}{z^2 + 1}$.

2. Establish a special case of l'Hôpital's rule. Suppose that $f(z)$ and $g(z)$ have formal power series about $z = a$, and

   $$f(a) = f'(a) = f''(a) = \cdots = f^{(k)}(a) = 0,$$

   $$g(a) = g'(a) = g''(a) = \cdots = g^{(k)}(a) = 0.$$

   If $f^{(k+1)}(a)$ and $g^{(k+1)}(a)$ are not simultaneously zero, show that

   $$\lim_{z \to a}\frac{f(z)}{g(z)} = \frac{f^{(k+1)}(a)}{g^{(k+1)}(a)}.$$

   What happens if $g^{(k+1)}(a) = 0$?

3. Where are the following functions differentiable?

$$\text{(a) } \sin z; \quad \text{(b) } \tan z; \quad \text{(c) } \frac{z-1}{z^2+1}; \quad \text{(d) } e^{1/z}; \quad \text{(e) } 2\bar{z}.$$

4. (a) If $|g(z)| \le M$, $M > 0$, for all z in a neighborhood of $z = z_0$, show that if $\lim_{z \to z_0} f(z) = 0$, then

$$\lim_{z \to z_0} f(z)g(z) = 0.$$

   (b) Using the definition of limit establish that $\lim_{z \to -i} z^3 = i$.
   Hint: Use: $|z^3 - i| = |z+i||z^2 - iz - 1| = |z+i||(z+i-i)^2 - i(z+i-i) - 1| = |z+i||(z+i)^2 - 2i(z+i) - 1 - i(z+i) - 2)| < \delta(\delta^2 + 3\delta + 3)$.

5. Show that the functions $\operatorname{Re} z$ and $\operatorname{Im} z$ are nowhere differentiable.

6. Let $f(z)$ be a continuous function for all z. Show that if $f(z_0) \ne 0$, then there must be a neighborhood of z_0 in which $f(z) \ne 0$.

7. Let $f(z)$ be a continuous function where $\lim_{z \to 0} f(z) = 0$. Show that $\lim_{z \to 0}(e^{f(z)} - 1) = 0$. What can be said about the $\lim_{z \to 0}\left((e^{f(z)} - 1)/z\right)$?

8. Let two polynomials $f(z) = a_0 + a_1 z + \cdots + a_n z^n$ and $g(z) = b_0 + b_1 z + \cdots + b_m z^m$ be equal at all points z in a region R. Use the concept of a limit to show that $m = n$ and that all the coefficients $\{a_j\}_{j=0}^n$ and $\{b_j\}_{j=0}^n$ must be equal. Hint: Consider $\lim_{z \to 0}(f(z) - g(z))$, $\lim_{z \to 0}(f(z) - g(z))/(z)$, etc.

9. (a) Use the real Taylor series formula,

$$e^x = 1 + x + O(x^2), \qquad \cos x = 1 + O(x^2),$$
$$\sin x = x(1 + O(x^2)),$$

where $O(x^2)$ means we are omitting terms proportional to power x^2 (i.e., $\lim_{x \to 0}(O(x^2))/(x^2) = C$, where C is a constant), to establish the following:

$$\lim_{z \to 0}\left(e^z - (1 + z)\right) = \lim_{r \to 0}\left(e^{r\cos\theta}e^{ir\sin\theta} - (1 + r(\cos\theta + i\sin\theta))\right) = 0.$$

   (b) Use the above Taylor expansions to show that (cf. Eq. (1.3.18))

$$\lim_{\Delta z \to 0}\left(\frac{e^{\Delta z} - 1}{\Delta z}\right) = \lim_{r \to 0}\left\{\frac{\left(e^{r\cos\theta}\cos(r\sin\theta) - 1\right) + ie^{r\cos\theta}\sin(r\sin\theta)}{r(\cos\theta + i\sin\theta)}\right\}$$
$$= 1.$$

10. Let $z = x$ be real. Use the relationship $(d/dx)e^{ix} = ie^{ix}$ to find the standard derivative formulae for trigonometric functions:

$$\frac{d}{dx}\sin x = \cos x,$$
$$\frac{d}{dx}\cos x = -\sin x.$$

11. (a) Show that $\lim_{z \to 0}(|z|^2/z) = 0$.
    (b) Show that $\lim_{z \to 0}(|z|/z)$ does not exist.
12. Let $f(z)$ be continuous for all z. Show that
    (a) $g(z) = \underline{f(\overline{z})}$ is continuous for all z;
    (b) $g(z) = f^2(z)$ is continuous for all z.

1.4 Elementary Applications to Ordinary Differential Equations

An important topic in the application of complex variables is the study of differential equations. Later in this text we discuss differential equations in the complex plane in some detail, but in fact we are already in a position to see why complex variables can be useful. Many readers will have had a course in differential equations, but it is not really necessary for what we shall discuss. Linear homogeneous differential equations with constant coefficients take the following form:

$$L_n w = \frac{d^n w}{dt^n} + a_{n-1}\frac{d^{n-1} w}{dt^{n-1}} + \cdots + a_1\frac{dw}{dt} + a_0 w = 0, \qquad (1.4.1)$$

where $\{a_j\}_{j=0}^{n-1}$ are all constant, n is called the order of the equation, and (for our present purposes) t is real. We could (and do, later in Section 3.7) allow t to be complex, in which case the study of such differential equations becomes intimately connected with many of the topics studied later in this text, but for now we keep t real. Solutions to Eq. (1.4.1) can be sought of the form

$$w(t) = c e^{zt}, \qquad (1.4.2)$$

where c is a nonzero constant. Substitution of Eq. (1.4.2) into Eq. (1.4.1), and factoring $c e^{zt}$ from each term (note e^{zt} does not vanish), yields the following *algebraic* equation:

$$z^n + a_{n-1}z^{n-1} + \cdots + a_1 z + a_0 = 0. \qquad (1.4.3)$$

There are various subcases to consider, but we shall only discuss the prototypical one where there are n *distinct* solutions of Eq. (1.4.3) which we call $\{z_1, z_2, \ldots, z_n\}$. Each of these values, say z_j, yields a solution to Eq. (1.4.1) $w_j = c_j e^{z_j t}$, where c_j is an arbitrary constant. Because Eq. (1.4.1) is a linear equation, we have the more general solution

$$w(t) = \sum_{j=1}^{n} w_j = \sum_{j=1}^{n} c_j e^{z_j t}. \qquad (1.4.4)$$

In differential equation texts it is proven that Eq. (1.4.4) is, in fact, the most general solution. In applications, the differential equations Eq. (1.4.1) frequently have real coefficients $\{a_j\}_{j=0}^{n-1}$. The study of algebraic equations of the form of Eq. (1.4.3), discussed later in this text, shows that there are at most n solutions — precisely n solutions if we count multiplicity of solutions. In fact, when the coefficients are real, then the solutions are either real or come in complex conjugate pairs. Corresponding to complex conjugate pairs, a real solution $w(t)$ is found by taking complex conjugate constants c_j and $\overline{c_j}$ corresponding to each pair of complex conjugate roots z_j and $\overline{z_j}$. For example, consider one such real solution, call it w_p, corresponding to the pair $z, \overline{z}$:

$$w_p(t) = ce^{zt} + \overline{c}e^{\overline{z}t}. \qquad (1.4.5)$$

We can rewrite this in terms of trigonometric functions and real exponentials. Let $z = x + iy$:

$$
\begin{aligned}
w_p(t) &= ce^{(x+iy)t} + \overline{c}e^{(x-iy)t} \\
&= e^{xt}\left[c(\cos yt + i\sin yt) + \overline{c}(\cos yt - i\sin yt)\right] \\
&= (c + \overline{c})e^{xt}\cos yt + i(c - \overline{c})e^{xt}\sin yt. \qquad (1.4.6)
\end{aligned}
$$

Because $c + \overline{c} = A$ and $i(c - \overline{c}) = B$ are real, we have that this pair of solutions may be put in the *real* form

$$w_c(t) = Ae^{xt}\cos yt + Be^{xt}\sin yt. \qquad (1.4.7)$$

Two examples of these ideas are simple harmonic motion (SHM) and vibrations of beams:

$$\frac{d^2w}{dt^2} + \omega_0{}^2 w = 0 \qquad \text{(SHM)}, \qquad (1.4.8a)$$

$$\frac{d^4w}{dt^4} + k^4 w = 0 \qquad \text{(beams)}, \qquad (1.4.8b)$$

where $\omega_0{}^2$ and k^4 are real nonzero constants that depend on the parameters in the physical model. Looking for solutions of the form of Eq. (1.4.2) leads to the equations

$$z^2 + \omega_0{}^2 = 0 \qquad \text{(SHM)}, \qquad (1.4.9a)$$

$$z^4 + k^4 = 0 \qquad \text{(beams)}, \qquad (1.4.9b)$$

which have solutions (see also Section 1.1)

$$z_1 = i\omega_0, \qquad z_2 = -i\omega_0 \qquad \text{(SHM)} \qquad (1.4.10a)$$

or

$$\left.\begin{array}{l} z_1 = ke^{i\pi/4} = \dfrac{k}{\sqrt{2}}(1+i) \\[2em] z_2 = ke^{3i\pi/4} = \dfrac{k}{\sqrt{2}}(-1+i) \\[2em] z_3 = ke^{5i\pi/4} = \dfrac{k}{\sqrt{2}}(-1-i) \\[2em] z_4 = ke^{7i\pi/4} = \dfrac{k}{\sqrt{2}}(1-i) \end{array}\right\} \quad \text{(beams).} \qquad (1.4.10b)$$

It follows from the above discussion that the corresponding real solutions $w(t)$ have the form

$$w = A\cos\omega_0 t + B\sin\omega_0 t \qquad \text{(SHM)}, \qquad (1.4.11a)$$

or

$$w = e^{\frac{kt}{\sqrt{2}}}\left[A_1\cos\frac{kt}{\sqrt{2}} + B_1\sin\frac{kt}{\sqrt{2}}\right]$$
$$+ e^{-\frac{kt}{\sqrt{2}}}\left[A_2\cos\frac{kt}{\sqrt{2}} + B_2\sin\frac{kt}{\sqrt{2}}\right] \quad \text{(beams)}, \qquad (1.4.11b)$$

where A, B, A_1, A_2, B_1, B_2 are arbitrary constants.

Summary

In this chapter we have introduced and summarized the basic properties of complex numbers and elementary functions. We have seen that the theory of functions of a single real variable have so far motivated many of the notions of complex variables; though the two-dimensional character of complex numbers has already led to some significant differences. In subsequent chapters a number of entirely new and surprising results will be obtained, and the departure from real variables will become more apparent.

1.4.1 Problems for Section 1.4

1. Solve the differential equation

$$\frac{d^2w}{dt^2} + 2\alpha\frac{dw}{dt} + \omega_0^2 w = 0,$$

where $0 < \alpha < \omega_0$, α, ω_0, are constants. Put the answer in terms of real functions.

2. Suppose we are given the following differential equations:

$$\text{(a)} \quad \frac{d^3 w}{dt^3} - k^3 w = 0;$$

$$\text{(b)} \quad \frac{d^6 w}{dt^6} - k^6 w = 0,$$

where t is real and k is a real constant. Find the general real solution of the above equations. Write the solution in terms of *real* functions.

3. Consider the differential equation

$$x^2 \frac{d^2 w}{dx^2} + x \frac{dw}{dx} + w = 0,$$

where x is real.

(a) Show that the transformation $x = e^t$ implies that

$$x \frac{d}{dx} = \frac{d}{dt},$$

$$x^2 \frac{d^2}{dx^2} = \frac{d^2}{dt^2} - \frac{d}{dt}.$$

(b) Use these results to find that w also satisfies the differential equation

$$\frac{d^2 w}{dt^2} + w = 0.$$

(c) Use these results to establish that w has the real solution

$$w = C e^{i(\log x)} + \bar{C} e^{-i(\log x)}$$

or

$$w = A \cos(\log x) + B \sin(\log x).$$

4. Use the ideas of Problem 3 to find the real solution of the following equations (x is real and k is a real constant):

$$\text{(a)} \quad x^2 \frac{d^2 w}{dx^2} + k^2 w = 0, \qquad 4k^2 > 1;$$

$$\text{(b)} \quad x^3 \frac{d^3 w}{dx^3} + 3x^2 \frac{d^2 w}{dx^2} + x \frac{dw}{dx} + k^3 w = 0.$$

5. Suppose we are given

$$\frac{dw}{dx} + 2ixkw = x, \quad k > 0, \ w(x = 0) = 0.$$

Solve this differential equation and show the real and imaginary parts of the solution are:

$$\text{Re}\, w(x) = \frac{1}{2k} \sin kx^2, \qquad \text{Im}\, w(x) = \frac{1}{2k}(\cos kx^2 - 1).$$

2

Analytic Functions and Integration

In this chapter we study the notion of analytic functions and their properties. It will be shown that a complex function is differentiable if and only if there is an important compatibility relationship between its real and imaginary parts. In later chapters it will be shown that this relationship reflects a more fundamental characterization of analytic functions: $f(z)$ is analytic if and only if it is a function of z only as opposed to being a function of z and $\bar{z}$. The concepts of multivalued functions and complex integration are considered in some detail. The technique of integration in the complex plane is discussed and two very important results of complex analysis are derived: Cauchy's theorem and a corollary – Cauchy's integral formula.

2.1 Analytic Functions

2.1.1 The Cauchy–Riemann Equations

In Section 1.3 we defined the notion of complex differentiation. For convenience, we remind the reader of this definition here. The derivative of $f(z)$, denoted by $f'(z)$, is defined by the following limit:

$$f'(z) = \lim_{\Delta z \to 0} \frac{f(z + \Delta z) - f(z)}{\Delta z}. \tag{2.1.1}$$

We write the real and imaginary parts of $f(z)$, $f(z) = u(x, y) + iv(x, y)$, and compute Eq. (2.1.1) for: (a) $\Delta z = \Delta x$ real; and (b) $\Delta z = i\Delta y$ pure imaginary (i.e., we take the limit along the real and then along the imaginary axis). Then, for case (a),

$$f'(z) = \lim_{\Delta x \to 0} \left(\frac{u(x + \Delta x, y) - u(x, y)}{\Delta x} + i\frac{v(x + \Delta x, y) - v(x, y)}{\Delta x} \right)$$
$$= u_x(x, y) + iv_x(x, y). \tag{2.1.2}$$

We use the subscript notation for partial derivatives; that is, $u_x = \partial u/\partial x$ and $v_x = \partial v/\partial x$. For case (b),

$$f'(z) = \lim_{\Delta y \to 0} \frac{u(x, y + \Delta y) - u(x, y)}{i\Delta y} + \frac{i\,(v(x, y + \Delta y) - v(x, y))}{i\Delta y}$$

$$= -iu_y(x, y) + v_y(x, y). \tag{2.1.3}$$

Setting Eqs. (2.1.2) and (2.1.3) equal yields

$$u_x = v_y, \qquad v_x = -u_y. \tag{2.1.4}$$

Equations (2.1.4) are called the Cauchy–Riemann conditions.

Equations (2.1.4) form a system of partial differential equations that are necessarily satisfied if $f(z)$ has a derivative at the point z. This is in stark contrast to real analysis where differentiability of a function $f(x)$ is only a mild smoothness condition on the function. We also note that if u, v have second derivatives, then we will show that they satisfy the equations $u_{xx} + u_{yy} = 0$ and $v_{xx} + v_{yy} = 0$ (cf. Eqs. (2.1.11a, 2.1.11b)).

Equation (2.1.4) is a **necessary condition** that must hold if $f(z)$ is differentiable. On the other hand, it turns out that if the partial derivatives of $u(x, y)$ and $v(x, y)$ do exist, satisfy Eq. (2.1.4), and are continuous, then $f(z) = u(x, y) + iv(x, y)$ must be differentiable at the point $z = x + iy$; that is, Eq. (2.1.4) is a **sufficient condition** as well. Namely, if Eq. (2.1.4) holds then $f'(z)$ exists and is given by Eqs. (2.1.1–2.1.2).

We discuss this latter point next. We use a well-known result of real analysis of two variables, namely, if u_x, u_y and v_x, v_y are continuous at the point (x, y), then

$$\Delta u = u_x \Delta x + u_y \Delta y + \epsilon_1 |\Delta z|,$$
$$\Delta v = v_x \Delta x + v_y \Delta y + \epsilon_2 |\Delta z|, \tag{2.1.5}$$

where $|\Delta z| = \sqrt{\Delta x^2 + \Delta y^2}$, $\lim_{\Delta z \to 0} \epsilon_1 = \lim_{\Delta z \to 0} \epsilon_2 = 0$, and

$$\Delta u = u(x + \Delta x, y + \Delta y) - u(x, y),$$
$$\Delta v = v(x + \Delta x, y + \Delta y) - v(x, y).$$

Calling $\Delta f = \Delta u + i\Delta v$, we have

$$\frac{\Delta f}{\Delta z} = \frac{\Delta u}{\Delta z} + i\frac{\Delta v}{\Delta z}$$

$$= \left(u_x \frac{\Delta x}{\Delta z} + u_y \frac{\Delta y}{\Delta z}\right) + i\left(v_x \frac{\Delta x}{\Delta z} + v_y \frac{\Delta y}{\Delta z}\right)$$

$$+ (\epsilon_1 + i\epsilon_2)\frac{\Delta z}{|\Delta z|}, \qquad |\Delta z| \neq 0. \tag{2.1.6}$$

Then, letting $\frac{\Delta z}{|\Delta z|} = e^{i\varphi}$ and using Eq. (2.1.4), Eq. (2.1.6) yields

$$\frac{\Delta f}{\Delta z} = (u_x + iv_x)\frac{\Delta x + i\Delta y}{\Delta z} + (\epsilon_1 + i\epsilon_2)e^{-i\varphi}$$
$$= f'(z) + (\epsilon_1 + i\epsilon_2)e^{-i\varphi}. \tag{2.1.7}$$

Taking the limit of $\Delta x, \Delta y$ approaching zero yields the desired result.

We state both of the above results as a theorem.

Theorem 2.1.1 *The function $f(z) = u(x, y) + iv(x, y)$ is differentiable at a point $z = x + iy$ of a region in the complex plane if and only if the partial derivatives u_x, u_y, v_x, v_y are continuous and satisfy the Cauchy–Riemann conditions, (2.1.4), at $z = x + iy$.*

A consequence of the Cauchy–Riemann conditions is that the "level" curves of u, that is, the curves $u(x, y) = c_1$ for constant c_1, are orthogonal to the level curves of v, where $v(x, y) = c_2$ for constant c_2, at all points where $f'(z)$ exists and is nonzero. From Eqs. (2.1.2) and (2.1.4) we have that

$$|f'(z)|^2 = \left(\frac{\partial u}{\partial x}\right)^2 + \left(\frac{\partial v}{\partial x}\right)^2 = \left(\frac{\partial u}{\partial x}\right)^2 + \left(\frac{\partial u}{\partial y}\right)^2 = \left(\frac{\partial v}{\partial x}\right)^2 + \left(\frac{\partial v}{\partial y}\right)^2,$$

hence the nontrivial two-dimensional vector gradients $\nabla u = (\frac{\partial u}{\partial x}, \frac{\partial u}{\partial y})$ and $\nabla v = (\frac{\partial v}{\partial x}, \frac{\partial v}{\partial y})$ are nonzero. We know from vector calculus that the gradient is orthogonal to its level curve (i.e., $du = \nabla u \cdot ds = 0$, where ds points in the direction of the tangent to the level curve), and from the Cauchy–Riemann condition (Eq. (2.1.4)) we see that the gradients $\nabla u, \nabla v$ are orthogonal because their vector dot product vanishes:

$$\nabla u \cdot \nabla v = \frac{\partial u}{\partial x}\frac{\partial v}{\partial x} + \frac{\partial u}{\partial y}\frac{\partial v}{\partial y} = -\frac{\partial u}{\partial x}\frac{\partial u}{\partial y} + \frac{\partial u}{\partial y}\frac{\partial u}{\partial x}$$
$$= 0.$$

Consequently, the two-dimensional level curves $u(x, y) = c_1$ and $v(x, y) = c_2$ are orthogonal.

For example the function $w = u + iv = z^2$ implies $u(x, y) = x^2 - y^2$ and $v(x, y) = 2xy$. The level curves $x^2 - y^2 = \text{constant}$ and $xy = \text{constant}$ are orthogonal for various constants: see Figure 2.1.

The Cauchy–Riemann conditions can be written in other coordinate systems, and it is frequently valuable to do so. Here we quote the result in polar coordinates:

$$\frac{\partial u}{\partial r} = \frac{1}{r}\frac{\partial v}{\partial \theta}, \qquad \frac{\partial v}{\partial r} = -\frac{1}{r}\frac{\partial u}{\partial \theta}. \tag{2.1.8}$$

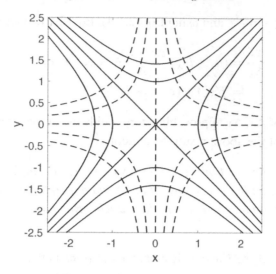

Figure 2.1 Contour Lines for $f(z) = z^2$

Equation (2.1.8) can be derived in the same manner as (2.1.4). An alternative derivation uses the following differential relationships:

$$\frac{\partial}{\partial x} = \cos\theta\frac{\partial}{\partial r} - \frac{\sin\theta}{r}\frac{\partial}{\partial\theta}, \qquad \frac{\partial}{\partial y} = \sin\theta\frac{\partial}{\partial r} + \frac{\cos\theta}{r}\frac{\partial}{\partial\theta}, \qquad (2.1.9)$$

which are derived from $x = r\cos\theta$ and $y = r\sin\theta$, $r^2 = x^2 + y^2$, $\tan\theta = y/x$.

Employing Eq. (2.1.9) in Eq. (2.1.4) yields

$$\cos\theta\frac{\partial u}{\partial r} - \frac{\sin\theta}{r}\frac{\partial u}{\partial\theta} = \sin\theta\frac{\partial v}{\partial r} + \frac{\cos\theta}{r}\frac{\partial v}{\partial\theta},$$

$$\sin\theta\frac{\partial u}{\partial r} + \frac{\cos\theta}{r}\frac{\partial u}{\partial\theta} = -\cos\theta\frac{\partial v}{\partial r} + \frac{\sin\theta}{r}\frac{\partial v}{\partial\theta}.$$

Multiplying the first of these equations by $\cos\theta$, the second by $\sin\theta$, and adding yields the first of the Eqs. (2.1.8). Similarly, multiplying the first by $\sin\theta$, the second by $-\cos\theta$, and adding yields the second of Eqs. (2.1.8).

Similarly, using the first relation of Eq. (2.1.9) in $f'(z) = \partial u/\partial x + i\partial v/\partial x$ yields

$$f'(z) = \cos\theta\frac{\partial u}{\partial r} - \frac{\sin\theta}{r}\frac{\partial u}{\partial\theta} + i\cos\theta\frac{\partial v}{\partial r} - i\frac{\sin\theta}{r}\frac{\partial v}{\partial\theta}$$

$$= (\cos\theta - i\sin\theta)\left(\frac{\partial u}{\partial r} + i\frac{\partial v}{\partial r}\right),$$

hence

$$f'(z) = e^{-i\theta}\left(\frac{\partial u}{\partial r} + i\frac{\partial v}{\partial r}\right). \qquad (2.1.10)$$

Example 2.1.2 Let $f(z) = e^z = e^{x+iy} = e^x e^{iy} = e^x(\cos y + i \sin y)$. Verify Eq. (2.1.4) for all x and y, and then show that $f'(z) = e^z$.

First,

$$u = e^x \cos y, \qquad v = e^x \sin y;$$

$$\frac{\partial u}{\partial x} = e^x \cos y = \frac{\partial v}{\partial y};$$

$$\frac{\partial u}{\partial y} = -e^x \sin y = -\frac{\partial v}{\partial x}.$$

Next,

$$f'(z) = \frac{\partial u}{\partial x} + i\frac{\partial v}{\partial x} = e^x(\cos y + i \sin y)$$
$$= e^x e^{iy} = e^{x+iy} = e^z.$$

We have therefore established the fact that $f(z) = e^z$ is differentiable for all finite values of z. Consequently standard functions like $\sin z$ and $\cos z$, which are linear combinations of the exponential function e^{iz} (see Eqs. (1.2.9–1.2.10)) are also seen to be differentiable functions of z for all finite values of z. It should be noted that these functions do not behave like their real counterparts. For example, the function $\sin x$ oscillates and is bounded between ± 1 for all real x. However, we have

$$\sin z = \sin(x + iy) = \sin x \cos iy + \cos x \sin iy$$
$$= \sin x \cosh y + i \cos x \sinh y.$$

Because $|\sinh y|$ and $|\cosh y|$ tend to infinity as y tends to infinity, we see that the real and imaginary parts of $\sin z$ grow without bound.

Example 2.1.3 Let $f(z) = \bar{z} = x - iy$, so that $u(x, y) = x$ and $v(x, y) = -y$. Because $\partial u/\partial x = 1$ while $\partial v/\partial y = -1$, condition (2.1.4) implies $f'(z)$ does not exist anywhere (see also Section 1.3).

Example 2.1.4 Let $f(z) = z^n = r^n e^{in\theta} = r^n(\cos n\theta + i \sin n\theta)$, for integer n, so that $u(r, \theta) = r^n \cos n\theta$ and $v(r, \theta) = r^n \sin n\theta$. Verify Eq. (2.1.8) and show that $f'(z) = nz^{n-1}$, with $z \neq 0$ if $n < 0$. By differentiation, we have that

$$\frac{\partial u}{\partial r} = nr^{n-1} \cos n\theta = \frac{1}{r}\frac{\partial v}{\partial \theta},$$

$$\frac{\partial v}{\partial r} = nr^{n-1} \sin n\theta = -\frac{1}{r}\frac{\partial u}{\partial \theta}.$$

From Eq. (2.1.10), we find

$$f'(z) = e^{-i\theta}(nr^{n-1})(\cos n\theta + i \sin n\theta)$$
$$= nr^{n-1}e^{-i\theta}e^{in\theta} = nr^{n-1}e^{i(n-1)\theta}$$
$$= nz^{n-1},$$

where $z \neq 0$ if $n < 0$.

Example 2.1.5   If a function is differentiable and has constant modulus, show that the function itself is constant. We may write f in terms of real, imaginary or complex form where

$$f = u + iv = Re^{i\Theta},$$
$$R^2 = u^2 + v^2, \qquad \tan \Theta = \frac{v}{u},$$
$$R = \text{constant}.$$

From Eq. (2.1.8) we have

$$u\frac{\partial u}{\partial r} + v\frac{\partial v}{\partial r} = \frac{1}{r}\left(u\frac{\partial v}{\partial \theta} - v\frac{\partial u}{\partial \theta}\right) = \frac{u^2}{r}\frac{\partial}{\partial \theta}\left(\frac{v}{u}\right),$$

so

$$\frac{\partial}{\partial r}\left(u^2 + v^2\right) = \frac{2u^2}{r}\frac{\partial}{\partial \theta}\left(\frac{v}{u}\right).$$

Thus, $\partial(v/u)/\partial\theta = 0$, because $R^2 = u^2 + v^2$ is a constant.

Similarly, using Eq. (2.1.8),

$$u^2\frac{\partial}{\partial r}\left(\frac{v}{u}\right) = \left(u\frac{\partial v}{\partial r} - v\frac{\partial u}{\partial r}\right)$$
$$= -\frac{1}{r}\left(u\frac{\partial u}{\partial \theta} + v\frac{\partial v}{\partial \theta}\right) = -\frac{1}{2r}\frac{\partial}{\partial \theta}\left(u^2 + v^2\right) = 0.$$

Thus if $u \neq 0$, $v/u = \text{constant}$, which implies Θ is constant, and hence so is f. If $u = 0$, the Cauchy–Riemann equations imply that v and therefore f are constant.

We have observed that the system of partial differential equations (PDEs), Eq. (2.1.4), that is, the Cauchy–Riemann equations, must hold at every point where $f'(z)$ exists. However, PDEs are really of interest when they hold not only at one point, but rather in a neighborhood or region containing the point. Hence we give the following definition.

Definition 2.1.6   A function $f(z)$ is said to be **analytic** at a point z_0 if $f(z)$ is differentiable in a neighborhood of z_0. The function $f(z)$ is said to be **analytic in a region** if it is analytic at every point in the region.

Of the previous examples, $f(z) = e^z$ is analytic in the entire finite z-plane whereas $f(z) = \bar{z}$ is analytic *nowhere*. The function $f(z) = 1/z^2$ (Example 2.1.4, with $n = -2$) is analytic for all finite $z \neq 0$ (the "punctured" z-plane).

Example 2.1.7   Determine where $f(z)$ is analytic when $f(z) = (x+\alpha y)^2 + 2i(x - \alpha y)$, for α real and constant. Write

$$u(x, y) = (x + \alpha y)^2, \qquad v(x, y) = 2(x - \alpha y),$$

$$\frac{\partial u}{\partial x} = 2(x + \alpha y), \qquad \frac{\partial v}{\partial y} = -2\alpha,$$

$$\frac{\partial u}{\partial y} = 2\alpha(x + \alpha y), \qquad \frac{\partial v}{\partial x} = 2.$$

The Cauchy–Riemann equations are satisfied only if $\alpha^2 = 1$ and only on the lines $x \pm y = \mp 1$. Because the derivative $f'(z)$ exists only on these lines, $f(z)$ is not analytic *anywhere* since it is not analytic in the neighborhood of these lines. As noted earlier an analytic function cannot depend on $\bar{z}$. Using $z = x + iy, \bar{z} = x - iy$, we find

$$x = \frac{z + \bar{z}}{2}, \quad y = \frac{z - \bar{z}}{2i}.$$

Then replacing x and y in the expression $u + iv$ the $\bar{z}$ dependence of the function takes the form

$$\frac{(1 + i\alpha)^2 \bar{z}^2}{4} + \frac{(1 + \alpha^2)z\bar{z}}{2} + i(1 - i\alpha)\bar{z}.$$

In order for the function to be independent of $\bar{z}$, we see that α must be equal to *both* i and $-i$. Hence the function f is nowhere analytic.

If we say that $f(z)$ is analytic in a region, such as $|z| \leq R$, we mean that $f(z)$ is analytic in a domain containing the circle because $f'(z)$ must exist in a neighborhood of every point on $|z| = R$. We also note that some authors use the term **holomorphic** instead of analytic.

An **entire** function is a function that is analytic at each point in the "entire" finite plane. As mentioned above, $f(z) = e^z$ is entire, as is $\sin z$ and $\cos z$. So also is $f(z) = z^n$, for integer $n \geq 0$, and therefore, any polynomial.

A **singular point** z_0 is a point where f fails to be analytic. Thus $f(z) = 1/z^2$ has $z = 0$ as a singular point. On the other hand, $f(z) = \bar{z}$ is analytic nowhere and has singular points everywhere in the complex plane. If any region $\mathcal{R}$ exists such that $f(z)$ is analytic in $\mathcal{R}$, we frequently speak of the function as being an **analytic function**. A further and more detailed discussion of singular points appears in Section 3.5.

As we have seen from our examples and from Section 1.3, the standard differentiation formulae of real variables hold for functions of a complex variable. Namely, if two functions are analytic in a domain D, their sum, product, and quotient are analytic in D provided the denominator of the quotient does not vanish at any point in D. Similarly, the composition of two analytic functions is also analytic.

We shall see later, in Section 2.6.1, that an analytic function has derivatives of all orders in the region of analyticity and that the real and imaginary parts have continuous derivatives of all orders as well. From Eq. (2.1.4), because $\partial^2 v/\partial x \partial y = \partial^2 v/\partial y \partial x$, we have

$$\frac{\partial^2 u}{\partial x^2} = \frac{\partial^2 v}{\partial x \partial y}, \qquad \frac{\partial^2 v}{\partial y \partial x} = -\frac{\partial^2 u}{\partial y^2},$$

hence

$$\nabla^2 u \equiv \frac{\partial^2 u}{\partial x^2} + \frac{\partial^2 u}{\partial y^2} = 0, \tag{2.1.11a}$$

and similarly,

$$\nabla^2 v \equiv \frac{\partial^2 v}{\partial x^2} + \frac{\partial^2 v}{\partial y^2} = 0. \tag{2.1.11b}$$

Equations (2.1.11a–2.1.11b) demonstrate that u and v satisfy the same uncoupled PDE. The equation $\nabla^2 w = 0$ is called **Laplace's equation**. It has wide applicability and plays a central role in the study of classical partial differential equations. The function $w(x, y)$ satisfying Laplace's equation in a domain D is called an **harmonic function** in D. The two functions $u(x, y)$ and $v(x, y)$, which are respectively the real and imaginary parts of an analytic function in D, both satisfy Laplace's equation in D. That is, they are **harmonic functions** in D, and v is referred to as the **harmonic conjugate** of u (and vice versa). The function v may be obtained from u via the Cauchy–Riemann conditions. It is clear from the derivation of Eqs. (2.1.11a,b) that $f(z) = u(x, y) + iv(x, y)$ is an analytic function if and only if u and v satisfy Eqs. (2.1.11a,b) and v is the harmonic conjugate of u.

The following example illustrates how, given $u(x, y)$, it is possible to obtain the harmonic conjugate $v(x, y)$ as well as the analytic function $f(z)$.

Example 2.1.8 Suppose we are given $u(x, y) = y^2 - x^2$ in the entire z-plane. Find its harmonic conjugate as well as $f(z)$. We have

$$\frac{\partial u}{\partial x} = -2x = \frac{\partial v}{\partial y} \quad \Rightarrow \quad v = -2xy + \phi(x),$$

$$\frac{\partial u}{\partial y} = 2y = -\frac{\partial v}{\partial x} \quad \Rightarrow \quad v = -2xy + \psi(y),$$

where $\phi(x)$, $\psi(y)$ are arbitrary functions of x, and y, respectively. Taking the difference of both expressions for v implies $\phi(x) - \psi(y) = 0$, which can only be satisfied by $\phi = \psi = c$, a constant; thus

$$f(z) = y^2 - x^2 - 2ixy + ic = -(x^2 - y^2 + 2ixy) + ic$$
$$= -z^2 + ic.$$

It follows from the remark following Theorem 2.1.1, that the two level curves $u = y^2 - x^2 = c_1$ and $v = -2xy = c_2$ are orthogonal to each other at each point (x, y). These are two orthogonal sets of hyperbolae.

In what follows we derive explicit formulas for computing an analytic function in terms of its real or imaginary parts. This provides an alternative to our earlier method which used the Cauchy–Riemann conditions, but now without differentiation or integration. This result is a direct consequence of the understanding that an analytic function is only a function of z as opposed to z and $\bar{z}$. We state the result as a theorem.

Theorem 2.1.9 *Let the complex-valued function $f(z) = u(x, y) + iv(x, y)$ be analytic in the neighborhood of a point z_0. The following formulae hold:*

$$f(z) = 2u\left(\frac{z + \bar{z}_0}{2}, \frac{z - \bar{z}_0}{2i}\right) - \overline{f(z_0)}, \tag{2.1.12}$$

$$f(z) = 2iv\left(\frac{z + \bar{z}_0}{2}, \frac{z - \bar{z}_0}{2i}\right) + \overline{f(z_0)}. \tag{2.1.13}$$

Proof To establish these expressions we will repeatedly use the important property of the analytic function $f(z)$ that it must be independent of $\bar{z}$. Indeed we have shown earlier that the function $\bar{z}$ is nowhere analytic, hence any function of $\bar{z}$ is nowhere analytic. We begin by expressing x and y in terms of z and $\bar{z}$ using $x = \frac{z+\bar{z}}{2}, y = \frac{z-\bar{z}}{2i}$ to find

$$f(z) = u\left(\frac{z + \bar{z}}{2}, \frac{z - \bar{z}}{2i}\right) + iv\left(\frac{z + \bar{z}}{2}, \frac{z - \bar{z}}{2i}\right). \tag{2.1.14}$$

We have shown that the functions $u(x, y)$ and $v(x, y)$ are harmonic, thus they are valid for any value of both variables x and y near the point z_0 in the analytic domain. Evaluating the expression (2.1.14) at $z = z_0$ we find

$$f(z_0) = u\left(\frac{z_0 + \bar{z}}{2}, \frac{z_0 - \bar{z}}{2i}\right) + iv\left(\frac{z_0 + \bar{z}}{2}, \frac{z_0 - \bar{z}}{2i}\right). \tag{2.1.15}$$

In the above expression we have *not* set $\bar{z} = \bar{z}_0$, as might have been expected from our choice of $z = z_0$, because the analytic function $f(z)$ must be independent of $\bar{z}$. So we can keep the general value of this variable, namely $\bar{z}$, instead of fixing it to be $\bar{z}_0$. In other words, in the expression of $f(z)$ that involves both z and $\bar{z}$, we treat z as the complex variable and $\bar{z}$ as a constant, for the moment unrelated to z. Computing the complex conjugate of the equation (2.1.15) we find

$$\overline{f(z_0)} = u\left(\frac{\bar{z}_0 + z}{2}, \frac{z - \bar{z}_0}{2i}\right) - iv\left(\frac{\bar{z}_0 + z}{2}, \frac{z - \bar{z}_0}{2i}\right). \tag{2.1.16}$$

Using again the fact that the analytic function $f(z)$ is independent of $\bar{z}$, we can now replace $\bar{z}$ by $\bar{z}_0$ in the right-hand side of equation (2.1.14), which leads to

$$f(z) = u\left(\frac{z + \bar{z}_0}{2}, \frac{z - \bar{z}_0}{2i}\right) + iv\left(\frac{z + \bar{z}_0}{2}, \frac{z - \bar{z}_0}{2i}\right). \tag{2.1.17}$$

Adding or subtracting this last equation to equation (2.1.16), we find the desired results; namely equations (2.1.12)–(2.1.13). □

These expressions are particularly useful in determining the analytic function $f(z)$ when we know only its real or imaginary part, *without having to do any differentiation or integration*. We demonstrate this below with two examples.

Example 2.1.10   Find the most general analytic function whose real part is given by

$$u(x, y) = \frac{x}{x^2 + y^2}. \tag{2.1.18}$$

We use equation (2.1.12):

$$f(z) = 2\frac{(z + \bar{z}_0)/2}{(z + \bar{z}_0)^2/4 - (z - \bar{z}_0)^2/4} - \overline{f(z_0)} \implies f(z) + \overline{f(z_0)} = \frac{z + \bar{z}_0}{z\bar{z}_0}$$

$$= \frac{1}{z} + \frac{1}{\bar{z}_0}. \tag{2.1.19}$$

From this last equation it follows that if we choose

$$f(z) = \frac{1}{z} + ia, \quad \text{for } a \text{ real}, \tag{2.1.20}$$

this function satisfies equation (2.1.18) and is the general solution we are seeking.

Example 2.1.11   Find the most general analytic function whose imaginary part is given by

$$v(x, y) = \sin(x)e^{-y}. \tag{2.1.21}$$

We use equation (2.1.13):

$$f(z) = 2i \sin\left(\frac{z + \bar{z}_0}{2}\right) e^{-\left(\frac{z - \bar{z}_0}{2i}\right)} + \overline{f(z_0)}. \tag{2.1.22}$$

Using the formula for $\sin z$ in terms of exponentials yields

$$f(z) - \overline{f(z_0)} = \left[e^{i\left(\frac{z + \bar{z}_0}{2}\right)} - e^{-i\left(\frac{z + \bar{z}_0}{2}\right)}\right] e^{i\left(\frac{z - \bar{z}_0}{2}\right)} = e^{iz} - e^{-i\bar{z}_0}. \tag{2.1.23}$$

From the last expression it follows that the function

$$f(z) = e^{iz} + c, \quad \text{for } c \text{ real} \tag{2.1.24}$$

satisfies equation (2.1.19); this is the general solution we are looking for.

Finally, we remark that Laplace's equation arises frequently in the study of physical phenomena. Applications include the study of two-dimensional ideal fluid flow, steady state heat conduction, electrostatics, and many others. In these applications we are usually interested in solving Laplace's equation $\nabla^2 w = 0$ in a domain D with boundary conditions, typically of the form

$$\alpha w + \beta \frac{\partial w}{\partial n} = \gamma \qquad \text{on } C, \tag{2.1.25}$$

where $\partial w/\partial n$ denotes the outward normal derivative of w on the boundary of D denoted by C; α, β, and γ are given functions on the boundary. We refer to the solution of Laplace's equation when $\beta = 0$ as the Dirichlet problem, and when $\alpha = 0$ the Neumann problem. The general case is usually called the mixed problem.

2.1.2 Ideal Fluid Flow

Two-dimensional **ideal fluid flow** is one of the prototypical examples of Laplace's equations and complex variable techniques. The corresponding flow configurations are usually easy to conceptualize. **Ideal fluid motion** refers to fluid motion which is steady (time independent), nonviscous (zero friction; usually called inviscid), incompressible (in this case, constant density), and irrotational (no local rotation of fluid "particles"). The two-dimensional equations of motion reduce to a system of two PDEs (see also the discussion in Section 5.4, Example 5.4.1):

(a) incompressibility (divergence of the velocity vanishes):

$$\frac{\partial v_1}{\partial x} + \frac{\partial v_2}{\partial y} = 0, \tag{2.1.26a}$$

where v_1 and v_2 are the horizontal and vertical components of the two-dimensional vector v; that is, $v = (v_1, v_2)$; and

(b) irrotationality (curl of the velocity vanishes),

$$\frac{\partial v_2}{\partial x} - \frac{\partial v_1}{\partial y} = 0. \tag{2.1.26b}$$

A simplification of these equations is found via the following substitutions:

$$v_1 = \frac{\partial \phi}{\partial x} = \frac{\partial \psi}{\partial y}, \qquad v_2 = \frac{\partial \phi}{\partial y} = -\frac{\partial \psi}{\partial x}. \tag{2.1.27}$$

In vector form: $\mathbf{v} = \nabla \phi$. We call ϕ the **velocity potential**, and ψ the **stream function**. Equations (2.1.26a,b) and (2.1.27) show that ϕ and ψ satisfy Laplace's equation. Because the Cauchy–Riemann conditions are satisfied for the functions ϕ and ψ, we have, quite naturally, an associated **complex velocity potential** $\Omega(z)$:

$$\Omega(z) = \phi(x, y) + i\psi(x, y). \tag{2.1.28}$$

The derivative of $\Omega(z)$ is usually called the **complex velocity**:

$$\Omega'(z) = \frac{\partial \phi}{\partial x} + i\frac{\partial \psi}{\partial x} = \frac{\partial \phi}{\partial x} - i\frac{\partial \phi}{\partial y} = v_1 - iv_2. \tag{2.1.29}$$

The complex conjugate, $\overline{\Omega'(z)} = \partial\phi/\partial x + i\partial\phi/\partial y = v_1 + iv_2$, is analogous to the usual velocity vector in two dimensions.

The associated boundary conditions are as follows. The normal derivative of ϕ

(i.e., the normal velocity) must vanish on a rigid boundary of an ideal fluid. Because we have shown that the level sets $\phi(x, y) =$ constant and $\psi(x, y) =$ constant are mutually orthogonal at any point (x, y), we conclude that the level sets of the stream function ψ follow the direction of the flow field; namely, they follow the direction of the gradients of ϕ, which are themselves orthogonal to the level sets of ϕ. Consequently, boundary conditions in an ideal flow problem at a boundary can be specified by either giving vanishing conditions on the normal derivative of ϕ at a boundary (no flow through the boundary) or by specifying that $\psi(x, y)$ is constant on a boundary, thereby making the boundary a streamline. The expression

$$\partial\varphi/\partial n = \nabla\varphi \cdot \hat{n},$$

where $\hat{n}$ is the unit normal, implies that $\nabla\varphi$ points in the direction of the tangent to the boundary. For problems with an infinite domain, some type of boundary condition – usually a boundedness condition – must be given at infinity. We usually specify that the velocity is uniform (constant) at infinity.

Briefly in this section, and more fully in subsequent sections and Chapter 5 (see Section 5.4), we shall discuss examples of fluid flows corresponding to various complex potentials. Upon considering boundary conditions, functions $\Omega(z)$ that are analytic in suitable regions may frequently be associated with two-dimensional fluid flows, though we also need to be concerned with locations of nonanalyticity of $\Omega(z)$. Some examples will clarify the situation.

Example 2.1.12   The simplest example is that of **uniform flow**,

$$\Omega(z) = v_0 e^{-i\theta_0} z = v_0 (\cos\theta_0 - i\sin\theta_0)(x + iy), \qquad (2.1.30)$$

where v_0 and θ_0 are positive real constants. Using Eqs. (2.1.28) and (2.1.29), the corresponding velocity potential and velocity field are given by

$$\phi(x, y) = v_0(\cos\theta_0 x + \sin\theta_0 y),$$

$$v_1 = \frac{\partial\phi}{\partial x} = v_0\cos\theta_0, \qquad v_2 = \frac{\partial\phi}{\partial y} = v_0\sin\theta_0,$$

which is identified with uniform flow making an angle θ_0 with the x-axis, as in Figure 2.2. Alternatively, the stream function $\psi(x, y) = v_0(\cos\theta_0 y - \sin\theta_0 x) =$ constant reveals the same flow field.

Example 2.1.13   A somewhat more complicated flow configuration, flow around a cylinder, corresponds to the complex velocity potential

$$\Omega(z) = v_0\left(z + \frac{a^2}{z}\right), \qquad (2.1.31)$$

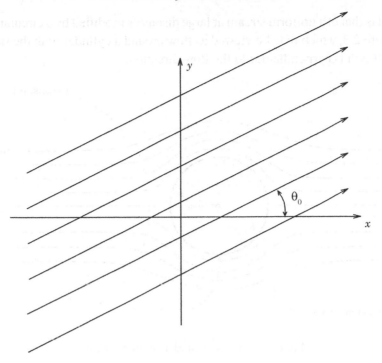

Figure 2.2 Uniform flow

where v_0 and a are positive real constants and $|z| > a$. The corresponding velocity potential and stream function are given by

$$\phi = v_0 \left(r + \frac{a^2}{r} \right) \cos \theta, \tag{2.1.32a}$$

$$\psi = v_0 \left(r - \frac{a^2}{r} \right) \sin \theta, \tag{2.1.32b}$$

and for the complex velocity we have

$$\Omega'(z) = v_0 \left(1 - \frac{a^2}{z^2} \right) = v_0 \left(1 - \frac{a^2 e^{-2i\theta}}{r^2} \right), \tag{2.1.33}$$

whereby from Eq. (2.1.29) the horizontal and vertical components of the velocity are given by

$$v_1 = v_0 \left(1 - \frac{a^2 \cos 2\theta}{r^2} \right), \qquad v_2 = -v_0 \frac{a^2 \sin 2\theta}{r^2}. \tag{2.1.34}$$

The circle $r = a$ is a streamline ($\psi = 0$) as is $\theta = 0$ and $\theta = \pi$. As $r \to \infty$, the limiting velocity is uniform in the x direction ($v_1 \to v_0, v_2 \to 0$). The corresponding flow field is that of a uniform stream at large distances modified by a circular barrier, as in Figure 2.3, which may be viewed as flow around a cylinder with the same flow field at all points perpendicular to the flow direction.

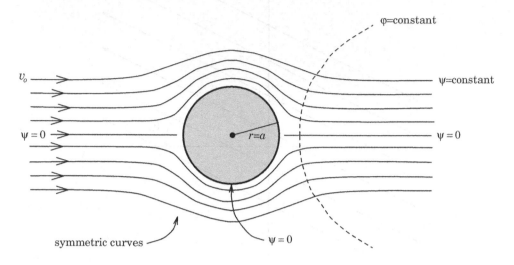

Figure 2.3 Flow around a circular barrier

Note that the velocity vanishes at $r = a$, $\theta = 0$, and $\theta = \pi$. These points are called **stagnation points** of the flow. On the circle $r = a$, which corresponds to the streamline $\psi = 0$, the normal velocity is zero because the corresponding velocity must be in the tangent direction to the circle. Another way to see this is to compute the normal velocity from ϕ using the gradient in two-dimensional polar coordinates:

$$\mathbf{v} = \nabla\phi = \frac{\partial\phi}{\partial r}\hat{\boldsymbol{u}}_r + \frac{1}{r}\frac{\partial\phi}{\partial\theta}\hat{\boldsymbol{u}}_\theta,$$

where $\hat{\boldsymbol{u}}_r$ and $\hat{\boldsymbol{u}}_\theta$ are the unit normal and tangential vectors. Thus, the velocity in the radial direction is $v_r = \frac{\partial\phi}{\partial r}$ and the velocity in the circumferential direction is $v_\theta = \frac{1}{r}\frac{\partial\phi}{\partial\theta}$. So, the radial velocity at any point (r, θ) is given by

$$\frac{\partial\phi}{\partial r} = v_0\left(1 - \frac{a^2}{r^2}\right)\cos\theta,$$

which vanishes when $r = a$. As mentioned earlier, as $r \to \infty$ the flow becomes uniform:

$$\phi \longrightarrow v_0 r\cos\theta = v_0 x,$$
$$\psi \longrightarrow v_0 r\sin\theta = v_0 y.$$

So, for large r and correspondingly large y, the curves $y = $ constant are streamlines as expected.

2.1.3 Problems for Section 2.1

1. Which of the following satisfy the Cauchy–Riemann (C–R) equations? If they satisfy the C–R equations, give the analytic function of z.

$$\text{(a) } f(x, y) = x - iy + 1;$$
$$\text{(b) } f(x, y) = y^3 - 3x^2y + i(x^3 - 3xy^2 + 2);$$
$$\text{(c) } f(x, y) = e^y(\cos x + i \sin y).$$

2. In the following we are given the real part of an analytic function of z. Find the imaginary part and the function of z.

$$\text{(a) } 3x^2y - y^3; \qquad \text{(b) } 2x(c - y), \quad c = \text{constant};$$
$$\text{(c) } \frac{y}{x^2 + y^2}; \qquad \text{(d) } \cos x \cosh y.$$

3. Determine whether the following functions are analytic. Discuss whether they have any singular points or if they are entire.

$$\text{(a) } \tan z; \qquad \text{(b) } e^{\sin z}; \qquad \text{(c) } e^{1/(z-1)}; \qquad \text{(d) } e^{\bar{z}};$$
$$\text{(e) } \frac{z}{z^4 + 1}; \qquad \text{(f) } \cos x \cosh y - i \sin x \sinh y.$$

4. Show that the real and imaginary parts of a twice-differentiable function $f(\bar{z})$ satisfy Laplace's equation. Show that $f(\bar{z})$ is nowhere analytic unless it is constant.

5. Let $f(z)$ be analytic in some domain. Show that $f(z)$ is necessarily a constant if either the function $\overline{f(z)}$ is analytic or $f(z)$ assumes only pure imaginary values in the domain.

6. Consider the following complex potential

$$\Omega(z) = -\frac{k}{2\pi}\frac{1}{z}, \qquad \text{for real } k,$$

referred to as a "doublet." Calculate the corresponding velocity potential, stream function, and velocity field. Sketch the stream function. The value of k is usually called the strength of the doublet. See also Problem 4 of Section 2.3.1, in which we obtain this complex potential via a limiting procedure of two elementary flows, referred to as a "source" and a "sink."

7. Consider the complex analytic function, $\Omega(z) = \phi(x, y) + i\psi(x, y)$, in a domain D. Let us transform from z to w using $w = f(z)$, $w = u + iv$, where $f(z)$ is

analytic in D, with the corresponding domain in the w-plane, D'. Assuming necessary differentiability, establish the following:

$$\frac{\partial \phi}{\partial x} = \frac{\partial u}{\partial x}\frac{\partial \phi}{\partial u} + \frac{\partial v}{\partial x}\frac{\partial \phi}{\partial v},$$

$$\frac{\partial^2 \phi}{\partial x^2} = \frac{\partial^2 u}{\partial x^2}\frac{\partial \phi}{\partial u} - \frac{\partial^2 u}{\partial x \partial y}\frac{\partial \phi}{\partial v} + \left(\frac{\partial u}{\partial x}\right)^2 \frac{\partial^2 \phi}{\partial u^2} - 2\frac{\partial u}{\partial x}\frac{\partial u}{\partial y}\frac{\partial^2 \phi}{\partial u \partial v}$$

$$+ \left(\frac{\partial u}{\partial y}\right)^2 \frac{\partial^2 \phi}{\partial v^2}.$$

Also find the corresponding formulae for $\partial \phi / \partial y$ and $\partial^2 \phi / \partial y^2$. Recall that $f'(z) = \frac{\partial u}{\partial x} - i\frac{\partial u}{\partial y}$, and $u(x, y)$ satisfies Laplace's equation in the domain D. Show that

$$\nabla^2_{x,y}\phi = \frac{\partial^2 \phi}{\partial x^2} + \frac{\partial^2 \phi}{\partial y^2} = \left(u_x^2 + u_y^2\right)\left(\frac{\partial^2 \phi}{\partial u^2} + \frac{\partial^2 \phi}{\partial v^2}\right)$$

$$= |f'(z)|^2 \, \nabla^2_{u,v}\phi.$$

Consequently, we find that if ϕ satisfies Laplace's equation $\nabla^2_{x,y}\phi = 0$ in the domain D, then as long as $f'(z) \neq 0$ in D, it also satisfies Laplace's equation $\nabla^2_{u,v}\phi = 0$ in the domain D'.

8. Given the complex analytic function $\Omega(z) = z^2$, show that the real part of Ω, $\phi(x, y) = \mathrm{Re}\,\Omega(z)$, satisfies Laplace's equation, $\nabla^2_{x,y}\phi = 0$. Let $z = (1 - w)/(1 + w)$, where $w = u + iv$. Show that $\phi(u, v) = \mathrm{Re}\,\Omega(w)$ satisfies Laplace's equation $\nabla^2_{u,v}\phi = 0$.

9. Let

$$P(z) = (z - z_1)(z - z_2)\cdots(z - z_n)$$

be a polynomial where $z_1, z_2, \ldots, z_n$ are the distinct roots of $P(z)$.

(a) Show that

$$\frac{P'(z)}{P(z)} = \frac{1}{z - z_1} + \frac{1}{z - z_2} + \cdots + \frac{1}{z - z_n}, \qquad z \neq z_j, \quad j = 1, 2, \ldots, n.$$

(b) Suppose $\mathrm{Re}\, z_j < 0$ for all $j = 1, 2, \ldots, n$, and $\mathrm{Re}\, z \geq 0$. Use $\frac{1}{\beta} = \frac{\bar{\beta}}{|\beta|^2}$, for any complex number β, to establish that $\mathrm{Re}\left(\frac{1}{z - z_j}\right) > 0$.

(c) Use part (b) to show that for $\mathrm{Re}\, z > 0$ it follows that $\mathrm{Re}\, P'(z) > 0$. Hence conclude that all the zeroes of $P'(z)$ must lie in the left half plane $\mathrm{Re}\, z < 0$.

2.2 Multivalued Functions

A single-valued function $w = f(z)$ yields one value w for a given complex number z. We say a 'multivalued function' admits more than one value w for a given z.

Indeed the nomenclature is at first glance strange since earlier in Chapter 1 we stated that a function has a unique value. Here we add the term 'multivalued' to distinguish this new function. This is the standard usage in the literature so we will continue using it. Such functions are more complicated and frequently require a great deal of care. Indeed it wasn't until Riemann's efforts that there was a deep understanding of them.

Multivalued functions are naturally introduced as the inverse of single-valued functions. The simplest one is the square root function. If we consider $z = w^2$, the inverse is written as

$$w = z^{1/2}. \tag{2.2.1}$$

From real variables we know that $x^{1/2}$ has two values, often written as $\pm\sqrt{x}$ where $\sqrt{x} \geq 0$. For the complex function (Eq. (2.2.1)) and from $w^2 = z$, we can ascertain the multivaluedness by letting $z = re^{i\theta}$, and $\theta = \theta_p + 2\pi n$, where, say, $0 \leq \theta_p < 2\pi$:

$$w = r^{1/2}e^{i\theta_p/2}e^{n\pi i}, \tag{2.2.2}$$

where $r^{1/2} \equiv \sqrt{r} \geq 0$ and n is an integer. (See also the discussion in Section 1.1.) For a given value z, the function $w(z)$ is multivalued; for the same value of z it takes two possible values corresponding to n even and n odd, namely

$$\sqrt{r}e^{i\theta_p/2} \quad \text{and} \quad \sqrt{r}e^{i\theta_p/2}e^{i\pi} = -\sqrt{r}e^{i\theta_p/2}.$$

An important consequence of the multivaluedness of w is that as z traverses a small circuit around $z = 0$, w does not return to its original value. Indeed, suppose we start at $z = \epsilon$ for real $\epsilon > 0$. Let us see what happens to w as we return to this point after going around a circle with radius ϵ. Let $n = 0$. When we start, $\theta_p = 0$ and $w = \sqrt{\epsilon}$; when we return to $z = \epsilon$, $\theta_p = 2\pi$ and $w = \sqrt{\epsilon}e^{\frac{2i\pi}{2}} = -\sqrt{\epsilon}$. We note that the value $-\sqrt{\epsilon}$ can also be obtained from $\theta_p = 0$ provided we take $n = 1$. In other words, we started with a value w corresponding to $n = 0$ and ended up with a value w corresponding to $n = 1$! (Any even/odd values of n suffice for this argument.) The point $z = 0$ is called a **branch point**. A point is a branch point if the multivalued function $w(z)$ is discontinuous upon traversing a small circuit around this point. It should be noted that the point $z = \infty$ is also a branch point. This is seen by using the transformation $z = 1/t$, which maps $z = \infty$ to $t = 0$. Using arguments such as that above, it follows that $t = 0$ is a branch point of the function $t^{-1/2}$, and hence $z = \infty$ is a branch point of the function $z^{1/2}$. The points $z = 0$ and $z = \infty$ are the only branch points of the function $z^{1/2}$. Indeed, if we take a closed circuit C (see Figure 2.4) which does *not* enclose $z = 0$ or $z = \infty$, then $z^{1/2}$ returns to its original value as z traverses C. Along C the phase θ will vary continuously between $\theta = \theta_R$ and $\theta = \theta_L$. So if we begin at $z_R = r_R e^{i\theta_R}$ and follow the curve C, the value z will return to exactly its previous value with no phase change. Hence $z^{1/2}$ will not have a jump as the curve C is traversed.

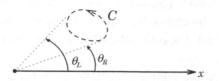

Figure 2.4 Closed circuit away from branch cut

The analytic study of multivalued functions usually is best effected by expressing the multivalued function in terms of a single-valued function. One method of doing this is to consider the multivalued function in a restricted region of the plane, and choose a value at every point such that the resulting function is single-valued and continuous. A continuous function obtained from a multivalued function in this way is called a **branch** of the multivalued function. For the function $w = z^{1/2}$ we can carry out this procedure by taking $n = 0$ and restricting the region of z to be the open or cut plane in Figure 2.5. For this purpose the real positive axis in the z-plane is cut out. The values of $z = 0$ and $z = \infty$ are also deleted. The function $w = z^{1/2}$ is now continuous in the cut plane which is an open region. The semiaxis Re $z > 0$ is referred to as a **branch cut**. Across the branch cut the function $w = z^{1/2}$ is discontinuous; hence each point on Re $z > 0$ would be a singular point.

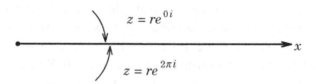

Figure 2.5 Cut plane, $z^{1/2}$

It should be noted that the location of the branch cut is arbitrary save that it ends at branch points. If we restrict θ_p to $-\pi \le \theta_p < \pi$, $n = 0$ in the polar representation of $z = re^{i\theta}$, $\theta = \theta_p + 2n\pi$, then the branch cut would naturally be on the negative real axis. More complicated curves (e.g. spirals) could equally well be chosen as branch cuts but rarely do we do so because a cut is chosen for convenience. The simplest choice (sometimes motivated by physical application) is generally satisfactory. We reiterate that the main purpose of a branch cut is to artificially create a region in which the function is single-valued and continuous. In fact, when we consider integrals with multivalued functions, such as those that arise in solutions of partial differential equations via transform methods (cf. Section 4.6), we introduce branch cuts.

On the other hand we note that if we took a closed circuit that didn't enclose the branch point $z = 0$, then the function $z^{1/2}$ would return to its same value. We depicted in Figure 2.4 a typical closed circuit C not enclosing the origin, with the choice of branch cut ($z = re^{i\theta}, 0 \le \theta < 2\pi$) on the positive real axis.

Note that if we had chosen $w = (z - z_0)^{1/2}$ as our prototype example, a (finite) branch point would have been at $z = z_0$. Similarly, if we had investigated $w = (az + b)^{1/2}$ a, b constants, $a \ne 0$, then a (finite) branch point would have been at $-b/a$. (In either case, $z = \infty$ would be another branch point.) We could deduce these facts by translating to a new origin in our coordinate system and investigating the change upon a circuit around the branch point, namely letting $z = z_0 + re^{i\theta}$, $0 \le \theta < 2\pi$. We shall see that multivalued functions can be considerably more exotic than the ones described above.

A somewhat more complicated situation is illustrated by the inverse of the exponential function; that is, the logarithm (see Figure 2.6). Consider

$$z = e^w. \tag{2.2.3}$$

Let $w = u + iv$. Using the properties of the exponential function, we have

$$z = e^{u+iv} = e^u e^{iv} = e^u(\cos v + i \sin v); \tag{2.2.4a}$$

in polar coordinates, $z = re^{i\theta_p}$ for $0 \le \theta_p < 2\pi$, so

$$r = e^u, \qquad v = \theta_p + 2\pi n, \quad \text{for integer } n. \tag{2.2.4b}$$

From the properties of real variables,

$$u = \log r.$$

Thus, in analogy with real variables, we write $w = \log z$, which is

$$w = \log z = \log r + i\theta_p + 2n\pi i, \tag{2.2.4c}$$

where $n = 0, \pm 1, \pm 2, \ldots$, and where θ_p takes on values in a particular range of 2π. Here we take

$$0 \le \theta_p < 2\pi.$$

When $n = 0$, Eq. (2.2.4a) is frequently referred to as the principal branch of the logarithm; the corresponding value of the function is referred to as the **principal value**. From (2.2.4a–c) we see that, as opposed to the square root example, the function is infinitely valued; that is, n takes on an infinite number of integer values. For example, if $z = i$, then $|z| = r = 1$, $\theta_p = \pi/2$; hence

$$\log i = \log 1 + i\left(\frac{\pi}{2} + 2n\pi\right) \qquad n = 0, \pm 1, \pm 2, \ldots. \tag{2.2.5}$$

Similarly if $z = x$, a real positive quantity, $|z| = r = |x|$, then

$$\log z = \log |x| + 2n\pi i \qquad n = 0, \pm 1, \pm 2, \ldots . \qquad (2.2.6)$$

The complex logarithm function differs from the real logarithm by additive multiples of $2\pi i$. If z is real and positive, we normally take $n = 0$ so that the principal branch of the complex logarithm function agrees with the usual one for real variables.

Suppose we consider a given point $z = x_0$, x_0 real and positive, and fix a branch of $\log z$, $n = 0$. So $\log z = \log |x_0|$. Let us now allow z to vary on a circle about $z = 0$: $z = |x_0|e^{i\theta}$. As θ varies from $\theta = 0$ to $\theta = 2\pi$ the value of $\log z$ varies from $\log |x_0|$ to $\log |x_0| + 2\pi i$. Thus we see that $z = 0$ is a branch point: a small circuit (x_0 can be as small as we wish) about the origin results in a change in $\log z$. Indeed we see that after one circuit we come to the $n = 1$ branch of $\log z$. The next circuit would put us on the $n = 2$ branch of $\log z$ and so on. The function $\log z$ is thus seen to be infinitely branched, and the line Re $z > 0$ is a branch cut (see Figure 2.6).

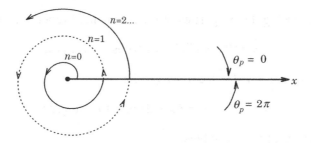

Figure 2.6 Logarithm function and branch cut

We reiterate that the branch cut Re $z > 0$ is arbitrarily chosen, although in a physical problem a particular choice might be indicated. Had we defined $\log z$ as

$$\log z = \log |x_0| + i(\theta_p + 2n\pi), \qquad -\pi \leq \theta_p < \pi, \qquad (2.2.7)$$

this would be naturally related to values of $\log z$ that have a jump on the negative real axis. So $n = 0$, $\theta_p = -\pi$ corresponds to $\log z = \log |x_0| - i\pi$. A full circuit in the counterclockwise direction puts us on the first branch $\log z = \log |x_0| + i\pi$ (see Figure 2.7). As opposed to taking $0 \leq \theta < 2\pi$, sometimes the principal branch of the logarithm is defined by equation (2.2.7) with $n = 0$.

It should be noted that the point $z = \infty$ is also a branch point for $\log z$. As we have seen, the point at infinity is easily understood via the transformation $z = 1/t$, so that t near zero corresponds to z near ∞. The above arguments, which are used to establish whether a point is in fact a branch point, apply at $t = 0$. The use of this transformation and the properties of $\log |z|$ yields $\log z = \log 1/t = -\log t$. We establish $t = 0$ as a branch point by letting $t = re^{i\theta}$, varying θ by 2π, and noting that this function does not return to its original values.

It is convenient to visualize the branch cut as joining the two branch points $z = 0$ and $z = \infty$. For those who studied the stereographic projection (Section 1.2.2), this branch cut is a (great circle) curve joining the south ($z = 0$) and the north ($z = \infty$) poles (see Figure 2.8).

The analyticity of $\log z$ ($z \neq 0$) in the cut plane can be established using the Cauchy–Riemann conditions. We shall also show the important relationship $d/dz(\log z) = 1/z$. Using (2.2.3) and (2.2.4a,b,c), we see that for $z = x + iy$, $w = \log z$, $w = u + iv$,

$$e^{2u} = x^2 + y^2, \qquad \tan v = \frac{y}{x}. \qquad (2.2.8)$$

Note that in deriving Eq. (2.2.8) we use $w = \log[|z|e^{i \arg z}]$, $|z| = (x^2 + y^2)^{1/2}$, and $\theta = \arg z = \tan^{-1} y/x$. A branch is fixed by assigning suitable values for the *real* functions u and v. The function u is given by

$$u = \frac{1}{2} \log(x^2 + y^2). \qquad (2.2.9)$$

To fix the branch of v corresponding to the inverse tangent of y/x is more subtle. Suppose we fix $\tan^{-1}(y/x)$ to be the standard real-valued function taking values between $-\pi/2$ and $\pi/2$; that is,

$$\frac{-\pi}{2} \leq \tan^{-1}(y/x) < \frac{\pi}{2}.$$

Thus the value of v will have a jump whenever x passes through zero (e.g. a jump of π when we pass from the first to the second quadrant).

Alternatively we could have written

$$v = \tan^{-1}\left(\frac{y}{x}\right) + C_i, \qquad (2.2.10)$$

with $C_1 = 0, C_2 = C_3 = \pi, C_4 = 2\pi$, where the values of the constant C_i correspond to suitable values in each of the four quadrants.

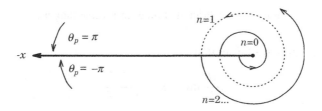

Figure 2.7 Logarithm function and alternative branch cut

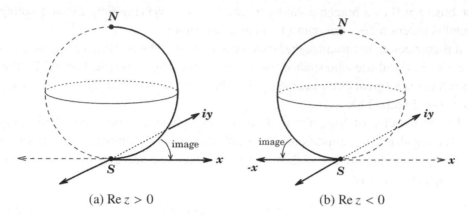

(a) Re $z > 0$ (b) Re $z < 0$

Figure 2.8 Branch cuts, stereographic projection

It can be verified that v is continuous in the z-plane apart from Re $z > 0$ where there is a jump of 2π across the Re $z > 0$ axis. Figure 2.9 depicts the choice of $v = \tan^{-1}(y/x)$ which will make $\log z$ continuous off the positive real axis. In this way v can be associated with the angle θ:

$$w = \log z = \log re^{i\theta} = \log r + i\theta, \qquad 0 \le \theta \le 2\pi.$$

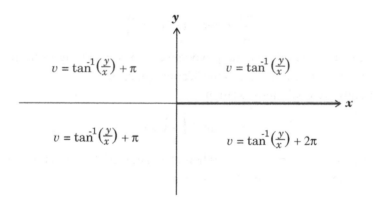

Figure 2.9 A branch choice for inverse tangent

From real variables we know that

$$\frac{d}{ds} \tan^{-1} s = \frac{1}{1 + s^2} \tag{2.2.11}$$

and with this, from Eq. (2.2.10), we can verify that the Cauchy–Riemann conditions are satisfied for Eq. (2.2.8). The partial derivatives of u and v are given by

$$u_x = \frac{x}{x^2 + y^2}, \qquad u_y = \frac{y}{x^2 + y^2}, \qquad (2.2.12a)$$

$$v_x = \frac{-y}{x^2 + y^2}, \qquad v_y = \frac{x}{x^2 + y^2}, \qquad (2.2.12b)$$

hence the Cauchy–Riemann conditions $u_x = v_y$ and $u_y = -v_x$ are satisfied and the function $\log z$ is analytic in the cut plane $\operatorname{Re} z > 0$ (as implied by the properties of the inverse tangent function). Alternatively we could have used $u = \log r$, $v = \theta$ and Eq. (2.1.8).

Because $\log z$ is analytic in the cut plane, its derivative can be easily calculated. We need only to calculate the derivative along the x direction

$$\frac{d}{dz} \log z = \frac{\partial u}{\partial x} + i\frac{\partial v}{\partial x} = \frac{x - iy}{x^2 + y^2} = \frac{1}{x + iy} = \frac{1}{z}. \qquad (2.2.13)$$

Hence the expected result is obtained for the derivative of $\log z$ in a cut plane. Indeed, this development can be carried out for any of the branches (suitable cut planes) of $\log z$. Alternatively, from (2.1.10) we have

$$f'(z) = e^{-i\theta}\frac{\partial}{\partial r}(\log r) = \frac{1}{re^{i\theta}} = \frac{1}{z}.$$

In Section 2.3 we discuss multisheeted Riemann surfaces where the $\log z$ is analytic everywhere except at the branch points $z = 0, \infty$.

The generalized power function is defined in terms of the logarithm

$$z^a = e^{a \log z}, \qquad (2.2.14)$$

for any complex constant a. When $a = m$, an integer, the power function is simply z^m. Using Eq. (2.2.4a) and $e^{2k\pi i} = \cos 2k\pi + i \sin 2k\pi = 1$, where k is an integer, we have

$$z^m = e^{m[\log r + i(\theta_p + 2\pi n)]} = e^{m \log r}e^{mi\theta_p} = (re^{i\theta_p})^m,$$

whereupon we have the usual integer power function with no branching and no branch points. If, however, a is a rational number, i.e.,

$$a = \frac{m}{l},$$

where m and l are integers with no common factor, then we have

$$z^{m/l} = \exp\left[\frac{m}{l}(\log r + i(\theta_p + 2\pi n))\right]$$

$$= \exp\left[\frac{m}{l}(\log r + i\theta_p)\right]\exp\left[2\pi i\left(\frac{mn}{l}\right)\right]. \qquad (2.2.15)$$

It is evident that when $n = 0, 1, \ldots, (l - 1)$, the expression (2.2.15) takes on different values corresponding to the term $e^{2\pi i(mn/l)}$. Thus $z^{(m/l)}$ takes on l different values. If n increases beyond $n = l-1$, say $n = l, (l+1), \ldots, (2l-1)$, the above values are correspondingly repeated, and so on. The formula (2.2.15) yields l branches for the function $z^{m/l}$. The function $z^{m/l}$ has branch points at $z = 0$ and $z = \infty$. Similar considerations apply to the function $w = (z - z_0)^{m/l}$ with a (finite) branch point now being located at $z = z_0$. A cut plane can be fixed by choosing θ_p appropriately. Hence a branch cut on Re $z > 0$ is fixed by requiring $0 \leq \theta_p < 2\pi$. Similarly a cut for Re $z < 0$ is fixed by assigning $-\pi \leq \theta_p < \pi$. Thus, if $m = 1, l = 4$, the formula (2.2.15) yields four branches of the function $z^{1/4}$.

Values of a which are neither integer nor rational result in functions that are infinitely branched with branch points at $z = 0$, $z = \infty$. Branch cuts can be defined via choices of θ_p as above. For any suitable branch, standard differentiation formulae give

$$\frac{d}{dz} z^a = \frac{d}{dz} e^{a \log z} = z^a \left(\frac{a}{z}\right) = a z^{a-1}. \tag{2.2.16}$$

From Eq. (2.2.4a–c) we also have

$$\begin{aligned} \log(z_1 z_2) &= \log \left(r_1 e^{i\theta_1} r_2 e^{i\theta_2}\right) \\ &= \log r_1 r_2 + i(\theta_{1p} + \theta_{2p}) + 2n\pi i \\ &= \log r_1 + i(\theta_{1p} + 2n_1\pi) + \log r_2 + i(\theta_{2p} + 2n_2\pi) \\ &= \log z_1 + \log z_2, \end{aligned}$$

where $n_1 + n_2 = n$. The other standard algebraic properties of the complex logarithm, which are analogous to the real logarithm, follow in a similar manner.

The inverse of trigonometric and hyperbolic functions can be computed via logarithms. It is another step in complication regarding multivalued functions. For example,

$$w = \cos^{-1} z \tag{2.2.17}$$

satisfies

$$\cos w = z = \frac{e^{iw} + e^{-iw}}{2}.$$

Thus,

$$e^{2iw} - 2z e^{iw} + 1 = 0. \tag{2.2.18}$$

Hence solving this quadratic equation for e^{iw} yields

$$e^{iw} = z + (z^2 - 1)^{\frac{1}{2}} = z + i(1 - z^2)^{\frac{1}{2}}$$

and then

$$w(z) = -i \log \left(z + i(1 - z^2)^{\frac{1}{2}}\right). \tag{2.2.19}$$

This function $w(z)$ has two sources of multivaluedness; one due to the logarithm, the other due to $f(z) = (1 - z^2)^{\frac{1}{2}}$. The function $f(z)$ has two branches and two branch points, at $z = \pm 1$. We can deduce that $z = \pm 1$ are branch points of $f(z)$ by investigating the local behavior of $f(z)$ near the points $z = \pm 1$. Namely, use $z = 1 + r_1 e^{i\theta_1}$ and $z = -1 + r_2 e^{i\theta_2}$ for small values of r_1 and r_2. For, say, $z = -1$, we have $f(z) \approx (2r_2)^{1/2} e^{i\theta_2/2}$ (dropping r_2^2 terms as much smaller than r_2), which certainly has a discontinuity as θ_2 changes by 2π. The function $f(z)$ has two branches. The log function has an infinite number of branches, hence so does w; sometimes we say that $w(z)$ is doubly infinite because for each of the infinity of branches of the log we also have two branches of $f(z)$. In the finite plane the only branch points of $w(z)$ are at $z = \pm 1$ because the function $g(z) = z + i(1 - z^2)^{1/2}$ has no solutions of $g(z) = 0$. (Equating both sides, $z = -i(1 - z^2)^{1/2}$ leads to a contradiction.) The branch structure of $w(z)$ in (2.2.19) is discussed further in Section 2.3 (cf. Eq. (2.3.8)).

Because the log function is determined up to additive multiples of $2\pi i$, it follows that for a *fixed* value of $(1 - z^2)^{1/2}$, and a particular branch of the log function, $w = \cos^{-1} z$ is determined only to within multiples of 2π. Namely, if we write $w_1 = -i \log \left(z + i(1 - z^2)^{1/2} \right)$ for a particular branch, then the general form for w satisfies

$$w = -i \log \left(z + i(1 - z^2)^{1/2} \right) + 2n\pi$$

or $w = w_1 + 2n\pi$, with n integer, which expresses the periodicity of the cosine function. Similarly, from the quadratic equation (2.2.18) we have that the product of the two roots e^{iw_1} and e^{iw_2} satisfy

$$e^{iw_1} e^{iw_2} = 1; \tag{2.2.20}$$

or, by taking the logarithm of Eq. (2.2.20) with $1 = e^{i0}$ or $1 = e^{2\pi i}$, we see that the two solutions of Eq. (2.2.18) are simply related:

$$w_1 + w_2 = 0 \qquad \text{or} \qquad w_1 + w_2 = 2\pi, \quad \text{etc.,} \tag{2.2.21}$$

which reflects the fact that the cosine of an angle, say α, equals the cosine of $-\alpha$ or the cosine of $2\pi - \alpha$, etc.

Differentiation establishes the relationship

$$\begin{aligned}
\frac{d}{dz} \cos^{-1} z &= \frac{-i}{z + i(1 - z^2)^{1/2}} \left(1 - \frac{iz}{(1 - z^2)^{1/2}} \right) \\
&= \frac{-i}{z + i(1 - z^2)^{1/2}} (-i) \frac{\left(z + i(1 - z^2)^{1/2} \right)}{(1 - z^2)^{1/2}} \\
&= \frac{-1}{(1 - z^2)^{1/2}}, \tag{2.2.22}
\end{aligned}$$

for $z^2 \neq 1$. Formulae for the other inverse trigonometric and hyperbolic functions can be established in a similar manner. For reference we list some of them below:

$$\sin^{-1} z = -i \log \left(iz + (1 - z^2)^{1/2} \right), \tag{2.2.23a}$$

$$\tan^{-1} z = \frac{1}{2i} \log \frac{i - z}{i + z}, \tag{2.2.23b}$$

$$\sinh^{-1} z = \log \left(z + (1 + z^2)^{1/2} \right), \tag{2.2.23c}$$

$$\cosh^{-1} z = \log \left(z + (z^2 - 1)^{1/2} \right), \tag{2.2.23d}$$

$$\tanh^{-1} z = \frac{1}{2} \log \frac{1 + z}{1 - z}. \tag{2.2.23e}$$

In the following section we will examine a deeper issue: Riemann surfaces. Multivalued functions when viewed on a suitable Riemann surface are analytic everywhere except at their branch points. For example, the functions $z^{1/2}, \log z$ on their Riemann surfaces are analytic everywhere except at the branch points: $z = 0, \infty$. We will also discuss the branch structure of more complicated functions such as $((z - a)(z - b))^{1/2}$ and $\cos^{-1} z$.

In Section 2.1 we mentioned that the real and imaginary parts of an analytic function in a domain D satisfy Laplace's equation in D. In fact, some simple complex functions yield fundamental and physically important solutions to Laplace's equation.

For example, consider the function

$$\Omega(z) = A \log z + iB \tag{2.2.24}$$

where A and B are real and we take the branch cut of the logarithm along the real axis with $z = re^{i\theta}$, $0 \leq \theta < 2\pi$. The imaginary part of $\Omega(z)$, where $\Omega(z) = \phi(x, y) + i\psi(x, y)$, satisfies Laplace's equation:

$$\nabla^2 \psi = \frac{\partial^2 \psi}{\partial x^2} + \frac{\partial^2 \psi}{\partial y^2} = 0 \tag{2.2.25}$$

in the upper half plane, i.e., in $-\infty < x < \infty$, where $y > 0$. From Eq. (2.2.24) a solution of Laplace's equation is

$$\psi(x, y) = A\theta + B$$

$$= A \tan^{-1} \left(\frac{y}{x} \right) + B, \tag{2.2.26}$$

where $\tan^{-1}(y/x)$ stands for the identifications in Eq. (2.2.10) (see Figure 2.9). Thus, for $y > 0$, we have $0 < \tan^{-1}(y/x) < \pi$.

Note that as $y \to 0^+$, then $\theta = \tan^{-1}(y/x) \to 0$ for $x > 0$, and $\to \pi$ for $x < 0$. Taking $B = 1$ and $A = -1/\pi$, we have that

$$\psi(x, y) = 1 - \frac{1}{\pi} \tan^{-1} \left(\frac{y}{x} \right) \tag{2.2.27}$$

is the solution of Laplace's equation in the upper half plane bounded at infinity, corresponding to the boundary conditions

$$\psi(x,0) = \begin{cases} 1 & \text{for } x > 0, \\ 0 & \text{for } x < 0. \end{cases} \tag{2.2.28}$$

Physically speaking, Eq. (2.2.27) corresponds to the steady-state heat distribution of a plate with the prescribed temperature distribution, Eq. (2.2.28), on the bottom of the plate (steady-state heat flow satisfies Laplace's equation).

We also mention briefly that in many applications it is useful to employ suitable transformations which have the effect of transforming Laplace's equation in a complicated domain to a "simple" one, that is, one for which Laplace's equation can be easily solved such as in a half plane or inside a circle. In terms of two-dimensional ideal fluid flow, this means that a flow in a complicated domain would be converted to one in a simpler domain under the appropriate transformation of variables. (A number of physical applications are discussed in Chapter 5.)

The essential idea is the following. Suppose we are given a complex analytic function in a domain D:

$$\Omega(z) = \phi(x, y) + i\psi(x, y),$$

where ϕ and ψ satisfy Laplace's equation in D. Let us transform to a new independent complex variable w, where $w = u + iv$, via the transformation

$$z = F(w), \tag{2.2.29}$$

where $F(w)$ is analytic in the corresponding domain D' in the (u, v)-plane. Then $\Omega(F(w))$, which we shall call $\Omega(w)$, where

$$\Omega(w) = \phi(u, v) + i\psi(u, v)$$

is also analytic in D'. Hence the function ϕ and ψ will satisfy Laplace's equation in D'. (A direct verification of this statement is included in the problems section; see Problem 7 in Section 2.1.3.) For this transformation to be useful, D' must be a simplified domain in which Laplace's equation is easily solved.

The complication inherent in this procedure is that of returning back from the w-plane to the z-plane in order to obtain the required solution $\Omega(z)$, or $\phi(x, y)$ and $\psi(x, y)$. We must invert Eq. (2.2.29) to find w as a function of z. In general, this introduces multivaluedness, which we shall discuss in Section 2.3. From a general point of view we can deduce where the "difficulties" in the transformation occur by examining the derivative of the function $\Omega(w)$. We denote the inverse of the transformation (2.2.29) by

$$w = f(z), \tag{2.2.30}$$

where $f(z)$ is assumed to be analytic in D. By the chain rule, we find that

$$\frac{d\Omega}{dw} = \frac{d\Omega}{dz}\frac{dz}{dw} = \frac{d\Omega}{dz} \bigg/ \frac{dw}{dz} = \frac{d\Omega}{dz} \bigg/ \frac{df(z)}{dz}. \tag{2.2.31}$$

Consequently, $\Omega(w)$ will be an analytic function of w in D' as long as there are no points in the w-plane which correspond to points in the z-plane via Eq. (2.2.30), where $df/dz = 0$.

In Chapter 5 we shall discuss in considerable detail transformations or mappings of the form of Eqs. (2.2.29)–(2.2.30). There it will be shown that if two curves intersect at a point z_0, then their angle of intersection is preserved by the mapping (i.e., the angle of intersection in the z-plane equals the angle between the corresponding images of the intersecting curves in the w-plane) as long as $f'(z_0) \neq 0$. Such mappings are referred to as **conformal mappings**, and as mentioned above they are important for applications.

A simple example where a multivalued function arises is ideal fluid flow discussed in Section 2.1. Consider the complex flow potential given by

$$\Omega(z) = z^2. \tag{2.2.32}$$

As discussed in Section 2.1, the streamlines correspond to the imaginary part of $\Omega(z) = \phi + i\psi$, hence

$$\psi = r^2 \sin 2\theta = 2xy. \tag{2.2.33}$$

Clearly, the streamline $\psi = 0$ corresponds to the edges of the quarter plane, $\theta = 0$ and $\theta = \pi/2$ (see Figure 2.10) and the streamlines of the flow inside the quarter plane are the hyperbolae $xy = $ const.

On the other hand, we can introduce the transformation

$$z = w^{1/2}, \tag{2.2.34}$$

which converts the flow configuration $\Omega = z^2$ to the "standard" problem

$$\Omega(z(w)) = w \tag{2.2.35}$$

discussed in Section 2.1. This equation corresponds to uniform straight line flow (see Eq. (2.1.30) with $v_0 = 1$ and $\theta_0 = 0$). Equation (2.2.35) may be viewed as describing a uniform flow over a flat plate with $w = u + iv$, with the boundary streamline $v = 0$ (see Figure 2.11). The speed of the flow is $|\Omega'(z)| = 2|z| = 2r$, which can also be obtained from Eq. (2.2.35) via $\dfrac{d\Omega}{dz} = \dfrac{d\Omega}{dw}\dfrac{dw}{dz}$.

The transformation (2.2.34) is an elementary example of conformal mapping. In Chapter 5 we will discuss conformal mappings in detail.

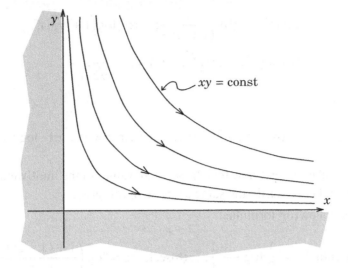

Figure 2.10 Flow configuration corresponding to $\Omega(z) = z^2$

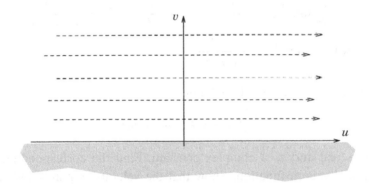

Figure 2.11 Uniform flow

2.2.1 Problems for Section 2.2

1. Find the location of the branch points and discuss possible branch cuts for the following functions.

$$\text{(a)} \; \frac{1}{(z-1)^{1/2}}; \quad \text{(b)} \; (z+1-2i)^{1/4}; \quad \text{(c)} \; 2\log z^2; \quad \text{(d)} \; z^{\sqrt{2}}.$$

2. Determine all possible values and give the principal value of the following numbers (put in the form $x + iy$).

(a) $i^{1/2}$;   (b) $\dfrac{1}{(1+i)^{1/2}}$;   (c) $\log(1 + \sqrt{3}i)$;

(d) $\log i^3$;   (e) $i^{\sqrt{3}}$;   (f) $\sin^{-1} \dfrac{1}{\sqrt{2}}$.

3. Solve for z:

(a) $z^5 = 1$;   (b) $3 + 2e^{(z-i)} = 1$;   (c) $\tan z = 1$;   (d) $\log z = i\dfrac{\pi}{2}$.

4. Let α be a real number. Show that the set of all values of the multivalued function $\log(z^\alpha)$ is not necessarily the same as those of $\alpha \log(z)$.

5. Derive the following formulae:

$$\text{(a) } \coth^{-1} z = \frac{1}{2} \log \frac{z+1}{z-1}; \quad \text{(b)} \operatorname{sech}^{-1} z = \log \left(\frac{1 + (1 - z^2)^{1/2}}{z} \right).$$

6. Deduce the following derivative formulae:

$$\text{(a) } \frac{d}{dz} \tan^{-1} z = \frac{1}{1 + z^2}; \quad \text{(b) } \frac{d}{dz} \sin^{-1} z = \frac{1}{(1 - z^2)^{1/2}};$$

$$\text{(c) } \frac{d}{dz} \sinh^{-1} z = \frac{1}{(1 + z^2)^{1/2}}.$$

7. Consider the complex velocity potential

$$\Omega(z) = k \log(z - z_0),$$

where k is real and z_0 a complex constant. Find the corresponding velocity potential and stream function. Show that the velocity is purely radial relative to the point $z = z_0$, and sketch the flow configuration. Such a flow is called a "source" if $k > 0$ and a "sink" if $k < 0$. The strength M is defined as the outward rate of flow of fluid, with unit density, across a circle enclosing $z = z_0$: $M = \oint_C V_r ds$, where V_r is the radial velocity, and ds the increment of arc length in the direction tangent to the circle C. Show that $M = 2\pi k$. (See also Section 2.1.2.)

8. Consider the complex velocity potential $\Omega(z) = -ik \log(z - z_0)$, where k is real. Find the corresponding velocity potential and stream function. Show that the velocity is purely circumferential relative to the point $z = z_0$, being counterclockwise if $k > 0$. Sketch the flow configuration. The strength of this flow, called a **point vortex**, is defined to be $M = \oint_C V_\theta ds$, where V_θ is the velocity in the circumferential direction, and ds the increment of arc length in the direction tangent to the circle C. Show that $M = 2\pi k$. (See also Section 2.1.2.)

9. (a) Show that the solution to Laplace's equation $\nabla^2 T = \partial^2 T/\partial u^2 + \partial^2 T/\partial v^2 = 0$
in the region $-\infty < u < \infty$, $v > 0$, with the boundary conditions $T(u,0) = T_0$
if $u > 0$ and $T(u,0) = -T_0$ if $u < 0$, is given by

$$T(u,v) = T_0\left(1 - \frac{2}{\pi}\tan^{-1}\frac{v}{u}\right).$$

(b) We shall use the result of part (a) to solve Laplace's equation inside a circle
of radius $r = 1$ with the boundary conditions

$$T(r,\theta) = \begin{cases} T_0 & \text{on } r = 1, \quad 0 < \theta < \pi; \\ -T_0 & \text{on } r = 1, \quad \pi < \theta < 2\pi. \end{cases}$$

Show that the transformation

$$w = i\left(\frac{1-z}{1+z}\right) \quad \text{or} \quad z = \frac{i-w}{i+w},$$

where $w = u + iv$, maps the interior of the circle $|z| = 1$ onto the upper half
of the w-plane $(-\infty < u < \infty$, $v > 0)$, and maps the boundary conditions
$r = 1$, $0 < \theta < \pi$ onto $0 < u < \infty$, $v = 0$, and $r = 1$, $\pi < \theta < 2\pi$ onto
$-\infty < u < 0$, $v = 0$. (See Problem 7, Section 2.1.3, which explains the
relationship between Laplace's equation in parts (a), (b).)

(c) Use the result of part (b) and the mapping function to show that the solution
of the boundary value problem in the circle is given by

$$T(x,y) = T_0\left(1 - \frac{2}{\pi}\cot^{-1}\left(\frac{2y}{1-(x^2+y^2)}\right)\right)$$

$$= T_0\left(1 - \frac{2}{\pi}\tan^{-1}\left(\frac{1-(x^2+y^2)}{2y}\right)\right),$$

or, in polar coordinates,

$$T(r,\theta) = T_0\left(1 - \frac{2}{\pi}\cot^{-1}\left(\frac{2r\sin\theta}{1-r^2}\right)\right)$$

$$= T_0\left(1 - \frac{2}{\pi}\tan^{-1}\left(\frac{1-r^2}{2r\sin\theta}\right)\right).$$

*2.3 More Complicated Multivalued Functions and Riemann Surfaces

We begin this section by discussing the branch structure associated with the
function

$$w = [(z-a)(z-b)]^{1/2}. \tag{2.3.1}$$

Functions such as Eq. (2.3.1) arise very frequently in applications. The function (2.3.1) is obviously the solution of the equation $w^2 = (z - a)(z - b)$, for real values a and b, $a < b$. Hence we expect square root type branch points at $z = a, b$. Indeed $z = a, b$ are branch points as can be verified by letting z be near, say, a, $z = a + \epsilon_1 e^{i\theta_1}$. Formula (2.3.1) implies that $w \approx q^{1/2} e^{i\theta_1/2}$ with $q = \epsilon_1(a - b)$ and as θ_1 varies between $\theta_1 = 0$ and $\theta_1 = 2\pi$, w jumps from $q^{1/2}$ to $-q^{1/2}$ (similarly near $z = b$). Perhaps surprising is the fact that $z = \infty$ is not a branch point. Letting $z = 1/t$, formula (2.3.1) yields

$$w = \frac{[(1 - ta)(1 - tb)]^{1/2}}{t} \tag{2.3.2}$$

and hence there is no jump near $t = 0$, because near $t = 0$, $w \approx 1/t$, which is single-valued. This is a consequence of the fact that for large z, $w \approx z$.

We can fix a branch cut for Eq. (2.3.1) as follows. We define the local polar coordinates

$$z - b = r_1 e^{i\theta_1}, \qquad z - a = r_2 e^{i\theta_2}, \qquad 0 \le \theta_1, \theta_2 < 2\pi. \tag{2.3.3}$$

Note that the magnitudes r_1 and r_2 are fixed uniquely by the location of the point z: $r_1 = |z - b|$, $r_2 = |z - a|$. However, there is freedom in the choice of angles. In Eq. (2.3.3) we have taken $0 \le \theta_1, \theta_2 < 2\pi$, but another branch could be specified by choosing θ_1 and θ_2 differently; as we discuss below.

Then Eq. (2.3.1) yields

$$w = (r_1 r_2)^{1/2} e^{i(\theta_1 + \theta_2)/2}. \tag{2.3.4}$$

In Figure 2.12 we denote values of the function w and the respective phases θ_1, θ_2 in those regions where a jump could be expected, that is, on the Re $z = x$ axis. (A heavy solid line denotes a branch cut.) We denote $\Theta = (\theta_1 + \theta_2)/2$.

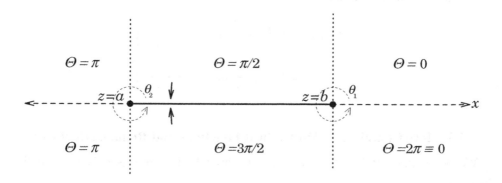

Figure 2.12 A branch cut for $w = (z - a)^{1/2}(z - b)^{1/2}$

For the above choice of angles θ_1, θ_2, the only jump of w (which depends on $(\theta_1 + \theta_2)/2$) occurs on the real axis between a and b, $a \le \mathrm{Re}\, z \le b$. Hence the branch cut is located on the $\mathrm{Re}\, z = x$ axis between (a, b). The points $z = a, b$ are square root branch points. Increasing θ_1, θ_2 to 4π, 6π, etc., would only put us on either side of the two branches of Eq. (2.3.1). Sometimes the branch depicted in Figure 2.12 is referred to as the one for which $w(z)$ is real and positive for $z = x$, $x > a, b$.

Other branches can be obtained by taking different choices of the angles θ_1, θ_2. For example, if we choose θ_1, θ_2 as follows, $0 \le \theta_1 < 2\pi$, $-\pi \le \theta_2 < \pi$, we would have a branch cut in the region $(-\infty, a) \cup (b, \infty)$ whereas the function is continuous in the region (a, b). In Figure 2.13 we give the phase angles in the respective regions which indicate why the branch cut is in the above-mentioned location.

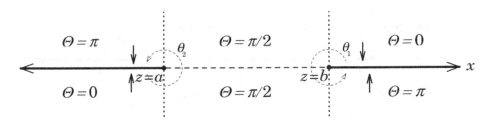

Figure 2.13 Another branch cut for $w = (z - a)^{1/2}(z - b)^{1/2}$

The branch cut in this latter case is best thought of as passing from $z = a$ to $z = b$ through the point at infinity. As mentioned earlier (see Eq. (2.3.2)), infinity is not a branch point. An alternative and useful view follows from the stereographic projection. The stereographic projections of the plane to a Riemann sphere corresponding to the branch cuts of Figures 2.12 and 2.13 are depicted in Figure 2.14.

More complicated functions are handled in similar ways. For example, consider the function

$$w = ((z - x_1)(z - x_2)(z - x_3))^{1/2} \qquad \text{with} \quad x_k \text{ real and } x_1 < x_2 < x_3. \quad (2.3.5)$$

If we let

$$z - x_k = r_k e^{i\theta_k}, \qquad 0 \le \theta_k < 2\pi \qquad (2.3.6)$$

then

$$w = \sqrt{r_1 r_2 r_3}\, e^{i(\theta_1 + \theta_2 + \theta_3)/2}. \qquad (2.3.7)$$

Defining $\Theta = (\theta_1 + \theta_2 + \theta_3)/2$, the phase diagram is given in Eq. (2.3.4). From the choices of phase (see Figure 2.15) it is clear that the branch cuts lie in the region $\{x_1 < \mathrm{Re}\, z < x_2\} \cup \{\mathrm{Re}\, z > x_3\}$.

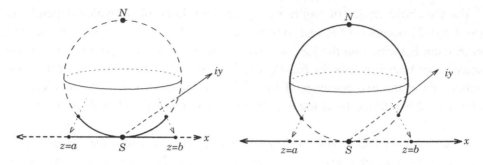

Figure 2.14 Projection of w onto Riemann sphere

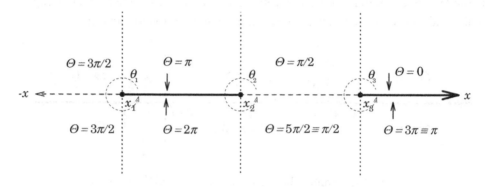

Figure 2.15 Triple choice of phase angles

A somewhat more complicated example is given by Eq. (2.2.19)

$$w = \cos^{-1} z = -i \log(z + i(1 - z^2)^{1/2})$$
$$= -i \log(z + (z^2 - 1)^{1/2}). \tag{2.3.8}$$

It is clear from the previous discussion that the points $z = \pm1$ are square root branch points. However, $z = \infty$ is a logarithmic branch point. Letting $z = 1/t$, we have

$$w = -i \log \left(\frac{1 + i(t^2 - 1)^{1/2}}{t} \right) = -i[\log(1 + i(t^2 - 1)^{1/2}) - \log t],$$

which demonstrates the logarithmic branch point behavior near $t = 0$. (We assume the branch of the square root is such that the first logarithm does not have a vanishing modulus, with the other sign of of the square root more work is required.) There are no other branch points because $z + i(1 - z^2)^{1/2}$ never vanishes in the finite z-plane. It should also be noted that owing to the fact that $(1 - z^2)^{1/2}$ has two branches, and the logarithm has an infinite number of branches, the function $\cos^{-1}$ can be thought of as having a "double infinity" of branches.

A particular branch of this function can be obtained by first taking

$$z + 1 = r_1 e^{i\theta_1}, \qquad z - 1 = r_2 e^{i\theta_2}, \qquad 0 \le \theta_i < 2\pi, \qquad i = 1, 2.$$

Then, by adding the above relations, we get

$$z = \left(r_1 e^{i\theta_1} + r_2 e^{i\theta_2} \right) /2,$$

and the function $q(z) = z + (z^2 - 1)^{1/2}$ is given by

$$q(z) = (r_1 e^{i\theta_1} + r_2 e^{i\theta_2})/2 + \sqrt{r_1 r_2} e^{i(\theta_1 + \theta_2)/2} \qquad (2.3.9)$$

whereupon

$$q(z) = \frac{r_1 e^{i\theta_1}}{2} \left(1 + \frac{r_2}{r_1} e^{i(\theta_2 - \theta_1)} + 2\sqrt{\frac{r_2}{r_1}} e^{i(\theta_2 - \theta_1)/2} \right). \qquad (2.3.10)$$

We further make the choice

$$1 + \frac{r_2}{r_1} e^{i(\theta_2 - \theta_1)} + 2\sqrt{\frac{r_2}{r_1}} e^{i(\theta_2 - \theta_1)/2} = R e^{i\Theta}, \qquad 0 \le \Theta < 2\pi.$$

We can choose Θ to be any interval of length 2π, which determines the particular branch of the logarithm. Here we made a convenient choice: $0 \le \Theta < 2\pi$.

With these choices of phase angle it is immediately clear that the function (2.3.8), $\log q(z)$, has a branch cut for $\operatorname{Re} z > -1$. In this regard we note that $\log(R e^{i\Theta})$ has no jump for $\operatorname{Re} z < -1$, nor does $\log(r_1 e^{i\theta_1})$, but for $\operatorname{Re} z > -1$, $\log(r_1 e^{i\theta_1})$ does have a jump.

In what follows we give a brief description of the concept of a Riemann surface. Actually, for the applications in this book, the preceding discussion of branch cuts and branch points is sufficient. Nevertheless, the notion of a Riemann surface for a multivalued function is helpful, and arises in applications. By a Riemann surface we mean an extension of the ordinary complex plane to a surface which has more than one "sheet." The multivalued function will have only one value corresponding to each point on the Riemann surface. In this way the function is single valued, and standard theory applies.

For example, consider again the square root function

$$w = z^{1/2}. \qquad (2.3.11)$$

Rather than considering the normal complex plane for z, it is useful to consider the two-sheeted surface depicted in Figure 2.16. This is the Riemann surface for Eq. (2.3.11).

Referring to Figure 2.16 we have double copies **I** and **II** of the z-plane with a cut along the positive x-axis. Each copy of the z-plane has identical coordinates z placed one on top of the other. Along the cut plane we have the planes joined in

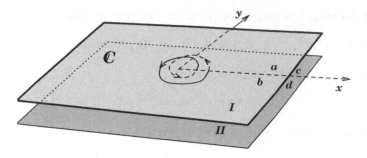

Figure 2.16 Two-sheeted Riemann surface

the following way. The cut along **Ib** is joined with the cut on **IIc**, while **Ia** is joined with the cut on **IId**. In this way, we produce a continuous one-to-one map from the Riemann surface for the function $z^{1/2}$ onto the w-plane; that is, the set of values $w = u + iv = z^{1/2}$. If we follow the curve C in Figure 2.16, we begin on sheet **Ia**, wind around the origin (the branch point) to **Ib**; we then *go through the cut and come out on* **IIc**. We again wind around the origin to **IId**, *go through the cut and come out on* **Ia**. The process obviously repeats after this. On this two-sheeted surface the function $f(z) = z^{1/2}$ is analytic everywhere except at $z = 0, z = \infty$.

In a similar manner we can construct an n-sheeted Riemann surface for the function $w = z^{m/n}$, where m and n are integers with no common factors. This would contain n identical sheets stacked one on top of the other with a cut on the positive x-axis and each successive sheet connected in the same way that **Ia** is connected to **IIc** in Figure 2.16. The nth sheet would be connected to the first in the same manner as **IId** is connected to **Ia** in Figure 2.16.

The logarithmic function is infinitely multivalued, as discussed in Section 2.2. The corresponding Riemann surface is infinitely sheeted. For example, Figure 2.17 depicts an infinitely-sheeted Riemann surface on which each sheet connects smoothly.

Each sheet is labeled $n = 0, n = 1, n = 2, \ldots$, corresponding to the branch of the log function (2.3.12):

$$w = \log z = \log |z| + i(\theta_p + 2n\pi), \qquad 0 \le \theta_p < 2\pi. \qquad (2.3.12)$$

The branch $n = 0$ is connected to $n = 1$, the branch $n = 1$ to $n = 2$, the branch $n = 2$ to $n = 3$, etc., in the same fashion that **Ib** is connected to **IIc** in Figure 2.18. A continuous closed circuit around the branch point $z = 0$ continuing on all the sheets $n = 0$ to $n = 1$ to $n = 2$, and so on, resembles an "infinite" spiral staircase. The main point here is that because the logarithmic function is infinitely branched (we say it has a branch point of infinite order) it has an infinitely-sheeted Riemann surface.

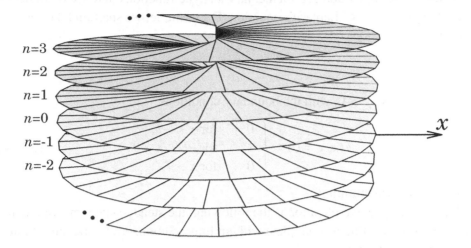

Figure 2.17 Infinitely sheeted Riemann surface

This beautiful geometric description, while useful, will be of far less importance for our purposes than the analytical understanding of how to specify particular branches, and how to work with these multivalued functions in examples and concrete applications.

Finally we remark that more complicated multivalued functions can have very complicated Riemann surfaces. For example, the function given by formula (2.3.1) with local coordinates given by Eq. (2.3.3) has a two-sheeted Riemann surface depicted in Figure 2.18.

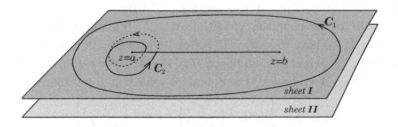

Figure 2.18 Riemann surface of two sheets

A closed circuit, for example, C_1 in Figure 2.18, enclosing both branch points $z = a$ and $z = b$, stays on the same sheet. However, a circuit enclosing either branch point, for example, the $z = a$ circuit C_2 in Figure 2.18, would start on sheet **I**; then after encircling the branch point would go through the cut onto sheet **II** and encircling the branch point again would end up on sheet **I**, and so on.

As described in Section 2.2, elementary analytic functions may yield physically interesting solutions of Laplace's equation. For example, we shall find the solution to Laplace's equation

$$\frac{\partial^2 \psi}{\partial x^2} + \frac{\partial^2 \psi}{\partial y^2} = 0, \tag{2.3.13}$$

for $-\infty < x < \infty$, $y > 0$, with the boundary conditions

$$\psi(x, y = 0) = \begin{cases} 0 & \text{for } x < -\ell, \\ 1 & \text{for } -\ell < x < \ell, \\ 0 & \text{for } x > \ell, \end{cases} \tag{2.3.14}$$

which is bounded at infinity.

A typical physical application is the following: the steady-state temperature distribution of a two-dimensional plate with an imposed nonzero temperature (unity) on a portion of the bottom of the plate.

Consider the function

$$\Omega(z) = A \log(z + \ell) + B \log(z - \ell) + iC, \qquad A, B, C \text{ are real constants} \quad (2.3.15)$$

with branch cuts taken by choosing $z + l = r_1 e^{i\theta_1}$ and $z - l = r_2 e^{i\theta_2}$, where $0 \le \theta_i < 2\pi$ for $i = 1, 2$. The function (2.3.15) is therefore analytic in the upper half plane, and consequently, we know that the imaginary part ψ of $\Omega(z) = \phi + i\psi$ satisfies Laplace's equation. This solution is given by

$$\psi(x, y) = A\theta_1 + B\theta_2 + C$$
$$= A \tan^{-1}\left(\frac{y}{x + \ell}\right) + B \tan^{-1}\left(\frac{y}{x - \ell}\right) + C, \tag{2.3.16}$$

where we are taking $0 < \tan^{-1}\alpha < \pi$; see Eq. (2.2.10).

It remains to fix the boundary conditions on $y = 0$ given by Eq. (2.3.14). For $x > \ell$ and $y = 0$, we have $\theta_1 = \theta_2 = 0$; hence we take $C = 0$. For $-\ell < x < \ell$ and $y = 0$ we have $\theta_1 = 0$ and $\theta_2 = \pi$; hence $B = 1/\pi$. For $x < -\ell$ and $y = 0$ we have $\theta_1 = \theta_2 = \pi$; hence $A + 1/\pi = 0$. The boundary value solution is therefore given by

$$\psi(x, y) = \frac{1}{\pi}\left[\tan^{-1}\left(\frac{y}{x - \ell}\right) - \tan^{-1}\left(\frac{y}{x + \ell}\right)\right]. \tag{2.3.17}$$

2.3.1 Problems for Section 2.3

1. Find the location of the branch points and discuss the branch cut structure of the following functions:

$$\text{(a) } (z^2 + 1)^{1/2}; \quad \text{(b) } ((z + 1)(z - 2))^{1/3}.$$

2. Find the location of the branch points and discuss the branch cuts associated with the following functions;

(a) $\log((z-1)(z-2))$;   (b) $\left(\dfrac{z-1}{z}\right)^{1/2}$;

(c) $\coth^{-1}\dfrac{z}{a} = \dfrac{1}{2}\log\dfrac{z+a}{z-a}$, $a > 0$.

(d) Related to part (c), show that when n is an integer

$$\coth^{-1}\frac{z}{a} = \frac{1}{4}\log\frac{(x+a)^2+y^2}{(x-a)^2+y^2}$$

$$+ \frac{i}{2}\tan^{-1}\left(\left(\frac{2ay}{a^2-x^2-y^2}\right)+2n\pi\right).$$

3. Given the function

$$\log(z-(z^2+1)^{1/2})$$

discuss the branch point/branch cut structure and where this function is analytic.

4. Consider the complex velocity potential

$$\Omega(z,z_0) = \frac{M}{2\pi}[\log(z-z_0)-\log z]$$

for $M > 0$, which corresponds to a source at $z = z_0$ and a sink at $z = 0$. (See also Exercise 6 in Section 2.1.3, and Exercises 7 and 8 of Section 2.2.1.) Find the corresponding velocity potential and stream function. Let $M = k/|z_0|$, $z_0 = |z_0|e^{i\theta_0}$, and show that

$$\Omega(z,z_0) = -\frac{k}{2\pi}\left(\frac{\log z - \log(z-z_0)}{z_0}\right)\frac{z_0}{|z_0|}.$$

Take the limit as $z_0 \to 0$ to obtain

$$\Omega(z) = \lim_{z\to 0}\Omega(z,z_0) = -\frac{ke^{i\theta_0}}{2\pi}\frac{1}{z}.$$

This is called a "doublet" with strength k. The angle θ_0 specifies the direction along which the source/sink coalesces. Find the velocity potential and the stream function of the "doublet," and sketch the flow.

5. Consider the complex velocity potential

$$\Omega(w) = -\frac{i\Gamma}{2\pi}\log w, \qquad \text{with } \Gamma \text{ real.}$$

(a) Show that the transformation $z = \frac{1}{2}(w+\frac{1}{w})$ transforms the complex velocity potential to

$$\Omega(z) = -\frac{i\Gamma}{2\pi}\log\left(z+(z^2-1)^{1/2}\right).$$

(b) Choose a branch of $(z^2-1)^{1/2}$ as follows:

$$(z^2-1)^{1/2} = (r_1r_2)^{1/2}e^{i(\theta_1+\theta_2)/2},$$

where $0 \leq \theta_i < 2\pi$, $i = 1, 2$, so that there is a branch cut on the x-axis, $-1 < x < 1$, for $(z^2 - 1)^{1/2}$. Show that a positive circuit around a closed curve enclosing $z = -1$ and $z = +1$ increases Ω by Γ (we say the circulation increases by Γ).

(c) Establish that the velocity field $v = (v_1, v_2)$ satisfies

$$v_1 = -\frac{\Gamma}{2\pi\sqrt{1 - x^2}} \qquad \text{on } y = 0^+ \text{ for } -1 < x < 1,$$

and

$$v_2 = \begin{cases} \dfrac{\Gamma}{2\pi\sqrt{x^2 - 1}} & \text{for } x > 1, \, y = 0; \\[4mm] -\dfrac{\Gamma}{2\pi\sqrt{x^2 - 1}} & \text{for } x < -1, \, y = 0. \end{cases}$$

6. Consider the transformation (see also Problem 5 above) $z = \frac{1}{2}(w + \frac{1}{w})$. Show that $T(x, y) = -\operatorname{Im}\Omega(z)$, where $\Omega = 1/w$ satisfies Laplace's equation and satisfies the following conditions:

$$T(x, y = 0^+) = \sqrt{1 - x^2} \qquad \text{for } |x| \leq 1;$$
$$T(x, y = 0^-) = -\sqrt{1 - x^2} \qquad \text{for } |x| \leq 1;$$
$$T(x, y = 0) = 0 \qquad \text{for } |x| \geq 1;$$

and

$$T(x = 0, y) = \begin{cases} \dfrac{1}{y + \sqrt{y^2 + 1}} & \text{for } y > 0, \\[4mm] -\dfrac{1}{-y + \sqrt{y^2 + 1}} & \text{for } y < 0. \end{cases}$$

2.4 Complex Integration

In this section we consider the evaluation of integrals of functions of a complex variable along appropriate curves in the complex plane. We shall see that some of the analysis bears a similarity to that of functions of real variables. However, for analytic functions, very important new results can be derived, namely Cauchy's Theorem (sometimes called the Cauchy–Goursat Theorem). Complex integration has wide applicability, and we shall describe some of the applications in this book.

We begin by considering a complex-valued function f of a real variable t on a fixed interval, $a \leq t \leq b$:

$$f(t) = u(t) + iv(t), \qquad (2.4.1)$$

where $u(t)$ and $v(t)$ are real valued. The function $f(t)$ is said to be **integrable** on the interval $[a, b]$ if the functions u and v are integrable. Then

$$\int_a^b f(t)\, dt = \int_a^b u(t)\, dt + i \int_a^b v(t)\, dt. \qquad (2.4.2)$$

The usual rules of integration for real functions apply; in particular, from the fundamental theorems of calculus, we have for continuous functions $f(t)$,

$$\frac{d}{dt} \int_a^t f(\tau)\, d\tau = f(t) \qquad (2.4.3\text{a})$$

and for $f'(t)$ continuous,

$$\int_a^b f'(t)\, dt = f(b) - f(a). \qquad (2.4.3\text{b})$$

Next we extend the notion of complex integration to integration on a curve in the complex plane. A curve in the complex plane can be described via the parameterization

$$z(t) = x(t) + iy(t), \qquad a \leq t \leq b. \qquad (2.4.4)$$

For each given t in $[a, b]$, there is a point $(x(t), y(t))$ that yields the image point $z(t)$. The image points $z(t)$ are ordered according to increasing t. The curve is said to be continuous if $x(t)$ and $y(t)$ are continuous functions of t. Similarly it is said to be differentiable if $x(t)$ and $y(t)$ are differentiable.

A curve or arc C is **simple** (sometimes called a **Jordan arc**) if it does not intersect itself: that is, $z(t_1) \neq z(t_2)$ if $t_1 \neq t_2$ for $t \in [a, b]$, except that $z(b) = z(a)$ is allowed; in the latter case we say that C is a **simple closed curve** (or **Jordan**). Examples are seen in Figure 2.19. Note also that a "figure 8" is an example of a nonsimple closed curve.

Next we shall discuss evaluation of integrals along curves. When the curve is closed, our convention shall be to take the positive direction to be one in which the interior remains to the left of C. Integrals along closed curves will be taken along the positive direction unless otherwise specified. The function $f(z)$ is said to be **continuous** on C if $f(z(t))$ is continuous for $a \leq t \leq b$, and f is said to be **piecewise continuous** on $[a, b]$ if $[a, b]$ can be broken up into a finite number of subintervals in which $f(z)$ is continuous. A **smooth arc** C is one for which $z'(t)$ is continuous. A **contour** is an arc consisting of a finite number of connected smooth arcs; that is,

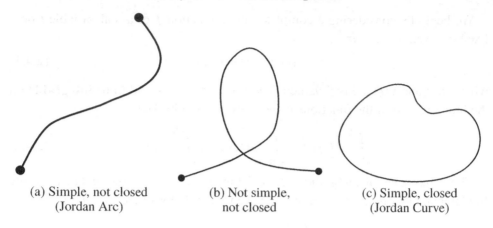

(a) Simple, not closed   (b) Not simple,   (c) Simple, closed
(Jordan Arc)   not closed   (Jordan Curve)

Figure 2.19 Examples of curves

a contour is a piecewise smooth arc. Thus, on a contour C, $z(t)$ is continuous and $z'(t)$ is piecewise continuous. Hereafter we shall only consider integrals along such contours unless otherwise specified. Asimple closed contour is frequently referred to as a **Jordan contour**.

The contour integral of a piecewise continuous function on a smooth contour C is defined to be

$$\int_C f(z)\, dz = \int_a^b f(z(t)) z'(t)\, dt, \tag{2.4.5}$$

where the right-hand side of Eq. (2.4.5) is obtained via the formal substitution $dz = z'(t)dt$. In general, Eq. (2.4.5) depends on $f(z)$ and the contour C. Thus, the integral (2.4.5) is really a line integral in the (x, y)-plane and is naturally related to the study of vector calculus in the plane. As mentioned earlier, the complex variable $z = x + iy$ can be thought of as a two-dimensional vector.

We remark that values of the above integrals are invariant if we redefine the parameter t appropriately. Namely, if we make the change of variables $t \to s$ by $t = T(s)$, where $T(s)$ maps the interval $A \leq s \leq B$ to the interval $a \leq t \leq b$, $T(s)$ is continuously differentiable, and $T'(s) > 0$ (needed to insure that t increases with s), then only the form the integrals take on is modified, but its value is invariant. The importance of this remark is that one can evaluate integrals by the most convenient choice of parameterization. Examples discussed later in this section will serve to illustrate this point.

The usual properties of integration apply. We have

$$\int_C [\alpha f(z) + \beta g(z)]\, dz = \alpha \int_C f(z)\, dz + \beta \int_C g(z)\, dz, \tag{2.4.6}$$

for constants α and β, and piecewise continuous functions f and g. The arc C traversed in the opposite direction – that is, from $t = b$ to $t = a$ – is denoted by $-C$. We then have

$$\int_{-C} f(z)\, dz = -\int_{C} f(z)\, dz, \tag{2.4.7}$$

because the left-hand side of Eq. (2.4.7) is equivalent to $\int_{b}^{a} f(z(t))z'(t)dt$. Similarly, if C consists of n connected contours with endpoints from z_1 to z_2 for C_1, from z_2 to z_3 for $C_2, \ldots$, from z_n to z_{n+1} for C_n, then we have

$$\int_{C} f = \sum_{j=1}^{n} \int_{C_j} f.$$

The fundamental theorem of calculus yields the following result.

Theorem 2.4.1 *Suppose $F(z)$ is an analytic function, and that $f(z) = F'(z)$ is continuous in a domain D. Then for a contour C lying in D with endpoints z_1 and z_2,*

$$\int_{C} f(z)\, dz = F(z_2) - F(z_1). \tag{2.4.8}$$

Proof Using the definition of the integral (2.4.5), the chain rule, and assuming for simplicity that $z'(t)$ is continuous (otherwise add integrals separately over smooth arcs) we have

$$\int_{C} f(z)\, dz = \int_{C} F'(z)\, dz = \int_{a}^{b} F'(z(t))z'(t)\, dt$$

$$= \int_{a}^{b} \frac{d}{dt} [F(z(t))]\, dt$$

$$= F(z(b)) - F(z(a))$$

$$= F(z_2) - F(z_1). \qquad \square$$

As a consequence of Theorem 2.4.1, for closed curves we have

$$\oint_{C} f(z)\, dz = \oint_{C} F'(z)\, dz = 0, \tag{2.4.9}$$

where $\oint_{C}$ denotes a closed contour C (that is, the endpoints are equal).

If the function $f(z)$ satisfies the hypothesis of Theorem 2.4.1, then, for *all* contours C lying in D beginning at z_1 and ending at z_2, we have Eq. (2.4.8). Hence the result demonstrates that the integral is independent of path. Indeed, Figure 2.20 illustrates this fact.

Referring to Figure 2.20 we have $\int_{C_1} f\, dz = \int_{C_2} f\, dz$ because

$$\oint_{C} f\, dz = \int_{C_1} f\, dz - \int_{C_2} f\, dz = 0, \tag{2.4.10}$$

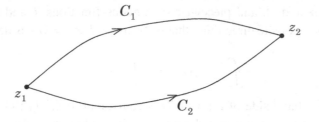

Figure 2.20 Independent paths forming a closed curve

where the closed curve $C = C_1 - C_2$.

The hypothesis in Theorem 2.4.1 requires the existence of $F(z)$ such that $f(z) = F'(z)$. Later in this chapter we shall show this for a large class of functions $f(z)$.

Sometimes it is convenient to evaluate the complex integral by reducing it to two real-line integrals in the (x, y)-plane. In the definition (2.4.5) we use $f(z) = u(x, y) + iv(x, y)$ and $dz = dx + i\,dy$ to obtain

$$\int_C f(z)\,dz = \int_C \left[(u\,dx - v\,dy) + i(v\,dx + u\,dy)\right] . \qquad (2.4.11)$$

This can be shown, via parameterization, to be equivalent to

$$\int_a^b f(z(t))z'(t)\,dt.$$

Later in this chapter we shall use Eq. (2.4.11) in order to derive one form of Cauchy's Theorem.

In the following examples we illustrate how line integrals may be calculated in prototypical cases.

Example 2.4.2   Evaluate $\int_C \bar{z}\,dz$ for: (a) $C = C_1$, a contour from $z = 0$ to $z = 1$ to $z = 1 + i$; (b) $C = C_2$, the line from $z = 0$ to $z = 1 + i$; and (c) $C = C_3$, the unit circle $|z| = 1$ (see Figure 2.21).

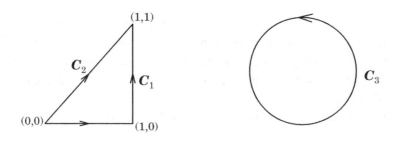

Figure 2.21 Contours C_1, C_2, and C_3

(a)
$$\int_{C_1} \bar{z}\, dz = \int_{C_1} (x - iy)(dx + i\,dy)$$

$$= \int_{x=0}^{1} x\, dx + \int_{y=0}^{1} (1 - iy)(i\, dy)$$

$$= \frac{1}{2} + i[y - iy^2/2]_0^1$$

$$= 1 + i.$$

Note, in the integral from $z = 0$ to $z = 1$, that $y = 0$, hence $dy = 0$. In the integral from $z = 1$ to $z = 1 + i$, similarly $x = 1$, hence $dx = 0$.

(b)
$$\int_{C_2} \bar{z}\, dz = \int_{x=0}^{1} (x - ix)(dx + i\,dx)$$

$$= (1 - i)(1 + i) \int_0^1 x\, dx$$

$$= 1.$$

Note that C_2 is the line $y = x$, hence $dy = dx$. Because $\bar{z}$ is *not* analytic we see that $\int_{C_2} \bar{z}\, dz$ and $\int_{C_1} \bar{z}\, dz$ need not be equal.

(c)
$$\int_{C_3} \bar{z}\, dz = \int_{\theta=0}^{2\pi} e^{-i\theta} i e^{i\theta}\, d\theta = 2\pi i.$$

Note that $z = e^{i\theta}$, $\bar{z} = e^{-i\theta}$, and $dz = ie^{i\theta} d\theta$; on the unit circle, $r = 1$.

Example 2.4.3 Evaluate $\int_C z\, dz$ along the three contours described above and as illustrated in Figure 2.21. Because z is analytic in the region containing z, and $z = (d/dz)(z^2/2)$, we immediately have, from Theorem 2.4.1

$$\int_{C_1} z\, dz = \int_{C_2} z\, dz = \frac{(1 + i)^2}{2} = i,$$

$$\int_{C_3} z\, dz = 0.$$

These results can be calculated directly via the line integral methods described above – which we will leave for the reader to verify.

Example 2.4.4 Evaluate $\int_C (1/z)\, dz$ for (a) any simple closed contour C not enclosing the origin, and; (b) any simple closed contour C enclosing the origin.

(a) Because $1/z$ is analytic for all $z \neq 0$, we immediately have from Theorem 2.4.1 and from $1/z = (d/dz)(\log z)$,

$$\int_C \frac{1}{z} \, dz = 0,$$

because $[\log z]_C = 0$ so long as C does not enclose the branch point of $\log z$ at $z = 0$ (see Figure 2.22a).

(b) Any simple closed contour around the origin, call it C_2, can be *deformed* into a small, but finite circle C_1 of radius r, as follows. Introduce a "crosscut" (L_1, L_2) as in Figure 2.22b. Then in the limit of r and the crosscut width tending to zero we have a closed contour: $C = C_2 + L_1 + L_2 - C_1$. (Note that for C_1 we take the positive counterclockwise orientation.) In Figure 2.22b we have taken care to distinguish the positive and negative directions of C_1 and C_2, respectively. From part (a) of this problem,

$$\int_C \frac{1}{z} \, dz = 0;$$

then, because $\int_{L_1} + \int_{L_2} = 0$, we have (using $z = re^{i\theta}$ and $dz = rie^{i\theta} \, d\theta$, and since the curve is simple we can take $0 \leq \theta < 2\pi$),

$$\int_{C_2} \frac{1}{z} \, dz = \int_{C_1} \frac{1}{z} \, dz = \int_0^{2\pi} r^{-1} e^{-i\theta} \, ie^{i\theta} \, r \, d\theta = 2\pi i.$$

Thus the integral of $1/z$ around any simple closed curve enclosing the origin is $2\pi i$. We also note that if we formally use the antiderivative of $1/z$ (i.e., $1/z = d/dz(\log z)$), we can also find $\int_{C_2} (1/z) \, dz = 2\pi i$. In this case, even though we enclose the branch point of $\log z$, the argument θ_p of $\log z = \log r + i(\theta_p + 2n\pi)$

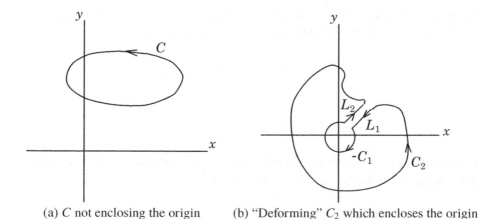

(a) C not enclosing the origin   (b) "Deforming" C_2 which encloses the origin

Figure 2.22 Integration contours in Example 2.4.4

increases by 2π as we enclose the origin. In this case, we need only select a convenient branch of $\log z$.

Example 2.4.5 Evaluate $\int_C z^n \, dz$ for integer n and some simple closed contour C that encloses the origin.

Using the crosscut segment as indicated in Figure 2.22, the integral in question is equal to that on C_1, a small, but finite circle of radius r. Thus

$$\int_{C_1} z^n \, dz = \int_0^{2\pi} r^{n+1} e^{in\theta} i e^{i\theta} \, d\theta$$

$$= i \int_0^{2\pi} r^{n+1} e^{i(n+1)\theta} \, d\theta = \begin{cases} 0 & n \neq -1, \\ 2\pi i & n = -1. \end{cases}$$

Hence, even though z^n is nonanalytic at $z = 0$ for $n < 0$, only the value $n = -1$ gives a nontrivial contribution. We remark that use of the antiderivative

$$z^n = \frac{d}{dz}\left(\frac{z^{n+1}}{n+1}\right), \qquad n \neq -1$$

yields the same results.

As mentioned earlier, complex line integrals arise in many physical applications. For example, in ideal fluid flow problems (see Section 2.1 where we briefly discussed ideal fluid flows), the real-line integrals

$$\Gamma = \int_C (\phi_x \, dx + \phi_y \, dy) = \int_C \mathbf{v} \cdot \hat{\mathbf{t}} \, ds, \qquad (2.4.12)$$

$$\mathcal{F} = \int_C (\phi_x \, dy - \phi_y \, dx) = \int_C \mathbf{v} \cdot \hat{\mathbf{n}} \, ds, \qquad (2.4.13)$$

where s is the arc length, $\mathbf{v} = (\phi_x, \phi_y)$ is the velocity vector, $\hat{\mathbf{t}} = (\frac{dx}{ds}, \frac{dy}{ds})$ is the unit tangent vector to C, and $\hat{\mathbf{n}} = (\frac{dy}{ds}, -\frac{dx}{ds})$ is the unit normal vector to C, represent (Γ) the circulation around the curve C (when C is closed), and ($\mathcal{F}$) the flux across the curve C. We note that in terms of analytic complex functions we have the simple equation

$$\Gamma + i\mathcal{F} = \int_C (\phi_x - i\phi_y)(dx + idy) = \int_C \Omega'(z) \, dz. \qquad (2.4.14)$$

Recall from (2.1.29) that the complex velocity is given by $\Omega'(z) = \phi_x + i\psi_x = \phi_x - i\phi_y$ (the latter follows from the Cauchy–Riemann conditions). Using complex function theory to evaluate Eq. (2.4.14) often provides an easy way to calculate the real-line integrals (2.4.12–2.4.13), which are the real and imaginary parts of the integral in Eq. (2.4.14). An example is discussed in the problem section.

Next we derive an important inequality which we shall use frequently.

Theorem 2.4.6   *Let $f(z)$ be continuous on a contour C. Then*

$$\left| \int_C f(z)\, dz \right| \leq ML, \tag{2.4.15}$$

where L is the length of C, and M is an upper bound for $|f|$ on C.

Proof   Write

$$I = \left| \int_C f(z)\, dz \right| = \left| \int_a^b f(z(t)) z'(t)\, dt \right|. \tag{2.4.16}$$

From real variables we know that, for $a \leq t \leq b$,

$$\left| \int_a^b G(t)\, dt \right| \leq \int_a^b |G(t)|\, dt.$$

The analogous result holds for complex integrals:

$$I \leq \int_a^b |f(z(t))|\, |z'(t)|\, dt.$$

(This can be shown by using Eq. (2.4.19) below, with the triangle inequality.) Since $|f|$ is bounded on C, i.e., $|f(z(t))| \leq M$ on C, where M is a constant, we have

$$I \leq M \int_a^b |z'(t)|\, dt.$$

However, because

$$|z'(t)|\, dt = |x'(t) + iy'(t)|\, dt$$
$$= \sqrt{(x'(t))^2 + (y'(t))^2}\, dt = ds, \tag{2.4.17}$$

where s represents arc length along C, we have Eq. (2.4.15).   □

For an alternative proof of Eq. (2.4.15) see Problem 12 in Section 2.4.1.

Example 2.4.7   Find an upper bound for the magnitude of the following integral

$$\int_C \frac{e^{ikz}}{z}\, dz, \qquad k > 0,$$

where C is an open semi-circle in the upper half plane radius R centered at $z = 0$.
   Taking the absolute value of the integral and using (2.4.15), we have

$$\left| \int_C \frac{e^{ikz}}{z}\, dz \right| \leq \int_C \frac{e^{-ky}}{|z|} |dz| \leq \int_C \frac{|dz|}{|z|} \leq \int_0^\pi \frac{R\, d\theta}{R} = \pi,$$

where we have used $z = x + iy = Re^{i\theta}$, with $y \geq 0$.

We also remark that the preceding developments of contour integration could also have been derived using limits of appropriate sums. This would be in analogy to the one-dimensional evaluation of integrals by Riemann sums. More specifically, given a contour C in the z-plane beginning at z_a and terminating at z_b, choose any ordered sequence $\{z_j\}$ of $n + 1$ points on C such that $z_0 = z_a$ and $z_n = z_b$. Define $\Delta z_j = z_{j+1} - z_j$ and form the sum

$$S_n = \sum_{j=1}^{n} f(\xi_j)\Delta z_j, \tag{2.4.18}$$

where ξ_j is any point on C between z_{j-1} and z_j. If $f(x)$ is piecewise continuous on C, then the limit of S_n as $n \to \infty$ and $|\Delta z_j| \to 0$ converges to the integral of $f(z)$; namely,

$$\int_C f(z)\, dz = \lim_{\substack{n \to \infty \\ |\Delta z_j| \to 0}} \sum_{j=1}^{n} f(\xi_j)\Delta z_j. \tag{2.4.19}$$

Finally, we define a **simply connected** domain D to be one for which every simple closed contour within it encloses only points of D (see Figure 2.23a). The points within a circle, square, and polygon are examples of a simply-connected domain. An annulus (doughnut) is not simply connected. A domain that is not simply connected is called multiply connected. An annulus is multiply connected, because a contour encircling the inner hole encloses points within and outside D (see Figure 2.23b).

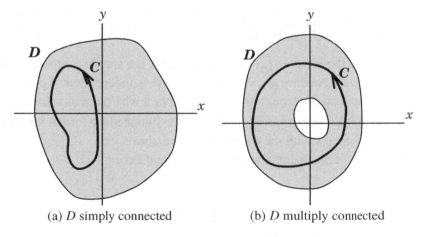

(a) D simply connected   (b) D multiply connected

Figure 2.23 Connectedness of domain D

2.4.1 Problems for Section 2.4

1. From the basic definition of complex integration, evaluate the integral $\oint_C f(z)\, dz$, where C is the parametrized unit circle enclosing the origin: $C : x(t) = \cos t,\ y(t) = \sin t$ or $z = e^{it}$, and where $f(z)$ is given by

$$\text{(a)}\ z^2; \qquad \text{(b)}\ \bar{z}^2; \qquad \text{(c)}\ \frac{z+1}{z^2}.$$

2. Evaluate the integral $\oint_C f(z)\, dz$, where C is the unit circle enclosing the origin, and $f(z)$ is given as follows:

$$\text{(a)}\ 1 + 2z + z^2; \qquad \text{(b)}\ 1/(z - 1/2)^2; \qquad \text{(c)}\ 1/\bar{z}; \qquad \text{(d)}\ z\bar{z}; \qquad \text{(e)}^*\ e^{\bar{z}}.$$

   *Hint: use (1.2.19).

3. Let C be a square with diagonal corners at $-1 - i$ and $1 + i$. Evaluate $\oint_C f(z)\, dz$, where $f(z)$ is given by the following:

$$\text{(a)}\ \sin z; \qquad \text{(b)}\ \frac{1}{2z + 1}; \qquad \text{(c)}\ \bar{z}; \qquad \text{(d)}\ \operatorname{Re} z.$$

4. Use the principal branch of $\log z$ and $z^{1/2}$ to evaluate

$$\text{(a)}\quad \int_{-1}^{1} \log z\, dz; \qquad \text{(b)}\quad \int_{-1}^{1} z^{1/2}\, dz.$$

5. (a) Consider the integral

$$\oint_{C_0} \frac{z}{z^4 + 1}\, dz$$

   where C_0 is the unit circle: centered at $z = 0$, radius one.
   Using Theorem 2.4.6 and the triangle inequality, show that

$$\left| \oint_{C_0} \frac{z}{z^4 + 4}\, dz \right| \le \oint_{C_0} \frac{|z|}{|z^4 + 4|}\, |dz| \le \oint_{C_0} \frac{|z|}{(4 - |z|^4)}\, |dz| \le \int_0^{2\pi} \frac{d\theta}{4 - 1} = \frac{2\pi}{3}.$$

   (b) Show that the integral $\int_C (1/z^2)\, dz$, where C is a simple contour beginning at $z = -a$ and ending at $z = b$, and $a, b > 0$, is independent if path as long as C does not go through the origin. Explain why the real-valued integral $\int_{-a}^{b} (1/x^2)\, dx$ doesn't exist, but the value obtained by formal substitution of limits agrees with the complex integral above.

6. Let $z^{1/2}$ have a branch cut along the positive real axis.
   (a) Evaluate

$$\oint_{C_0} z^{1/2}\, dz$$

   where C_0 is the unit circle: centered at $z = 0$, radius one.

(b) Evaluate $\int_0^b z^{1/2} dz$, where $b > 0$, obtained by integrating along the top half of the cut. Show that the value of this integral, obtained by integrating along the top half of the cut, is exactly minus that obtained by integrating along the bottom half of the cut. What is the difference between taking the principal versus the second branch of $z^{1/2}$?

7. Let C be an open (upper) semicircle of radius R with its center at the origin, and consider $\int_C f(z)\, dz$. Let $f(z) = 1/(z^2 + a^2)$, for real $a > 0$. Show that $|f(z)| \le 1/(R^2 - a^2)$, $R > a$, and

$$\left| \int_C f(z) dz \right| \le \frac{\pi R}{R^2 - a^2}, \qquad R > a.$$

8. Let C be an arc of the circle $|z| = R$ $(R > 1)$ of angle $\pi/3$. Show that

$$\left| \int_C \frac{dz}{z^3 + 1} \right| \le \frac{\pi}{3} \left(\frac{R}{R^3 - 1} \right),$$

and deduce $\lim_{R \to \infty} \int_C \frac{dz}{z^3 + 1} = 0$.

9. Consider $I_R = \int_{C_R} (e^{iz}/z^2)\, dz$ where C_R is the open semicircle with radius R in the upper half plane with endpoints $(-R, 0)$ and $(R, 0)$; note C_R is open: it does not include the x-axis. Show that $\lim_{R \to \infty} I_R = 0$.

10. Consider

$$I_\epsilon = \oint_{C_\epsilon} z^\alpha f(z)\, dz, \qquad \alpha > -1, \quad \alpha \text{ real},$$

where C_ϵ is a circle of radius ϵ centered at the origin, and $f(z)$ is analytic inside the circle. Show that $\lim_{\epsilon \to 0} I_\epsilon = 0$.

11. (a) Suppose we are given the complex flow field $\Omega(z) = -ik \log(z - z_0)$, where k is a real constant and z_0 a complex constant. Show that the circulation around a closed curve C_0 encircling $z = z_0$ is given by $\Gamma = 2\pi k$. (Hint: from Section 2.4, $\Gamma + i\mathcal{F} = \oint_{C_0} \Omega'(z)\, dz$.)

(b) Suppose $\Omega(z) = k \log(z - z_0)$. Find the circulation around C_0 and the flux through C_0.

12. In this problem we will give an alternative derivation of Eq. (2.4.15):

$$\left| \int_C f(z) dz \right| \le ML,$$

where $f(t)| \le M$ and L is the length of the curve C.

(a) Show that

$$\left|\int_a^b F(t)\, dt\right| = e^{-i\phi} \int_a^b F(t)\, dt = \int_a^b e^{-i\phi} F(t)\, dt,$$

where $\phi = \arg\left(\int_a^b F(t)\, dt\right)$.

(b) Show that, since $\left|\int_a^b F(t)\, dt\right|$ is real,

$$\left|\int_a^b F(t)\, dt\right| = \mathrm{Re}\left[\int_a^b e^{-i\phi} F(t)\, dt\right] = \int_a^b \mathrm{Re}\left[e^{-i\phi} F(t)\right] dt$$

$$\leq \int_a^b \left|e^{-i\phi} F(t)\, dt\right| = \int_a^b |F(t)|\, dt.$$

(c) Let $F(t) = f(t) z'(t)$ and show

$$\left|\int_a^b f(t) z'(t)\, dt\right| \leq \int_a^b |f(t) z'(t)|\, dt \leq M \int_a^b |z'(t)|\, dt = ML.$$

2.5 Cauchy's Theorem

In this section we study Cauchy's theorem, which is one of the most important theorems in complex analysis. In order to prove Cauchy's theorem in a convenient manner we will use a well-known result from vector analysis in real variables, known as Green's theorem in the plane, which can be found in advanced calculus texts; see, for example, Buck (1956).

Theorem 2.5.1 (Green)   *Let the real functions $u(x, y)$ and $v(x, y)$ along with their partial derivatives $\partial u/\partial x$, $\partial u/\partial y$, $\partial v/\partial x$, $\partial v/\partial y$, be continuous throughout a simply-connected region $\mathcal{R}$ consisting of points interior to and on a simple closed contour C in the (x, y)-plane. Let C be described in the positive (counterclockwise) direction. Then*

$$\oint_C (u\, dx + v\, dy) = \iint_R \left(\frac{\partial v}{\partial x} - \frac{\partial u}{\partial y}\right) dx\, dy. \tag{2.5.1}$$

We remark for those readers who may not recall or have not seen this formula, Eq. (2.5.1) is a two-dimensional version of the divergence theorem of vector calculus (taking the divergence of a vector $\mathbf{v} = (v, -u)$).

An elementary derivation of Eq. (2.5.1) can be given if we restrict the region $\mathcal{R}$ to be such that every vertical and horizontal line intersects the boundary of $\mathcal{R}$

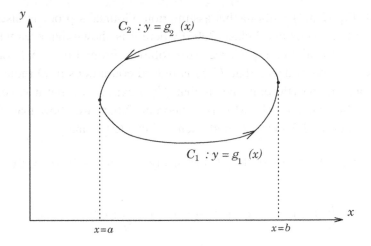

Figure 2.24 Deriving Equation 2.5.1 for region $\mathcal{R}$

in at most two points. Then if we call the "top" and "bottom" curves defining C, $y = g_2(x)$ and $y = g_1(x)$, respectively (see Figure 2.24)

$$-\iint_{\mathcal{R}} \frac{\partial u}{\partial y}\, dx\, dy = -\int_a^b \int_{g_1(x)}^{g_2(x)} \frac{\partial u}{\partial y}\, dy\, dx$$

$$= -\int_a^b \left[u(x, g_2(x)) - u(x, g_1(x)) \right]\, dx$$

$$= +\int_{C_2} u(x, y)\, dx + \int_{C_1} u(x, y)\, dx$$

$$= \oint_C u(x, y)\, dx.$$

Following the same line of thought we also find

$$\iint_{\mathcal{R}} \frac{\partial v}{\partial x}\, dx\, dy = \oint_C v\, dy.$$

From these relationships we obtain Eq. (2.5.1).

With Green's theorem we can give a simple proof of Cauchy's theorem as long as we make a certain extra assumption to be explained shortly.

Theorem 2.5.2 (Cauchy) *If a function f is analytic in a simply connected domain D, then along a simple closed contour C in D*

$$\oint_C f(z)\, dz = 0. \tag{2.5.2}$$

We remark that in the proof given here, we shall also *require* that $f'(z)$ be continuous in D. In fact, a more general proof owing to Goursat enables one to establish Eq. (2.5.2) without this assumption. Goursat's proof is discussed in Section 2.7 of Ablowitz and Fokas (2003) where it is shown that even when $f(z)$ is assumed only analytic, Eq. (2.5.2) still follows. From Eq. (2.5.2) one could then derive as a consequence that $f'(z)$ is indeed continuous in D (note so far in our development, analytic only means that $f'(z)$ exists, not that it is necessarily continuous). In a later result (Morera's theorem, 2.6.5) we show that if $f(z)$ is continuous and Eq. (2.5.2) is satisfied, then in fact $f(z)$ is analytic.

Proof   From the definition of $\oint_C f(z)dz$, using $f(z) = u + iv$, $dz = dx + i\,dy$, we have

$$\oint_C f(z)\,dz = \oint_C (u\,dx - v\,dy) + i \oint_C (u\,dy + v\,dx). \qquad (2.5.3)$$

Then, using $f'(z)$ continuous, we find that u and v have continuous partial derivatives, hence Theorem 2.5.1 holds, and each of the above line integrals can be converted to the following double integrals for points of D enclosed by C:

$$\oint_C f(z)\,dz = -\iint_D \left(\frac{\partial v}{\partial x} + \frac{\partial u}{\partial y}\right) dx\,dy + i \iint_D \left(\frac{\partial u}{\partial x} - \frac{\partial v}{\partial y}\right) dx\,dy. \quad (2.5.4)$$

Because $f(z)$ is analytic we have that the Cauchy–Riemann conditions (2.1.4) hold:

$$\frac{\partial u}{\partial y} = -\frac{\partial v}{\partial x} \qquad \text{and} \qquad \frac{\partial u}{\partial x} = \frac{\partial v}{\partial y},$$

hence we have $\oint_C f(z)\,dz = 0$.   □

We also note that Cauchy's Theorem can be alternatively stated as: If $f(z)$ is analytic everywhere interior to and on a simple closed contour C, then $\oint_C f(z)\,dz = 0$.

Knowing that $\oint_C f(z)\,dz = 0$ yields numerous results of interest. In particular we will see that this condition for continuous $f(z)$ yields an analytic antiderivative for f.

Theorem 2.5.3 *If $f(z)$ is continuous in a simply connected domain D and if $\oint_C f(z)\,dz = 0$ for every simple closed contour C lying in D, then there exists a function $F(z)$, analytic in D, such that $F'(z) = f(z)$.*

Proof   Consider three points within D: z_0, z and $z + h$. Define F by

$$F(z) = \int_{z_0}^{z} f(z')\,dz', \qquad (2.5.5)$$

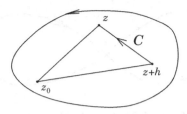

Figure 2.25 Three points lying in D

where the contour from z_0 to z lies within D (see Figure 2.25). Then from $\oint_C f(z)\, dz = 0$ we have

$$\int_{z_0}^{z+h} f(z')\, dz' + \int_{z+h}^{z} f(z')\, dz' + \int_{z}^{z_0} f(z')\, dz' = 0, \qquad (2.5.6)$$

where again all paths must lie within D. Although it may seem that choosing a contour in this way is special, shortly we will show that when $f(z)$ is analytic in D, the integral of $f(z)$ over the simple contour C enclosing the domain D is equivalent to any closed integral along a simple contour inside D.

Then, using Eq. (2.5.6) and reversing the order of integration of the last two terms gives

$$F(z+h) - F(z) - \left(\int_{z_0}^{z+h} - \int_{z_0}^{z} \right) f(z')\, dz' = \int_{z}^{z+h} f(z')\, dz',$$

hence

$$\frac{F(z+h) - F(z)}{h} = \frac{\int_{z}^{z+h} f(z')\, dz'}{h}. \qquad (2.5.7)$$

Because $f(z)$ is continuous, we find, from the definition of the derivative and the properties of real integration, that as $h \to 0$

$$F'(z) = f(z). \qquad \Box$$

We remark that any (nonsimple) contour that has self-intersections can be decomposed into a sequence of contours that are simple. This fact is illustrated in Figure 2.26, where the complete nonsimple contour ("figure eight" contour) can be decomposed into two simple closed contours corresponding to each "loop" of the nonsimple contour. A consequence of this observation is that Cauchy's Theorem can be applied to a nonsimple contour with a finite number of intersections.

In a multiply-connected domain with a function $f(z)$ analytic in this domain we can also apply Cauchy's theorem. The best way to see this is to introduce crosscuts, as mentioned earlier, such that Cauchy's theorem can be applied to a

Figure 2.26 Non-simple contour

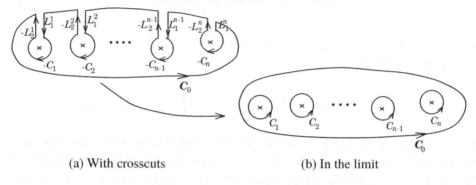

(a) With crosscuts   (b) In the limit

Figure 2.27 Non-simple contour

simple contour. Consider the multiply-connected region depicted in Figure 2.27b with outer boundary C_0 and n holes with boundaries $C_1, C_2, \ldots, C_n$, and introduce n crosscuts $L_1^1 L_2^1, L_1^2 L_2^2, \ldots, L_1^n L_2^n$, as in Figure 2.27a.

Then Cauchy's theorem applies to an analytic function in a domain D with the simple contour

$$\tilde{C} = C_0 - \sum_{j=1}^{n} C_j + \sum_{j=1}^{n} (L_1^j - L_2^j),$$

where we have used the convention that each closed contour is taken in the positive counterclockwise direction, and we take L_1^j, L_2^j in the same direction.

Because the integrals along the crosscuts vanish as the width between the cross-cuts vanishes (i.e., $\int_{L_1^j - L_2^j} f(z)\, dz \to 0$), we have

$$\oint_C f(z)\, dz = 0,$$

where $C = C_0 - \sum_{j=1}^{n} C_j = C_0 + \sum_{j=1}^{n} (-C_j)$. It is often best to interpret the integral

$$\oint_C = \oint_{C_0} + \sum_{j=1}^{n} \oint_{-C_j},$$

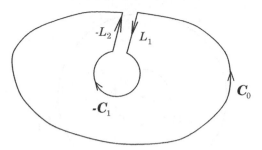

Figure 2.28 Non-intersecting closed curves C_0 and C_1

as one contour with the enclosed region bounded by C as that lying to the *left* of C_0 and to the *right* of the C_j (or to the *left* of $-C_j$). From $\oint_C f(z)dz = 0$ we have

$$\oint_{C_0} f(z)\,dz = \sum_{j=1}^{n} \oint_{C_j} f(z)\,dz, \qquad (2.5.8)$$

with all the contours taken in the counterclockwise direction as depicted in Figure 2.27b. We often say that the contour C_0 has been **deformed** into the contours C_j, $j = 1, \ldots, n$. A simple case is depicted in Figure 2.28.

This is an example of a **deformation** of the contour, deforming C_0 into C_1. By introducing crosscuts it is seen that

$$\oint_{C_0} f(z)\,dz = \oint_{C_1} f(z)\,dz, \qquad (2.5.9)$$

where C_0 and C_1 are two non-intersecting closed curves in which $f(z)$ is analytic on and in the region between C_0 and C_1. With respect to Eq. (2.5.9) we say that C_0 can be **deformed** into C_1, and for the purpose of this integration they are equivalent contours.

The process of introducing crosscuts, and deformation of the contour, effectively allows us to deal with multiply connected regions and closed contours which are not simple. That is, one can think of integrals along such contours as a sum of integrals along simple contours, as long as $f(z)$ is analytic in the relevant region.

Example 2.5.4 Evaluate

$$I = \oint_C \frac{e^z}{z\,(z^2 - 16)}\,dz,$$

where C is the boundary of the annulus between the circles $|z| = 1$, $|z| = 3$ (see Figure 2.29).

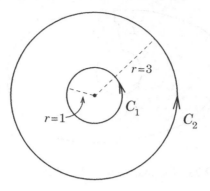

Figure 2.29 Annulus

We note that $C = C_2 + (-C_1)$, and in the region between C_1 and C_2 the function $f(z) = e^z/z(z^2 - 16)$ is analytic because its derivative $f'(z)$ exists and is continuous. The only nonanalytic points are at $z = 0$, $z = \pm 4$; hence, using the deformation of contours discussed above, $I = 0$.

Example 2.5.5   Evaluate

$$I = \frac{1}{2\pi i} \oint_C \frac{dz}{(z - a)^m}, \qquad m = 1, 2, \ldots, M,$$

where C is a simple closed contour.

The function $f(z) = 1/(z - a)^m$ is analytic for all $z \neq a$. Hence if C does not enclose $z = a$, then we have $I = 0$. If C encloses $z = a$ we use Cauchy's Theorem to *deform* the contour to C_a, a small circle of radius $r > 0$ centered at $z = a$ (see Figure 2.30). Namely,

$$\int_C f(z)\, dz - \int_{C_a} f(z)\, dz = 0, \qquad f(z) = 1/(z - a)^m.$$

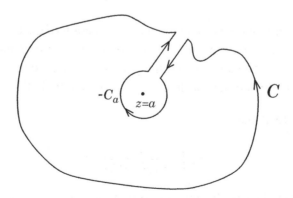

Figure 2.30 Deformed contour around $z = a$

We evaluate $\int_{C_a} f(z)\,dz$ by letting

$$z - a = re^{i\theta}, \qquad dz = ie^{i\theta}\,r\,d\theta,$$

in which case

$$I = \frac{1}{2\pi i} \oint_{C_a} \frac{1}{(z-a)^m}\,dz = \frac{1}{2\pi i} \int_0^{2\pi} \frac{1}{r^m e^{im\theta}} ie^{i\theta}\,r\,d\theta$$

$$= \frac{1}{2\pi i} \int_0^{2\pi} ie^{-i(m-1)\theta} r^{-m+1}\,d\theta = \delta_{m,1} = \begin{cases} 1 & \text{if } m = 1, \\ 0 & \text{otherwise.} \end{cases}$$

Thus,

$$I = \begin{cases} 0, & z = a \quad \text{outside } C, \\ 0, & z = a \quad \text{inside } C, \quad m \neq 1, \\ 1, & z = a \quad \text{inside } C, \quad m = 1. \end{cases}$$

By considering contour integrals over functions $f(z)$ that enclose many points in which $f(z)$ have the local behavior

$$\frac{g_j(z)}{\left(z - a_j\right)^m}, \qquad j = 1, 2, \ldots, N, \quad m = 1, 2, \ldots, M,$$

where $g_j(z)$ is analytic, numerous important results can be obtained. In Chapter 3 we discuss functions with this type of local behavior (we say $f(z)$ has a pole of order m at $z = a_j$). In Chapter 4 we discuss extensions of the crosscut concept and the methods described in Example 2.5.5 will be used to derive the well-known Cauchy Residue Theorem (Theorem 4.1.1). The following example is an application of these kinds of ideas.

Example 2.5.6 Let $P(z)$ be a polynomial of degree n, with n simple roots, none of which lie on a simple closed contour C. Evaluate

$$I = \frac{1}{2\pi i} \oint_C \frac{P'(z)}{P(z)}\,dz.$$

Because $P(z)$ is a polynomial with distinct roots, we can factor it as

$$P(z) = A(z - a_1)(z - a_2) \cdots (z - a_n),$$

where A is the coefficient of the term of highest degree. Because

$$\frac{P'(z)}{P(z)} = \frac{d}{dz}\left(\log P(z)\right)$$

$$= \frac{d}{dz} \log\left(A(z - a_1)(z - a_2) \cdots (z - a_n)\right),$$

it follows that

$$\frac{P'(z)}{P(z)} = \frac{1}{z - a_1} + \frac{1}{z - a_2} + \cdots + \frac{1}{z - a_n}.$$

Hence using the result from Example 2.5.5 above, we have

$$I = \frac{1}{2\pi i} \oint_C \frac{P'(z)}{P(z)} \, dz = \text{number of roots lying within } C.$$

Example 2.5.7 As an example where Cauchy's Theorem cannot be used, consider the integral of the nonanalytic function $f(z) = \bar{z}$ over a simple closed contour C. Breaking the integral into real and imaginary parts we have

$$\oint \bar{z} \, dz = \oint (x - iy)(dx + i \, dy) = \oint (x \, dx + y \, dy) + i \oint (-y \, dx + x \, dy).$$

Now we can use Green's theorem, Equation (2.5.1):

$$\oint (u(x, y) \, dx + v(x, y) \, dy) = \int \int_R \left(\frac{\partial v(x, y)}{\partial x} - \frac{\partial u(x, y)}{\partial y} \right) dx \, dy,$$

with $u(x, y) = x$, $v(x, y) = y$ in the real part and $u(x, y) = -y$, $v(x, y) = x$ in the imaginary part to find

$$\oint \bar{z} \, dz = \int \int_R 2i \, dx \, dy = 2i A, \tag{2.5.10}$$

where A is the enclosed area.

2.5.1 Problems for Section 2.5

1. Evaluate $\oint_C f(z) \, dz$, where C is the unit circle centered at the origin, and $f(z)$ is given by the following:

   (a) e^{iz};   (b) e^{z^2};   (c) $\dfrac{1}{z - \frac{1}{2}}$;   (d) $\dfrac{1}{z^2 - 4}$;

   (e) $\dfrac{1}{2z^2 + 1}$;   (f) $\sqrt{z - 4}$.

2. Use partial fractions to evaluate the following integrals $\oint_C f(z) \, dz$, where C is the unit circle centered at the origin, and $f(z)$ is given by the following:

   (a) $\dfrac{1}{z(z - 2)}$;   (b) $\dfrac{z}{z^2 - \frac{1}{9}}$;   (c) $\dfrac{1}{z(z + \frac{1}{2})(z - 2)}$.

3. Evaluate the following integral

$$\oint_C \frac{e^{iz}}{z(z-\pi)}\, dz$$

for each of the following four cases (all circles are centered at the origin; use Eq. (1.2.19) as necessary):

(a) C is the boundary of the annulus between circles of radius 1 and radius 3;
(b) C is the boundary of the annulus between circles of radius 1 and radius 4;
(c) C is a circle of radius R, where $R > \pi$;
(d) C is a circle of radius R, where $R < \pi$.

4. Evaluate

$$\oint_C \frac{e^{z^2}}{z^3}\, dz,$$

where C is a simple closed curve enclosing the origin; use (1.2.19) as necessary. Explain.

5. Show that $I = \int_C (z-z_0)^n\, dz = 0$ where C is a circle that does not pass through z_0 and $n \neq -1$, for n integer.

6. (a) Consider the integral

$$I = \oint_C \frac{z}{z^2 + 1}\, dz,$$

where C is a circle radius 2 enclosing the origin. Use

$$\frac{z}{z^2+1} = \frac{1}{2}\left(\frac{1}{z+i} + \frac{1}{z-i}\right)$$

to show that $I = \frac{2\pi i}{2} + \frac{2\pi i}{2} = 2\pi i$.

(b) Let $f(z) = \sum_{j=1}^{n} \frac{a_j}{z-z_j}$ where a_j, z_j are constants, to show that

$$\oint_C f(z)\, dz = 2\pi i \sum_{j=1}^{n} a_j,$$

where C is a closed contour enclosing all z_j.

7. (a) Suppose $F = (F_1, F_2) = \nabla\phi$ where F is continuously differentiable in a simply-connected region R. Show from Green's theorem that

$$\oint_C \nabla F \cdot d\mathbf{r} = 0,$$

where C is a simple closed curve enclosing R.

(b) Let $\Omega(z) = \phi + i\psi$ where ϕ, ψ satisfy the Cauchy–Riemann conditions:

$$\phi_x = \psi_y, \qquad \phi_y = -\psi_x.$$

Show that $\Omega'(z) = \phi_x + i\psi_x = \phi_x - i\phi_y$ and using the result of part (a)

$$\oint_C \Omega'(z)\, dz = 0,$$

where C is a simple closed contour enclosing a simply connected region R.

8. Evaluate

$$\oint_C \bar{z}^2\, dz,$$

where C is a circle with radius R centered at: a) $z = 0$ and b) $z = 1$.

9. We wish to evaluate the integral $I = \int_0^\infty e^{ix^2}\, dx$. Consider the contour $I_R = \oint_{C_{(R)}} e^{iz^2}\, dz$, where $C_{(R)}$ is the closed circular sector in the upper half plane with boundary points 0, R, and $Re^{i\pi/4}$. Show that $I_R = 0$ and that $\lim_{R\to\infty} \int_{C_{1(R)}} e^{iz^2}\, dz = 0$, where $C_{1(R)}$ is the line integral along the circular sector from R to $Re^{i\pi/4}$. (Hint: use $\sin x \geq \frac{2x}{\pi}$ on $0 \leq x \leq \frac{\pi}{2}$.) Then, breaking up the contour $C_{(R)}$ into three component parts, deduce

$$\lim_{R\to\infty} \left(\int_0^R e^{ix^2}\, dx - e^{i\pi/4} \int_0^R e^{-r^2}\, dr \right) = 0,$$

and from the well-known result of real integration, $\int_0^\infty e^{-x^2}\, dx = \sqrt{\pi}/2$, deduce that $I = e^{i\pi/4}\sqrt{\pi}/2$.

10. Consider the integral $I = \int_{-\infty}^\infty \dfrac{dx}{x^2 + 1}$. Show how to evaluate this integral by considering $\oint_{C_{(R)}} \dfrac{dz}{z^2 + 1}$, where $C_{(R)}$ is the closed semicircle in the upper half plane with endpoints at $(-R, 0)$ and $(R, 0)$ plus the x-axis. Hint: use

$$\frac{1}{(z^2 + 1)} = -\frac{1}{2i}\left(\frac{1}{z+i} - \frac{1}{z-i} \right),$$ and show that the integral along the open semicircle in the upper half plane vanishes as $R \to \infty$. Verify your answer by usual integration in real variables.

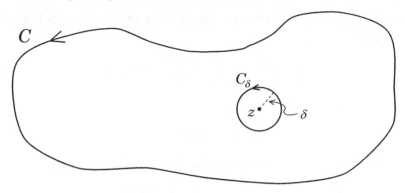

Figure 2.31 Circle C_δ inscribed in contour C

2.6 Cauchy's Integral Formula, its $\bar{\partial}$ Generalization, and Consequences

In this section we discuss a number of fundamental consequences and extensions of the ideas presented in earlier sections, especially Cauchy's Theorem. The section *2.6.3 is more difficult and can be skipped entirely or returned to when desired.

2.6.1 Cauchy's Integral Formula and its derivatives

An important result owing to Cauchy shows that the values of an analytic function f on the boundary of a closed contour C determine the values of f interior to C.

Theorem 2.6.1 *Let $f(z)$ be analytic interior to and on a simple closed contour C. Then at any interior point z,*

$$f(z) = \frac{1}{2\pi i} \oint_C \frac{f(\zeta)}{\zeta - z} \, d\zeta. \tag{2.6.1}$$

Equation (2.6.1) is referred to as Cauchy's Integral Formula.

Proof Inside the contour C, inscribe a small circle C_δ, radius δ with center at point z (see Figure 2.31).

From Cauchy's Theorem we can deform the contour C into C_δ:

$$\oint_C \frac{f(\zeta)}{\zeta - z} \, d\zeta = \oint_{C_\delta} \frac{f(\zeta)}{\zeta - z} \, d\zeta. \tag{2.6.2}$$

We rewrite the second integral as

$$\oint_{C_\delta} \frac{f(\zeta)}{\zeta - z} \, d\zeta = f(z) \oint_{C_\delta} \frac{d\zeta}{\zeta - z} + \oint_{C_\delta} \frac{f(\zeta) - f(z)}{\zeta - z} \, d\zeta. \tag{2.6.3}$$

Using polar coordinates, $\zeta = z + \delta e^{i\theta}$, the first integral on the right in Eq. (2.6.3) is computed to be

$$\oint_{C_\delta} \frac{d\zeta}{\zeta - z} = \int_0^{2\pi} \frac{i\delta e^{i\theta}}{\delta e^{i\theta}} \, d\theta = 2\pi i. \qquad (2.6.4)$$

Because $f(z)$ is continuous,

$$|f(\zeta) - f(z)| < \epsilon$$

for small enough $|z - \zeta| = \delta$. Then (see also the inequality (2.4.15))

$$\left| \oint_{C_\delta} \frac{f(\zeta) - f(z)}{\zeta - z} \, d\zeta \right| \le \oint_{C_\delta} \frac{|f(\zeta) - f(z)|}{|\zeta - z|} \, |d\zeta|$$

$$< \frac{\epsilon}{\delta} \int_{C_\delta} |d\zeta|$$

$$= 2\pi\epsilon.$$

Thus as $\epsilon \to 0$, the second integral in Eq. (2.6.3) vanishes. Hence Eqs. (2.6.3) and (2.6.4) yield Cauchy's Integral Formula, Eq. (2.6.1). ▫

A particularly simple example of Cauchy's integral formula is the following. If on the unit circle $|\zeta| = 1$ we are given $f(\zeta) = \zeta$, then by Eq. (2.6.1),

$$\frac{1}{2\pi i} \oint_C \frac{\zeta}{\zeta - z} \, d\zeta = z.$$

An alternative way to obtain this answer is as follows:

$$\frac{1}{2\pi i} \oint_C \frac{\zeta}{\zeta - z} \, d\zeta = \frac{1}{2\pi i} \oint_C \left(1 + \frac{z}{\zeta - z} \right) d\zeta$$

$$= \frac{1}{2\pi i} [\zeta + z \log(\zeta - z)]_C$$

$$= z,$$

where we use the notation $[\cdot]_C$ to denote the change around the unit circle, and we have selected some branch of the logarithm.

A corollary of Cauchy's theorem demonstrates that the derivatives of $f(z)$: $f'(z), f''(z), \ldots, f^{(n)}(z)$ all exist and there is a simple formula for them. Thus the analyticity of $f(z)$ implies the analyticity of all its derivatives.

Theorem 2.6.2 *If $f(z)$ is analytic interior to and on a simple closed contour C, then all the derivatives $f^{(k)}(z)$, $k = 1, 2, \ldots$ exist in the domain D interior to C, and*

$$f^{(k)}(z) = \frac{k!}{2\pi i} \oint_C \frac{f(\zeta)}{(\zeta - z)^{k+1}} \, d\zeta. \qquad (2.6.5)$$

Proof   Let z be any point in D. It will be shown that all the derivatives of $f(z)$ exist at z. Because z is arbitrary, this establishes the existence of all derivatives in D.

We begin by establishing Eq. (2.6.5) for $k = 1$. Consider the usual difference quotient:

$$\frac{f(z+h) - f(z)}{h} = \frac{1}{2\pi i} \frac{1}{h} \oint_C f(\zeta) \left(\frac{1}{\zeta - (z+h)} - \frac{1}{\zeta - z} \right) d\zeta$$

$$= \frac{1}{2\pi i} \oint_C \frac{f(\zeta)}{(\zeta - (z+h))(\zeta - z)} d\zeta$$

$$= \frac{1}{2\pi i} \oint_C \frac{f(\zeta)}{(\zeta - z)^2} d\zeta + R, \qquad (2.6.6)$$

where

$$R = \frac{h}{2\pi i} \oint_C \frac{f(\zeta)}{(\zeta - z)^2 (\zeta - z - h)} d\zeta. \qquad (2.6.7)$$

We shall put $\min |\zeta - z| = 2\delta > 0$. Then if $|h| < \delta$ for ζ on C, we have

$$|\zeta - (z+h)| \geq |\zeta - z| - |h| > 2\delta - \delta = \delta.$$

Because $|f(\zeta)| < M$ on C, then

$$|R| \leq \frac{|h|}{2\pi} \frac{M}{(2\delta)^2 \delta} L, \qquad (2.6.8)$$

where L is the length of the contour C. Because $|R| \to 0$ as $h \to 0$, we have established Eq. (2.6.5) for $k = 1$:

$$f'(z) = \frac{1}{2\pi i} \oint_C \frac{f(\zeta)}{(\zeta - z)^2} d\zeta. \qquad (2.6.9)$$

We may repeat the above argument beginning with Eq. (2.6.9) and thereby prove the existence of $f''(z)$; that is, Eq. (2.6.5) for $k = 2$. This shows that f' has a derivative f'', and so is itself analytic. Consequently we find that if $f(z)$ is analytic, so is $f'(z)$. Applying this argument to f' instead of f proves that f'' is analytic, and, more generally, the analyticity of $f^{(k)}$ implies the analyticity of $f^{(k+1)}$. By induction, we find that all the derivatives exist and hence are analytic. Because $f^{(k)}(z)$ is analytic, Eq. (2.6.1) gives

$$f^{(k)}(z) = \frac{1}{2\pi i} \oint_C \frac{f^{(k)}(\zeta)}{\zeta - z} d\zeta. \qquad (2.6.10)$$

Integration by parts (k) times (the boundary terms vanish) yields Eq. (2.6.5).   □

An immediate consequence of this result is the following.

Theorem 2.6.3 *All partial derivatives of u and v are continuous at any point where $f = u + iv$ is analytic.*

For example, the first derivative of $f(z)$, using the Cauchy–Riemann equations, is

$$f'(z) = u_x + iv_x = v_y - iu_y. \tag{2.6.11}$$

Because $f'(z)$ is analytic, it is certainly continuous. The continuity of $f'(z)$ ensures that u_x, v_y, v_x, and u_y are all continuous. Similar arguments are employed for the higher-order derivatives, $u_{xx}, u_{yy}, u_{xy}, \ldots$.

2.6.2 The Liouville, Morera, and Maximum-Modulus Theorems

First we establish a useful inequality. From

$$f^{(n)}(z) = \frac{n!}{2\pi i} \oint_C \frac{f(\zeta)}{(\zeta - z)^{n+1}} \, d\zeta, \tag{2.6.12}$$

where C is a circle, $|\zeta - z| = R$, and $|f(z)| < M$, we have

$$|f^{(n)}(z)| \leq \frac{n!}{2\pi} \oint_C \frac{|f(\zeta)|}{|\zeta - z|^{n+1}} \, |d\zeta|$$

$$\leq \frac{n!M}{2\pi R^{n+1}} \oint_C |d\zeta|$$

$$\leq \frac{n!M}{R^n}. \tag{2.6.13}$$

With Eq. (2.6.13) we can derive a result about functions that are everywhere analytic in the finite complex plane. Such functions are called **entire**.

Theorem 2.6.4 (Liouville) *If $f(z)$ is entire and bounded in the z-plane (including infinity), then $f(z)$ is a constant.*

Proof Using the inequality (2.6.13) with $n = 1$ we have

$$|f'(z)| \leq \frac{M}{R}.$$

Because this is true for any point z in the plane, we can make R arbitrarily large; hence $f'(z) = 0$ for any point z in the plane. Because

$$f(z) - f(0) = \int_0^z f'(\zeta) \, d\zeta = 0,$$

we have $f(z) = f(0) = $ constant, and the theorem is proven. □

Cauchy's theorem tells us that if $f(z)$ is analytic inside C, then $\oint_C f(z) \, dz = 0$. Now we prove that the converse is also true.

Theorem 2.6.5 (Morera) *If $f(z)$ is continuous in a domain D and if*

$$\oint_C f(z)\, dz = 0$$

for every simple closed contour C lying in D, then $f(z)$ is analytic in D.

Proof From Theorem 2.5.3 it follows that if the contour integral always vanishes, then there exists an analytic function $F(z)$ in D such that $F'(z) = f(z)$. Theorem 2.6.2 implies that $F'(z)$ is analytic if $F(z)$ is analytic, hence so is $f(z)$. □

A corollary to Liouville's theorem is the so-called Fundamental Theorem of Algebra; namely, any polynomial

$$P(z) = a_0 + a_1 z + \cdots + a_m z^m, \qquad a_m \neq 0, \tag{2.6.14}$$

with $m \geq 1$ an integer, has at least one point $z = \alpha$ such that $P(\alpha) = 0$; that is, $P(z)$ has at least one root.

We establish this statement by contradiction. If $P(z)$ does not vanish, then the function $Q(z) = 1/P(z)$ is analytic (has a derivative) in the finite z-plane. For $|z| \to \infty$, $P(z) \to \infty$; hence $Q(z)$ is bounded in the entire complex plane, including infinity. Liouville's Theorem then implies that $Q(z)$ and hence $P(z)$ is a constant, which violates $m \geq 1$ in Eq. (2.6.14) and thus contradicts the assumption that $P(z)$ does not vanish. In Section 4.4 it is shown that $P(z)$ has m and only m roots, including multiplicities.

There are a number of valuable statements that can be made about the maximum (minimum) modulus an analytic function can achieve, and certain mean value formulae can be ascertained.

For example, using Cauchy's integral formula (2.6.1) with C being a circle centered at z and radius r, we have $\zeta - z = re^{i\theta}$, and $d\zeta = ire^{i\theta}\, d\theta$; hence Eq. (2.6.1) becomes

$$f(z) = \frac{1}{2\pi} \int_0^{2\pi} f\left(z + re^{i\theta}\right) d\theta. \tag{2.6.15}$$

Equation (2.6.15) is a "mean-value" formula; that is, the value of an analytic function at any interior point is the "mean" of the function integrated over the circle centered at z. Similarly multiplying Eq. (2.6.15) by $r\, dr$, and integrating over a circle of radius R yields

$$f(z) \int_0^R r\, dr = \frac{1}{2\pi} \int_0^R \int_0^{2\pi} f\left(z + re^{i\theta}\right) r\, dr\, d\theta,$$

hence

$$f(z) = \frac{1}{\pi R^2} \iint_{D_0} f\left(z + re^{i\theta}\right) dA, \tag{2.6.16}$$

where D_0 is the region inside the circle C, radius R, center z.

Thus the value of $f(z)$ also equals its mean value over the area of a circle centered at z.

This result can be used to establish the following maximum-modulus theorem.

Theorem 2.6.6 (Maximium principles)   **(i)** *If $f(z)$ is analytic in a domain D, then $|f(z)|$ cannot have a maximum in D unless $f(z)$ is a constant.*

(ii) *If $f(z)$ is analytic in a bounded region D and $|f(z)|$ is continuous in the closed region $\overline{D}$, then $|f(z)|$ assumes its maximum on the boundary of the region.*

Proof   Equation (2.6.16) is a useful device for establishing this result. Suppose z is an interior point in the region such that $|f(\zeta)| \le |f(z)|$ for all points ζ in the region. Choose any circle center z radius R such that the circle lies entirely in the region. Calling $\zeta = z + re^{i\theta}$ for any point in the circle, we have, from Eq. (2.6.16),

$$|f(z)| \le \frac{1}{\pi R^2} \iint_{D_0} |f(\zeta)|\, dA. \tag{2.6.17}$$

Actually, the assumed inequality $|f(\zeta)| \le |f(z)|$ substituted into Eq. (2.6.17) implies that in fact $|f(\zeta)| = |f(z)|$ because if in any subregion equality, did not hold, Eq. (2.6.17) would imply $|f(z)| < |f(z)|$. Thus the modulus of $f(z)$ is constant. Use of the Cauchy–Riemann equations then shows that if $|f(z)|$ is constant, then $f(z)$ is also constant (see Example 2.1.5). This establishes the maximum principle **(i)** inside C.

Because $f(z)$ is analytic within and on the circle C, then $|f(z)|$ is continuous. A result of real variables states that a continuous function in a bounded region must assume a maximum somewhere in the closed bounded region, including the boundary. Hence the maximum for $|f(z)|$ must be achieved on the boundary of the circle C, and the maximum principle **(ii)** is established for the circle.

In order to extend these results to more general regions we may construct appropriate new circles centered at interior points of D and overlapping with the old ones. In this way, by using a sequence of such circles, the region can be filled and the above results follow.   □

We note that if $f(z)$ does not vanish at any point inside the contour, by considering $1/(f(z)) = g(z)$ it can be seen that $|g(z)|$ also attains its maximum value on the boundary, and hence $f(z)$ attains its minima on the boundary.

The real and imaginary parts, u and v, of an analytic function $f(z) = u(x, y) + iv(x, y)$, attain their maximum values on the boundary. This follows from the fact that $g(z) = \exp(f(z))$ is analytic, and hence it satisfies the maximum principle. Thus the modulus $|g(z)| = \exp u(x, y)$ must achieve its maximum   value on the boundary. Similar arguments for a function $g(z) = \exp(-if(z))$ yield analogous results for $v(x, y)$. Now, because $f(z)$ is analytic, we have from Theorem 2.6.2 and Eq. (2.6.11) that u and v are infinitely differentiable. Furthermore, from the Cauchy–Riemann conditions, u and v are harmonic functions, that is, they satisfy Laplace's equation

$$\nabla^2 u = 0, \qquad \nabla^2 v = 0 \tag{2.6.18}$$

(see Section 2.1, e.g. Eqs. 2.1.11a,b). Hence the maximum principle says that the harmonic functions $u = \mathrm{Re}\, f$ or $v = \mathrm{Im}\, f$, $f = u + iv$, achieve their maxima (and minima by a similar proof) on the boundary of the region.

*2.6.3 Generalized Cauchy formula and $\bar{\partial}$ derivatives

In previous sections we concentrated on analytic functions or functions that are analytic everywhere apart from isolated "singular" points where the function blows up or possesses branch points/cuts. On the other hand, as mentioned earlier (see, for example, Section 2.1, Example 2.1.3 and the subsequent discussion) there are functions which are nowhere analytic. For example, the Cauchy–Riemann conditions show that the function $f(z) = \bar{z}$ (and hence any function of $\bar{z}$) is nowhere analytic. The reader might mistakenly think that such functions are mathematical artifacts. However, mathematical formulations of physical phenomena are often described via such complicated nonanalytic functions. In fact, the main result, Theorem 2.6.7, described in this section, is used in an essential way to study the scattering and inverse scattering theory associated with certain problems arising in nonlinear wave propagation (Ablowitz, Bar Yaacov, and Fokas, 1983). Despite the fact that Cauchy's integral formula, (2.6.1), requires that $f(z)$ be an analytic function, there is nevertheless an important extension, which we shall develop below, that extends Cauchy's Integral Theorem to certain nonanalytic functions.

From the coordinate representation $z = x + iy, \bar{z} = x - iy$, we have $x = (z + \bar{z})/2$ and $y = (z - \bar{z})/2i$. Using the chain rule,

$$\frac{\partial}{\partial z} = \frac{\partial x}{\partial z}\frac{\partial}{\partial x} + \frac{\partial y}{\partial z}\frac{\partial}{\partial y},$$

we find

$$\frac{\partial}{\partial z} = \frac{1}{2}\left(\frac{\partial}{\partial x} - i\frac{\partial}{\partial y}\right), \tag{2.6.19a}$$

$$\frac{\partial}{\partial \bar{z}} = \frac{1}{2}\left(\frac{\partial}{\partial x} + i\frac{\partial}{\partial y}\right). \tag{2.6.19b}$$

Sometimes it is convenient to consider the function $f(x, y)$ as depending explicitly on both z and $\bar{z}$; that is, $f = f(z, \bar{z})$. For simplicity we still use the notation $f(z)$ to denote $f(z, \bar{z})$. If f is a differentiable function of z and $\bar{z}$, and

$$\frac{\partial f}{\partial \bar{z}} = 0 \tag{2.6.20}$$

then we say that $f = f(z)$. Moreover, any $f(z)$ satisfying Eq. (2.6.20) is an analytic function, because from Eq. (2.6.19b) and $f(z) = u + iv$, we find, from Eq. (2.6.20), that $\frac{\partial u}{\partial x} - \frac{\partial v}{\partial y} + i(\frac{\partial u}{\partial y} + \frac{\partial v}{\partial x}) = 0$, hence u and v satisfy the Cauchy–Riemann equations.

In what follows we shall use Green's Theorem, Eq. (2.5.1), in the following form:

$$\oint_C g \, d\zeta = 2i \iint_R \frac{\partial g}{\partial \bar{\zeta}}, dA(\zeta) \tag{2.6.21}$$

where $\zeta = \xi + i\eta$, $d\zeta = d\xi + i \, d\eta$, and $dA(\zeta) = d\xi \, d\eta$. Note in Eq. (2.5.1) use $u = g$, $v = ig$,

$$\frac{\partial g}{\partial \bar{\zeta}} = \frac{1}{2} \left(\frac{\partial g}{\partial \xi} + i \frac{\partial g}{\partial \eta} \right)$$

and replace x and y by ξ and η.

Next we establish the following:

Theorem 2.6.7 (Generalized Cauchy Formula)   *If $\partial f / \partial \bar{\zeta}$ exists and is continuous in a region R bounded by a simple closed contour C, and if f is continuous on C, then at any interior point z*

$$f(z) = \frac{1}{2\pi i} \oint_C \left(\frac{f(\zeta)}{\zeta - z} \right) d\zeta - \frac{1}{\pi} \iint_R \left(\frac{\partial f / \partial \bar{\zeta}}{\zeta - z} \right) dA(\zeta). \tag{2.6.22}$$

Proof   Consider Green's Theorem in the form of Eq. (2.6.21) in the region R_ϵ depicted in Figure 2.32, with $g = f(\zeta)/(\zeta - z)$ and the contour composed of two parts C and C_ϵ.

We have, from Eq. (2.6.21), noting that $\frac{1}{\zeta - z}$ is analytic in this region,

$$\oint_C \frac{f(\zeta)}{\zeta - z} d\zeta - \int_{C_\epsilon} \frac{f(\zeta)}{\zeta - z} d\zeta = 2i \iint_{R_\epsilon} \frac{\partial f / \partial \bar{\zeta}}{\zeta - z} dA. \tag{2.6.23}$$

Note that on C_ϵ we have $\zeta = z + \epsilon e^{i\theta}$

$$\oint_{C_\epsilon} \frac{f(\zeta)}{\zeta - z} d\zeta = \int_0^{2\pi} \frac{f(z + \epsilon e^{i\theta})}{\epsilon e^{i\theta}} i\epsilon e^{i\theta} \, d\theta$$

$$= \int_0^{2\pi} f\left(z + \epsilon e^{i\theta}\right) i \, d\theta$$

$$\xrightarrow[\epsilon \to 0]{} f(z)(2\pi i). \tag{2.6.24}$$

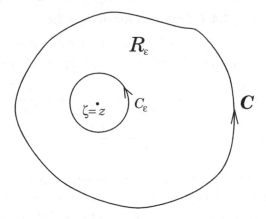

Figure 2.32 Generalized Cauchy formula in region R_ϵ

The limit result is due to the fact that $f(z)$ is assumed to be continuous, and from real variables we find that the limit $\epsilon \to 0$ and the integral of a continuous function over a bounded region can be interchanged. Similarly, because $1/(\zeta - z)$ is integrable over R_ϵ and $\partial f/\partial\bar{\zeta}$ is continuous, then the double integral over R_ϵ converges to the double integral over the whole region R, the difference tending to zero with ϵ: namely, using polar coordinates $\zeta = z + re^{i\theta}$

$$\left| \iint_{R-R_\epsilon} \left(\frac{\partial f/\partial\bar{\zeta}}{\zeta - z} \right) i \, dA \right| \leq \int_0^\epsilon \int_0^{2\pi} \frac{\left| \partial f/\partial\bar{\zeta} \right|}{r} r \, dr \, d\theta$$

$$\leq 2\pi M \epsilon. \tag{2.6.25}$$

Using the continuity of $\partial f/\partial\bar{\zeta}$ in a bounded region implies that

$$\left| \frac{\partial f}{\partial\bar{\zeta}} \right| \leq M.$$

Thus Eq. (2.6.23) yields in the limit $\epsilon \to 0$

$$\oint_C \left(\frac{f(\zeta)}{\zeta - z} \right) d\zeta - 2\pi i f(z) = 2i \iint_R \left(\frac{\partial f/\partial\bar{\zeta}}{\zeta - z} \right) i \, dA,$$

and hence the generalized Cauchy formula, (2.6.22), follows by manipulation. $\square$

We note that if $\partial f/\partial\bar{\zeta} = 0$, that is, if $f(z)$ is analytic inside R, then the generalized Cauchy formula reduces to the usual Cauchy Integral Formula, (2.6.1).

2.6.4 Problems for Section 2.6

1. Evaluate the integrals $\oint_C f(z)dz$, where C is the unit circle centered at the origin and $f(z)$ is given by the following (use Eq. (1.2.19) as necessary):

   (a) $\dfrac{\sin z}{z}$;　　(b) $\dfrac{1}{(2z-1)^2}$;　　(c) $\dfrac{1}{(2z-1)^3}$;

   (d) $\dfrac{e^z}{z}$;　　(e) $e^{z^2}\left(\dfrac{1}{z^2} - \dfrac{1}{z^3}\right)$.

2. Evaluate the integrals $\oint_C f(z)\,dz$ over a contour C, where C is the boundary of a square with diagonal opposite corners at $z = -(1+i)\,R$ and $z = (1+i)\,R$, where $R > a > 0$, and where $f(z)$ is given by the following (use Eq. (1.2.19) as necessary):

   (a) $\dfrac{e^z}{z - \frac{\pi i}{4}a}$;　　(b) $\dfrac{e^z}{\left(z - \frac{\pi i}{4}a\right)^2}$;　　(c) $\dfrac{z^2}{2z+a}$;

   (d) $\dfrac{\sin z}{z^2}$;　　(e) $\dfrac{\cosh z}{z}$.

3. Evaluate the integral

   $$\int_{-\infty}^{\infty} \frac{dx}{(x+i)^2}$$

   by considering $\oint_{C_{(R)}} (1/(z+i)^2)\,dz$, where $C_{(R)}$ is the closed semicircle in the upper half plane with corners at $z = -R$ and $z = R$, plus the x-axis. Hint: show that

   $$\lim_{R\to\infty} \int_{C_{1(R)}} \frac{1}{(z+i)^2}\,dz = 0,$$

   where $C_{1(R)}$ is the open semicircle in the upper half plane (not including the x-axis).

4. Show that

   $$\oint_{C_0} \frac{\zeta^3 + \zeta + 2}{(\zeta - z)^3}\,d\zeta = 6\pi i z$$

   where C_0 is the unit circle and z is a point interior to C_0.

5. Show that

   $$\oint_C \frac{f''(\zeta)}{(\zeta - z)}\,d\zeta = \oint_C \frac{f'(\zeta)}{(\zeta - z)^2}\,d\zeta,$$

   where the point z is interior to the simple closed contour C and $f(z)$ is analytic on and inside C.

6. Let $f(z)$ be analytic in a square containing a point w, and C be a circle with center ω and radius ρ inside the square. From Cauchy's Theorem show that

$$f(\omega) = \frac{1}{2\pi} \int_0^{2\pi} f\left(\omega + \rho e^{i\theta}\right) d\theta.$$

7. Consider two entire functions with no zeroes and having a ratio equal to unity at infinity. Use Liouville's theorem to show that they are in fact the same function.

8. Let $f(z)$ be analytic and nonzero in a region R. Show that $|f(z)|$ has a minimum value in R which occurs on the boundary. Hint: use the Maximum-Modulus Theorem for the function $1/f(z)$.

9. Let $f(z)$ be an entire function, with $|f(z)| \leq C|z|$ for all z, where C is a constant. Show that $f(z) = Az$, where A is a constant.

10. Find the $\bar{\partial}$ (dbar) derivative of the following functions:

$$\text{(a) } e^z; \qquad \text{(b) } z\bar{z} = r^2.$$

Verify the generalized Cauchy formula inside a circle of radius R for both of these functions. Hint: Reduce problem (b) to the verification of the following formula:

$$-\pi\bar{z} = \iint_A \frac{dA}{\zeta - z} = \iint_A \frac{d\xi \, d\eta}{\zeta - z} \equiv I,$$

where A is a circle of radius r. To establish this result, transform the integral I to polar coordinates, $\zeta - \xi + i\eta - re^{i\theta}$, and find

$$I = \int_0^{2\pi} \int_0^R \frac{r \, dr \, d\theta}{re^{i\theta} - z}.$$

In the θ integral, change variables to $u = e^{i\theta}$, and use $du = ie^{i\theta} d\theta$, $\int_0^{2\pi} d\theta = \frac{1}{i} \oint_{C_0} \frac{du}{u}$, where C_0 is the unit circle. The methods of Section 2.5 can be employed to calculate this integral. Show that we have

$$I = 2\pi \int_0^R r \, dr \left[-\frac{1}{z} + \frac{1}{z} H\left(1 - \frac{|z|}{r}\right) \right],$$

where $H(x) = \{1 \text{ if } x > 0, \ 0 \text{ if } x < 0\}$. Then show that $I = -\pi|z|^2/z = -\pi\bar{z}$ as required.

11. Use Morera's Theorem to verify that the following functions are indeed analytic inside a circle of radius R:

$$\text{(a) } z^n, \quad n \geq 0; \qquad \text{(b) } e^z.$$

From Morera's Theorem, what can be said about the following functions?

$$\text{(c)} \quad \frac{\sin z}{z}; \qquad \text{(d)} \quad \frac{e^z}{z}.$$

12. In Cauchy's integral formula, (2.6.1), take the contour to be a circle of unit radius centered at the origin. Let $\zeta = e^{i\theta}$ to deduce

$$f(z) = \frac{1}{2\pi} \int_0^{2\pi} \frac{f(\zeta)\zeta}{\zeta - z} d\theta,$$

where z lies inside the circle. Explain why we have

$$0 = \frac{1}{2\pi} \int_0^{2\pi} \frac{f(\zeta)\zeta}{\zeta - 1/\bar{z}} d\theta$$

and use $\zeta = 1/\bar{\zeta}$ to show

$$f(z) = \frac{1}{2\pi} \int_0^{2\pi} f(\zeta) \left(\frac{\zeta}{\zeta - z} \pm \frac{\bar{z}}{\bar{\zeta} - \bar{z}} \right) d\theta,$$

whereupon, using the plus sign

$$f(z) = \frac{1}{2\pi} \int_0^{2\pi} f(\zeta) \frac{\left(1 - |z|^2\right)}{|\zeta - z|^2} d\theta.$$

(a) Deduce the "Poisson formula" for the real part of $f(z)$: $u(r, \phi) = \text{Re } f$, $z = re^{i\phi}$;

$$u(r, \phi) = \frac{1}{2\pi} \int_0^{2\pi} u(\theta) \frac{1 - r^2}{\left[1 - 2r\cos(\phi - \theta) + r^2\right]} d\theta,$$

where $u(\theta) = u(1, \theta)$.

(b) If we use the minus sign in the formula for $f(z)$ above, show that

$$f(z) = \frac{1}{2\pi} \int_0^{2\pi} f(\zeta) \left[\frac{1 + r^2 - 2re^{i(\theta-\phi)}}{1 - 2r\cos(\phi - \theta) + r^2} \right] d\theta$$

and by taking the imaginary part

$$v(r, \phi) = C + \frac{1}{\pi} \int_0^{2\pi} u(\theta) \frac{r\sin(\phi - \theta)}{\left[1 - 2r\cos(\phi - \theta) + r^2\right]} d\theta,$$

where $C = \frac{1}{2\pi} \int_0^{2\pi} v(1, \theta) d\theta = v(r = 0)$. (This last relationship follows from Cauchy's Integral formula at $z = 0$ – see the Eq. (2.6.4).)

(c) Show that

$$\frac{2r\sin(\phi - \theta)}{1 - 2r\cos(\phi - \theta) + r^2} = \text{Im}\left[\frac{1 - r^2 + 2ir\sin(\phi - \theta)}{1 + r^2 - 2r\cos(\phi - \theta)}\right]$$

$$= \text{Im}\left[\frac{\zeta + z}{\zeta - z}\right]$$

and therefore the result for $v(r, \phi)$ from part (b) may be expressed as

$$v(r, \phi) = v(0) + \frac{1}{2\pi}\text{Im}\int_0^{2\pi} u(\theta)\frac{\zeta + z}{\zeta - z}\,d\theta.$$

This example illustrates that prescribing the real part of $f(z)$ on $|z| = 1$ determines (a) the real part of $f(z)$ everywhere inside the circle, and (b) the imaginary part of $f(z)$ inside the circle to within a constant. We *cannot* arbitrarily specify both the real and imaginary parts of an analytic function on $|z| = 1$.

13. The "complex delta function" possesses the following property;

$$\iint_A \delta(z - z_0)F(z)\,dA(z) = F(z_0)$$

or

$$\iint_A \delta(x - x_0)\delta(y - y_0)F(x, y)\,dA(x, y) = F(x_0, y_0),$$

where $z_0 = x_0 + iy_0$ is contained within the region A.

In (2.6.21) let $g(z) = F(z)/(z - z_0)$, where $F(z)$ is analytic in A. Show that

$$\oint_C \frac{F(z)}{z - z_0}\,dz = 2i\iint_A F(z)\frac{\partial}{\partial\bar{z}}\left(\frac{1}{z - z_0}\right)\,dA$$

where C is a simple closed curve enclosing the region A. Use $\oint_C F(z)/(z - z_0)\,dz = 2\pi iF(z_0)$ to establish

$$F(z_0) = \frac{1}{\pi}\iint_A F(z)\frac{\partial}{\partial\bar{z}}\left(\frac{1}{z - z_0}\right)\,dA$$

and therefore the action of $\partial/\partial\bar{z}\,(1/(z - z_0))$ is that of a complex delta function; that is, $\partial/\partial\bar{z}\,(1/(z - z_0)) = \pi\delta(z - z_0)$.

3

Sequences, Series and Singularities of Complex Functions

The representation of complex functions frequently requires the use of infinite series expansions. The best known are Taylor and Laurent series, which represent analytic functions in appropriate domains. Applications often require that we manipulate series by termwise differentiation and integration. These operations may be substantiated by employing the notion of uniform convergence. Series expansions break down at points or curves where the represented function is not analytic. Such locations are termed singular points or singularities of the function. The study of the singularities of analytic functions is vitally important in many applications including contour integration, differential equations in the complex plane, and conformal mappings.

3.1 Definitions of Complex Sequences, Series and their Basic Properties

Consider the following sequence of complex functions: $f_n(z)$ for $n = 1, 2, 3, \ldots$, defined in a region $\mathcal{R}$ of the complex plane. Usually, we denote the sequence of functions by $\{f_n(z)\}$, where $n = 1, 2, 3, \ldots$. The notion of convergence of a sequence is really the same as that of a limit. We say the sequence $f_n(z)$ **converges** to $f(z)$ on $\mathcal{R}$ or a suitable subset of $\mathcal{R}$, assuming that $f(z)$ exists and is finite, if

$$\lim_{n \to \infty} f_n(z) = f(z). \tag{3.1.1}$$

This means that for each z, given $\epsilon > 0$ there is an N depending on ϵ and z, such that whenever $n > N$ we have

$$|f_n(z) - f(z)| < \epsilon. \tag{3.1.2}$$

If the limit does not exist (or is infinite), we say the sequence **diverges** for those values of z.

An infinite series may be viewed as an infinite sequence, $\{s_n(z)\}, n = 1, 2, 3, \ldots$ by noting that a sequence of partial sums may be formed by

$$s_n(z) = \sum_{j=1}^{n} b_j(z) \tag{3.1.3}$$

and taking the infinite series as the infinite limit of partial sums:

$$S(z) = \lim_{n \to \infty} s_n(z) = \sum_{j=1}^{\infty} b_j(z). \tag{3.1.4}$$

Conversely, given the sequence of partial sums we may find the sequence of terms $b_j(z)$ via: $b_1(z) = s_1(z), b_j(z) = s_j(z) - s_{j-1}(z), j \geq 2$. With this correspondence, no real distinction exists between a series and a sequence.

A basic property of a convergent series such as Eq. (3.1.4) is:

$$\lim_{j \to \infty} b_j(z) = 0$$

because

$$\lim_{j \to \infty} b_j(z) = \lim_{j \to \infty} s_j(z) - \lim_{j \to \infty} s_{j-1}(z) = S - S = 0. \tag{3.1.5}$$

Thus a necessary condition for convergence is Eq. (3.1.5).

We say that the sequence of functions $s_n(z)$, defined for z in a region $\mathcal{R}$, **converges uniformly** in $\mathcal{R}$ if it is possible to choose N depending on ϵ only (and not z): $N = N(\epsilon)$ in Eq. (3.1.1). In other words, the same estimate for N holds for all z in the domain $\mathcal{R}$; that is, we may establish the validity of the limit process independently of which particular z we choose in $\mathcal{R}$.

For example, consider the sequence of functions

$$f_n(z) = \frac{1}{nz}, \qquad n = 1, 2, \ldots. \tag{3.1.6}$$

In the annular region $1 \leq |z| \leq 2$, the sequence of functions $\{f_n\}$ converges uniformly to zero. Namely, given $\epsilon > 0$ for n sufficiently large we have

$$|f_n(z) - f(z)| = \left| \frac{1}{nz} - 0 \right| = \frac{1}{n|z|} < \epsilon.$$

Thus, the estimate $1/n\,|z| < 1/n$ holds in the region $1 \leq |z| \leq 2$ for the first integer n such that $n > N(\epsilon) = 1/\epsilon$. The sequence is therefore uniformly convergent to **zero**.

On the other hand, $f_n(z)$ given in Eq. (3.1.6) converges to zero, but not uniformly, on the interval $0 < |z| \leq 1$; that is, $|f_n - f| < \epsilon$ only if $n > N(\epsilon, z) = 1/\epsilon\,|z|$ in the region $0 < |z| \leq 1$. Certainly $\lim_{n \to \infty} f_n(z) = f(z) = 0$, but irrespective of the choice of N there is a value of z (small) such that $|f_n - f| > \epsilon$.

Example 3.1.1   Discuss the uniform convergence of the sequence $\{f_n(z)\}$ where $f_n(z) = e^{-nz^2}$.

Let $z = re^{i\theta}$ hence $z^2 = r^2e^{2i\theta} = r^2(\cos 2\theta + i \sin 2\theta)$ and therefore $e^{-nz^2} = e^{-nr^2(\cos 2\theta + i \sin 2\theta)}$.

There are three possibilities:

(i) $\cos 2\theta > 0$;
(ii) $\cos 2\theta = 0$;
(iii) $\cos 2\theta < 0$.

The only case in which we have convergence is case (i). In case (ii), $2\theta = \pm\pi/2 + 2n\pi$, for integer n, and then $f_n(z)$ diverges due to rapid oscillations: $f_n(z) = e^{-nr^2(\pm i)}$. In case (iii) $|f_n(z)|$ grows without bound.

For case (i) $\lim_{n\to\infty} f_n(z) = 0$ which implies that we can find an N such that whenever $n > N$, $|e^{-nr^2 \cos 2\theta}| < \epsilon$. Hence we can take

$$0 < |e^{-nr^2 \cos 2\theta}| < |e^{-Nr^2 \cos 2\theta}| = \epsilon$$

so that

$$N = \frac{-\log \epsilon}{r^2 \cos 2\theta} = \frac{\log(1/\epsilon)}{r^2 \cos 2\theta}.$$

Note that the denominator can become arbitrarily small making N arbitrarily large. In other words N is a function of the point $z = re^{i\theta}$ and the convergence is not uniform for all z in the region r, θ defined by $\cos 2\theta > 0$ or $-\pi/4 < \theta < \pi/4 \cup 3\pi/4 < \theta < 5\pi/4$. If r, θ are restricted – for example $r > r_0, 0 < \theta < \pi/4 - \theta_0 -$ then we can take $N = \frac{\log(1/\epsilon)}{r_0^2 \cos 2\theta_0}$ in which case the convergence becomes uniform.

Further examples are given in the exercises at the end of the section.

Uniformly convergent sequences possess a number of important properties. In particular, we may employ the notion of uniform convergence to establish the following useful theorem.

Theorem 3.1.2   *Let the sequence of functions $f_n(z)$ be continuous for each integer n and let $f_n(z)$ converge to $f(z)$ uniformly in a region $\mathcal{R}$. Then $f(z)$ is continuous, and for any finite contour C inside $\mathcal{R}$,*

$$\lim_{n\to\infty} \int_C f_n(z)\, dz = \int_C f(z)\, dz. \tag{3.1.7}$$

Proof   **(a)** First we prove the continuity of $f(z)$. For z and z_0 in $\mathcal{R}$, we write

$$f(z) - f(z_0) = f_n(z) - f_n(z_0) + f_n(z_0) - f(z_0) + f(z) - f_n(z)$$

and hence

$$|f(z) - f(z_0)| \leq |f_n(z) - f_n(z_0)| + |f_n(z_0) - f(z_0)| + |f(z) - f_n(z)|.$$

Uniform convergence of $\{f_n(z)\}$ allows us to choose an N independent of z such that for $n > N$, N independent of z

$$|f_n(z_0) - f(z_0)| < \epsilon/3 \qquad \text{and} \qquad |f(z) - f_n(z)| < \epsilon/3.$$

Now take an $n > N$. Continuity of $f_n(z)$ allows us to choose $\delta > 0$ such that

$$|f_n(z) - f_n(z_0)| < \epsilon/3 \qquad \text{for} \quad |z - z_0| < \delta.$$

Thus for $|z - z_0| < \delta$ and $n > N$,

$$|f(z) - f(z_0)| < \epsilon,$$

which establishes the continuity of $f(z)$.

(b) Because the function $f(z)$ is continuous, it can be integrated by using the usual definition as described in Chapter 2. Given the continuity of $f(z)$ we shall prove Eq. (3.1.7); namely, for $\epsilon > 0$ we must find N such that when $n > N$

$$\left| \int_C f_n(z)\, dz - \int_C f(z)\, dz \right| < \epsilon. \tag{3.1.8}$$

But, for $n > N$,

$$\left| \int_C f_n\, dz - \int_C f\, dz \right| \leq \int_C |f_n - f|\, |dz| < \epsilon_1 L,$$

where the length of C is bounded by L and $|f_n - f| < \epsilon_1$ by uniform convergence of f_n. Taking $\epsilon_1 = \epsilon/L$ establishes Eq. (3.1.8) and hence Eq. (3.1.7). $\qquad\square$

Example 3.1.3 Consider the integral:

$$J(n; \alpha) = \int_0^\alpha \frac{n}{n^2 + z^2}\, dz,$$

where $\alpha > 0$, and $n \geq 1$, is an integer.

Take α to be finite and R to be a region containing the finite contour $C : (0, \alpha)$. Write $f_n(z) = \frac{1}{n^2 + z^2}$; it is analytic function of z for each n. Since, when $n > N$, we have $\frac{n}{n^2 + z^2} < \frac{n}{n^2} < \frac{1}{N} = \epsilon$ we can find $N = N(\epsilon) = 1/\epsilon$ such that $\lim_{n\to\infty} f_n(z) = 0$ uniformly. Thus we have that $\lim_{n\to\infty} J(n; \alpha) = 0$. Hence this is an example of Theorem 3.1.2. Alternatively

$$J(n; \alpha) = \frac{1}{n} \int_0^\alpha \frac{dz}{1 + (z/n)^2} = \int_0^{\alpha/n} \frac{du}{1 + u^2} = \tan^{-1} \frac{\alpha}{n},$$

which implies that $\lim_{n\to\infty} J(n; \alpha) = 0$ for finite α.

Note, when $\alpha = \infty$ we have that $\lim_{n\to\infty} J = \pi/2$. In this case the contour C is not finite and the hypotheses of Theorem 3.1.2 are no longer satisfied.

An immediate corollary of this theorem applies to series expansions. Namely, if the sequence of continuous partial sums converge uniformly, then we may integrate termwise; that is, for $b_j(z)$ continuous,

$$\sum_{j=1}^{\infty} \left(\int_C b_j(z)\, dz \right) = \int_C \left(\sum_{j=1}^{\infty} b_j(z) \right) dz. \qquad (3.1.9)$$

Equation (3.1.9) is important and we will use it extensively in our development of power series expansions of analytic functions.

We have already seen that uniformly convergent sequences and series have important and useful properties; for example, they allow the interchange of certain limit processes, such as interchanging infinite sums and integrals. In practice it is often unwieldy and frequently difficult to prove that particular series converge uniformly in a given region. Rather, we usually appeal to general theorems that provide conditions under which a series will converge uniformly. In what follows, we shall state one such important theorem, for which the proof is given in Section 3.4. The interested reader can follow the logical development by reading relevant portions of Section 3.4 at this point.

Theorem 3.1.4 ("Weierstrass M Test")   *Let $|b_j(z)| \leq M_j$ in a region $\mathcal{R}$, with M_j constant. If $\sum_{j=1}^{\infty} M_j$ converges, then the series $S(z) = \sum_{j=1}^{\infty} b_j(z)$ converges uniformly in $\mathcal{R}$.*

An immediate corollary to this theorem is the so-called **ratio test** for complex series. Namely, suppose $|b_1(z)|$ is bounded, and

$$\left| \frac{b_{j+1}(z)}{b_j(z)} \right| \leq M < 1, \qquad j > 1, \qquad (3.1.10)$$

for M constant. Then the series

$$S(z) = \sum_{j=1}^{\infty} b_j(z) \qquad (3.1.11)$$

is uniformly convergent.

In order to prove this statement, we write

$$b_n(z) = b_1(z) \frac{b_2(z)}{b_1(z)} \frac{b_3(z)}{b_2(z)} \cdots \frac{b_n(z)}{b_{n-1}(z)}. \qquad (3.1.12)$$

The boundedness of $b_1(z)$ implies

$$|b_1(z)| \leq B$$

hence

$$|b_n(z)| \leq BM^{n-1},$$

and therefore

$$\sum_{j=1}^{\infty} |b_j(z)| \le B \sum_{j=1}^{\infty} M^{j-1} = \left(\frac{B}{1-M}\right).$$

We see that the series $\sum_{j=1}^{\infty} |b_j(z)|$ is bounded by a series which converges and is independent of z. Consequently, we see that Theorem 3.1.4 (to be proved in Section 3.4), via the assertion (3.1.10), implies the uniform convergence of (3.1.11).

   We note in the above that if any finite number of terms do not satisfy the hypothesis, they can be added in separately; this will not affect the convergence results.

Example 3.1.5 Consider $\sum_{j=0}^{\infty} z^j$. The ratio test with $b_j(z) = z^j$:

$$\left|\frac{b_{j+1}}{b_j}\right| = \left|\frac{z^{j+1}}{z^j}\right| = |z|$$

implies that the series converges uniformly for $|z| < M < 1$.

   In the next section we will show that

$$\frac{1}{1-z} = \sum_{j=0}^{\infty} z^j.$$

Hence from the corollary to Theorem 3.1.2 we have that

$$\int_0^z \frac{dz'}{1-z'} = \int_0^z \sum_{j=0}^{\infty} (z')^j \, dz' = \sum_{j=0}^{\infty} \int_0^z (z')^j \, dz'$$

or

$$\log(1-z) = -\sum_{j=0}^{\infty} \frac{z^{j+1}}{j+1}.$$

3.1.1 Problems for Section 3.1

1. In the following we are given sequences. Discuss their limits and whether the convergence is uniform, in the region $\alpha \le |z| \le \beta$, for finite $\alpha, \beta > 0$.

   (a) $\left\{\dfrac{1}{nz^2}\right\}_{n=1}^{\infty}$;   (b) $\left\{\dfrac{1}{z^n}\right\}_{n=1}^{\infty}$;

   (c) $\left\{\sin \dfrac{z}{n}\right\}_{n=1}^{\infty}$;   (d) $\left\{\dfrac{1}{1+(nz)^2}\right\}_{n=1}^{\infty}$.

2. For each sequence in Problem 1, what can be said if

   (a) $\alpha = 0$;   (b) $\alpha > 0$, $\beta = \infty$.

3. Compute the integrals

$$\lim_{n\to\infty} \int_0^1 nz^{n-1}\,dz \qquad \text{and} \qquad \int_0^1 \lim_{n\to\infty} \left(nz^{n-1}\right) dz$$

and show that they are not equal. Explain why this is not a counterexample to Theorem 3.1.2.

4. In the following, let C denote the unit circle centered at the origin. Let $f(z) = \lim_{n\to\infty} f_n(z)$. Evaluate $\oint_C f(z)\,dz$ and the limit $\lim_{n\to\infty} \oint_C f_n(z)\,dz$, and discuss why they might or might not be equal:

(a) $f_n(z) = \dfrac{1}{z-n}$;   (b) $f_n(z) = \dfrac{1}{z-\left(1-\frac{1}{n}\right)}$.

5. Show that the following series converge uniformly in the given regions:

(a) $\displaystyle\sum_{n=1}^{\infty} z^n$, $0 \le |z| \le R,\ R < 1$;

(b) $\displaystyle\sum_{n=1}^{\infty} e^{-nz}$, $R \le \operatorname{Re} z \le 1,\ R > 0$;

(c) $\displaystyle\sum_{n=1}^{\infty} \operatorname{sech} nz$, $\operatorname{Re} z \ge 1$.

6. Show that the sequence $\{z^n\}_{n=1}^{\infty}$ converges uniformly inside $0 \le |z| \le R,\ R < 1$. (Hint: because $|z| < 1$, we find that $|z| \le R,\ R < 1$. Find $N(\epsilon, R)$ using the definition of uniform convergence.)

7. Show that if $\sum_{n=1}^{\infty} z_n = Z$ then $\sum_{n=1}^{\infty} \bar{z}_n = \bar{Z}$.

8. Consider the integral:

$$J(n) = \int_{-a}^{a} \frac{dz}{n(e^{z/n} + e^{-z/n})},$$

where $a > 0, n \ge 1$, integer.

(a) Let a be finite. Denoting $f_n(z) = \frac{1}{n(e^{z/n}+e^{-z/n})}$, show that both $\lim_{n\to\infty} f_n(z) = 0$ and $\lim_{n\to\infty} J(n) = 0$ and explain why Theorem 3.1.2 is satisfied.

(b) On the other hand when $a = \infty$ explain why the hypotheses of the theorem do not hold. Show that $\lim_{n\to\infty} J \ne 0$.

9. Establish that the sequence $\{\frac{1}{1+(nz)^2}\}_1^{\infty}$ converges uniformly to zero when $|z| > \delta > 0$.

   Hint: Recall that uniform convergence to zero means that there is an $N = N(\epsilon)$ such that $\left|\frac{1}{1+(nz)^2}\right| < \epsilon$. Use $1 + n^2|z|^2 > |1 + (nz)^2| > \frac{1}{\epsilon}$.

3.2 Taylor Series

In a similar manner to a function of a single real variable, as mentioned in Section 1.2, a **power series** about the point $z = z_0$ is defined as

$$f(z) = \sum_{j=0}^{\infty} b_j (z - z_0)^j, \tag{3.2.1}$$

where b_j, z_0 are constants or alternatively

$$f(z + z_0) = \sum_{j=0}^{\infty} b_j z^j. \tag{3.2.2}$$

Without loss of generality we shall simply work with the series

$$f(z) = \sum_{j=0}^{\infty} b_j z^j. \tag{3.2.3}$$

This corresponds to taking $z_0 = 0$. The general case can be obtained by replacing z by $(z - z_0)$.

We begin by establishing the uniform convergence of the above series.

Theorem 3.2.1 *If the series Eq. (3.2.3) converges for some $z_* \neq 0$, then it converges for all z in $|z| < |z_*|$. Moreover, it converges uniformly in $|z| \leq R$ for $R < |z_+|$.*

Proof For $j \geq J$, $|z| < |z_*|$,

$$\left| b_j z^j \right| = \left| b_j z_*^j \right| \left| \frac{z}{z_*} \right|^j < \left| \frac{z}{z_*} \right|^j \leq \left(\frac{R}{|z_*|} \right)^j.$$

This follows from the fact that $|b_j z_*^j| < 1$ for sufficiently large j owing to the assumed convergence of the series at $z = z_*$ (i.e., $\lim_{j \to \infty} b_j z_*^j = 0$). We now take

$$M = \frac{R}{|z_*|} < 1 \qquad \text{and} \qquad M_j \equiv M^j$$

in the Weierstrass M test for $j \geq J$. Thus, $\sum_{j=0}^{\infty} b_j z^j$ converges uniformly for $|z| \leq R$, $|R| < |z_*|$, because $\sum_{j=J}^{\infty} |b_j z^j| < \sum_{j=J}^{\infty} M^j = (M^J)/(1 - M)$. $\qquad \square$

We now establish the Taylor series for an analytic function.

Theorem 3.2.2 (Taylor Series) *Let $f(z)$ be analytic for $|z| \leq R$. Then*

$$f(z) = \sum_{j=0}^{\infty} b_j z^j, \tag{3.2.4}$$

where

$$b_j = \frac{f^{(j)}(0)}{j!},$$

converges uniformly in $|z| \leq R_1 < R$.

We note that this is the Taylor series about $z = 0$. If $z = 0$ is replaced by $z = z_0$ then the result of this theorem would state that $f(z) = \sum b_j(z - z_0)^j$, where $b_j = f^{(j)}(z_0)/j!$, converges uniformly in $|z - z_0| < R$. (If $f(z)$ is analytic for $|z| < R$ then (3.2.4) holds equally well.)

Proof   The proof is really an application of Cauchy's integral formula (Eq. (2.6.1) of Section 2.6). We write

$$f(z) = \frac{1}{2\pi i} \oint_C \frac{f(\zeta)}{\zeta - z}\, d\zeta = \frac{1}{2\pi i} \oint_C \frac{f(\zeta)}{\zeta}\left(1 - \frac{z}{\zeta}\right)^{-1} d\zeta, \qquad (3.2.5)$$

where C is a circle of radius R. We use the uniformly convergent expansion

$$(1 - z)^{-1} = \sum_{j=0}^{\infty} z^j. \qquad (3.2.6)$$

Equation (3.2.6) can be established directly. Consider $s_n(z) = \sum_{j=0}^{n} z^j$ for $|z| < 1$. Then

$$s_n(z) - z s_n(z) = 1 - z^{n+1}$$

hence $s_n(z) = (1 - z^{n+1})/(1 - z)$. Because $\lim_{n\to\infty} z^{n+1} = 0$, we have Eq. (3.2.6). Noting that $|z/\zeta| < 1$, we can replace z by z/ζ in Eq. (3.2.6). Using this expansion in Eq. (3.2.5) we deduce

$$f(z) = \frac{1}{2\pi i} \oint_C f(\zeta) \sum_{j=0}^{\infty} \left(\frac{z^j}{\zeta^{j+1}}\right) d\zeta. \qquad (3.2.7)$$

From Theorem 3.1.2 we may interchange $\oint_C$ and $\sum_{j=0}^{\infty}$ to obtain

$$f(z) = \sum_{j=0}^{\infty} b_j z^j, \qquad (3.2.8)$$

where

$$b_j = \frac{1}{2\pi i} \oint_C \frac{f(\zeta)}{\zeta^{j+1}}\, d\zeta = \frac{f^{(j)}(0)}{j!},$$

where the right-hand side of Eq. (3.2.8) follows from the corollary of Cauchy's Theorem (Theorem 2.6.2 of Section 2.6). The uniform convergence of the power series follows in the same way as discussed in Theorem 3.2.1 above. $\square$

Sometimes a Taylor series converges for all finite z. Then $f(z)$ is analytic for $|z| \le R$, for every R.

We note that:

(a) formula (3.2.4) is the same as that for functions of one real variable; and

(b) the Taylor series about the point $z = z_0$ is given by

$$f(z) = \sum_{j=0}^{\infty} b_j (z - z_0)^j, \qquad (3.2.9)$$

where

$$b_j = \frac{1}{2\pi i} \oint_C \frac{f(\zeta)}{(\zeta - z_0)^{j+1}} \, d\zeta = \frac{f^{(j)}(z_0)}{j!}.$$

Example 3.2.3

$$e^z = \sum_{j=0}^{\infty} \frac{z^j}{j!} \qquad \text{for} \quad |z| < \infty; \qquad (3.2.10)$$

$$e^{z^2} = \sum_{j=0}^{\infty} \frac{z^{2j}}{j!} \qquad \text{for} \quad |z| < \infty. \qquad (3.2.11)$$

The first of these formulae follows from Eq. (3.2.4) because $f^{(j)}(0) = 1$ for $f(z) = e^z$. Using the limit form of the "ratio test" discussed in Section 3.1, see (3.1.10), we get

$$\lim_{n \to \infty} \left| \frac{\frac{z^{n+1}}{(n+1)!}}{\frac{z^n}{n!}} \right| = \lim_{n \to \infty} \left| \frac{z}{n+1} \right| = 0.$$

The second of these formulae follow from the first by replacing z by z^2.

We see that convergence of Eq. (3.2.10) is obtained for all z. The largest number R for which the power series (3.2.3) converges inside the disc $|z| < R$ is called the **radius of convergence**. The value of R may be *zero* or *infinity* or a finite number. A value for the radius of convergence may be obtained via the usual absolute value tests of calculus such as the ratio test discussed in Section 3.1, Eq. (3.1.10), or more generally via the root test of calculus (i.e., $R = [\lim_{n \to \infty} (\sup_{m \ge n} |a_m|^{1/m})]^{-1}$ where supremum, sup, is the least upper bound). In Example 3.2.3 we say the radius of convergence is infinite. We leave it as an exercise for the reader to verify that formulae (1.2.19) of Section 1.2 are Taylor series representations (about $z = 0$) of the indicated functions.

Taylor series behave just like ordinary polynomials. We may integrate or differentiate Taylor series termwise. Integrating termwise inside its region of convergence, about $z = 0$,

$$\int f(z)\,dz = \int \left(\sum_{j=0}^{\infty} a_j z^j \right) dz = \sum_{j=0}^{\infty} \frac{a_j z^{j+1}}{j+1} + C, \qquad \text{where } C \text{ is a constant}$$

$$(3.2.12)$$

is justified by Eq. (3.1.9). Similarly, differentiation

$$f'(z) = \sum_{j=0}^{\infty} j\, a_j\, z^{j-1} \tag{3.2.13}$$

also follows. We formulate this result as a theorem.

Theorem 3.2.4 *Let $f(z)$ be analytic for $|z| \le R$. Then the series obtained by differentiating the Taylor series termwise converges uniformly to $f'(z)$ in $|z| \le R_1 < R$.*

Proof If $f(z)$ is analytic in $|z| \le R$, then from our previous results (e.g. Theorem 2.6.2, Section 2.6), $f'(z)$ is analytic in D for $|z| < R$. The Taylor series for $f'(z)$ is given by

$$f'(z) = \sum_{j=0}^{\infty} C_j z^j; \qquad C_j = \frac{f^{(j+1)}(0)}{j!}. \tag{3.2.14}$$

But the Taylor series for $f(z)$ is given by

$$f(z) = \sum_{j=0}^{\infty} \frac{f^{(j)}(0)}{j!} z^j, \tag{3.2.15}$$

hence formal differentiation termwise yields

$$f'(z) = \sum_{j=1}^{\infty} \frac{f^{(j)}(0)}{(j-1)!} z^{j-1} = \sum_{j=0}^{\infty} \frac{f^{(j+1)}(0)}{j!} z^j, \tag{3.2.16}$$

which is equivalent to Eq. (3.2.14). Moreover, the same argument as that presented in the proof of Theorem 3.2.2 holds for Eq. (3.2.14), which shows that the differentiated series converges uniformly in $|z| \le R_1 < R$. □

We remark that further differentiation for $f''(z), f'''(z), \ldots$, follows in the same manner by reapplying the arguments presented in Theorem 3.2.4.

The Taylor series representing the zero function is also clearly zero (because zero is an analytic function, (Eq. 3.2.4) applies). We easily deduce that Taylor series are unique; that is, there cannot be two Taylor series representations of a given function $f(z)$ because if there were two, say, $\sum a_n z^n$ and $\sum b_n z^n$, the difference $\sum c_n z^n$ (where $c_n = a_n - b_n$), must represent the zero function, which implies $a_n = b_n$.

Similarly, any convergent power series representation of an analytic function $f(z)$ must be the Taylor series representation of $f(z)$. In order to demonstrate this fact we first show that any convergent power series can be differentiated termwise.

Theorem 3.2.5 *If the power series (3.2.3) converges for $|z| \le R$, then it can be differentiated termwise to obtain a uniformly convergent series for $|z| \le R_1 < R$.*

Proof From Eq. (2.6.5) we have, for any closed contour C, $|z| = R_1 < R$,

$$f'(z) = \frac{1}{2\pi i} \oint_C \frac{f(\zeta)}{(\zeta - z)^2} \, d\zeta = \frac{1}{2\pi i} \oint_C \frac{\sum_{j=0}^{\infty} a_j \zeta^j}{(\zeta - z)^2} \, d\zeta. \tag{3.2.17}$$

Because the series in Eq. (3.2.17) is uniformly convergent we may interchange the sum and integral (see Eq. (3.1.9) above) to find

$$f'(z) = \sum_{j-0}^{\infty} a_j \left(\frac{1}{2\pi i} \oint_C \frac{\zeta^j}{(\zeta - z)^2} \, d\zeta \right) = \sum_{j=0}^{\infty} a_j \frac{d}{dz}(z^j)$$

$$= \sum_{j=0}^{\infty} j a_j z^{j-1}, \tag{3.2.18}$$

where we have employed Eq. (2.6.5) for the function $f(z) = z^j$. Uniform convergence follows in the same way as before. □

The formula $f(z) = \sum_{j=0}^{\infty} a_j z^j$ clearly may be differentiated over and over again for $|z| \le R_1 < R$. Thus, it immediately follows that

$$f(0) = a_0; \qquad f'(0) = a_1; \qquad f^{(j)}(0) = j! a_j. \tag{3.2.19}$$

Hence we have deduced that the power series of $f(z)$ is really the Taylor series of $f(z)$ (about $z = 0$).

The usual properties of series hold, namely, the sum/difference of a series are the sum/difference of the terms. Writing $g(z) = \sum_{j=0}^{\infty} b_j z^j$, we have

$$f(z) \pm g(z) = \sum_{j=0}^{\infty} a_j z^j \pm \sum_{j=0}^{\infty} b_j z^j = \sum_{j=0}^{\infty} (a_j \pm b_j) z^j, \tag{3.2.20}$$

and it also follows that the product of two convergent series may be written as

$$f(z)g(z) = \sum_{j=0}^{\infty} C_j z^j, \tag{3.2.21}$$

where

$$C_j = \sum_{k=0}^{j} b_k a_{j-k}. \tag{3.2.22}$$

An application of these ideas can be used to establish the form of Taylor series. If a series

$$\sum_0^\infty a_n z^n$$

converges to a function $f(z)$ at all points interior to some circle $R = |z - z_0|$ then it must be the Taylor expansion of $f(z)$. This follows from the differentiation of

$$f(z) = \sum_0^\infty a_n z^n.$$

Evaluating the nth derivative of the series at $z = z_0$ yields the Taylor coefficients

$$f^n(z = 0) = n! a_n \Rightarrow a_n = \frac{f^n(z = 0)}{n!}.$$

Another result similar to that of real analysis is the following.

Theorem 3.2.6 (Comparison test)   *Let the series $\sum_{j=0}^\infty a_j z^j$ converge for $|z| < R$. If $|b_j| \leq |a_j|$ for $j \geq J$, then the series $\sum_{j=0}^\infty b_j z^j$ also converges for $|z| < R$.*

Proof   For $j \geq J$ and $|z| < |z_*| < R$,

$$\left| b_j z^j \right| \leq \left| a_j z^j \right| = \left| a_j z_*^j \right| \left| \frac{z}{z_*} \right|^j < \left| \frac{z}{z_*} \right|^j < 1.$$

The latter inequalities follow because $\sum_{j=0}^\infty a_j z^j$ converges and we know that, for sufficiently large j, $|a_j z_*^j| < 1$. Convergence then follows, via the Weierstrass M test.   □

For example we know that $\sum_{n=0}^\infty (z^n/n!) = e^z$ converges for $|z| < \infty$. Thus, by the comparison test the series $\sum_{n=0}^\infty [z^n/(n!)^2]$ also converges for $|z| < \infty$ because $(n!)^2 \geq n!$ for all n.

Another example of a Taylor series (about $z = 0$) is given by

$$\frac{1}{1 + z} = \sum_{n=0}^\infty (-1)^n z^n \qquad \text{for } |z| < 1; \tag{3.2.23}$$

see also the remark below Eq. (3.2.7). Equation (3.2.23) is obtained by taking successive derivatives of the function $1/(1 + z)$, evaluating them at $z = 0$, and employing Eq. (3.2.4), or noting formula (3.2.6) and replacing $-z$ by z. The radius of convergence follows from the ratio test. Replacing z by z^2 yields

$$\frac{1}{1 + z^2} = \sum_{n=0}^\infty (-1)^n z^{2n}, \qquad \text{for } |z| < 1. \tag{3.2.24}$$

The divergence, for $|z| \geq 1$, of the series given in Eq. (3.2.24) is due to the zeroes of $1 + z^2 = 0$; that is, $z = \pm i$. In the case of real analysis it was not really clear why the series (3.2.24) with z replaced by x, diverges; only when we examine the series in the context of complex analysis do we understand the origins of divergence. Because the function $1/(1 + z^2)$ is nonanalytic only at $z = \pm i$, it is natural to ask whether there is another series representation valid for $|z| > 1$. In fact there is such a representation which is part of a more general series expansion (Laurent series) to be taken up shortly.

Having the ability to represent a function by a series, such as a Taylor series, allows us to analyze and work with a much wider class of functions than the usual elementary functions (e.g. polynomials, rational functions, exponentials, logarithms).

Example 3.2.7 Consider the "error function," erf(z):

$$\mathrm{erf}(z) = \frac{2}{\sqrt{\pi}} \int_0^z e^{-t^2}\, dt. \tag{3.2.25}$$

Using Eq. (3.2.11) with z^2 replaced by $-t^2$ and integrating termwise, we have the Taylor series representation

$$\mathrm{erf}(z) = \frac{2}{\sqrt{\pi}} \int_0^z \left(\sum_{n=0}^\infty \frac{(-t^2)^n}{n!} \right) dt = \frac{2}{\sqrt{\pi}} \sum_{n=0}^\infty \frac{(-1)^n z^{2n+1}}{(2n+1)n!}$$

$$= \frac{2}{\sqrt{\pi}} \left[z - \frac{z^3}{3} + \frac{z^5}{5 \cdot 2!} - \frac{z^7}{7 \cdot 3!} + \cdots \right]. \tag{3.2.26}$$

Because we are integrating an exponential function, which is entire, it follows that the error function is also entire.

With the results of this section we see that the notion of analyticity of a function $f(z)$ in a region $\mathcal{R}$ may now be broadened. Namely, if $f(z)$ is analytic, then by Definition 2.1.6, $f'(z)$ exists in $\mathcal{R}$. We have seen that this implies that:

(a) $f(z)$ has derivatives of all orders in $\mathcal{R}$ (an extension of Cauchy's formula, Theorem 2.6.2); and that

(b) $f(z)$ has a Taylor series representation in the neighborhood of all points of $\mathcal{R}$ (Theorem 3.2.2).

On the other hand, if $f(z)$ has a convergent power series expansion; that is

$$f(z) = \sum_{n=0}^\infty a_n (z - z_0)^n,$$

then integrating $f(z)$ over any simple closed contour C implies $\oint_C f(z)\, dz = 0$ because $\oint_C (z - z_0)^n\, dz = 0$. Hence, by Theorem 2.6.5, $f(z)$ is analytic inside C.

This is consistent with point (b) above because we have already shown that the power series is equivalent to the Taylor series representation.

3.2.1 Analytic Continuation

In later chapters we study analytic functions that coincide in a domain or on a curve or that are zero at distinct points. The theorems below will be useful; we will only sketch the proofs.

Theorem 3.2.8   *Let each of two function $f(z)$ and $g(z)$ be analytic in a common domain D. If $f(z)$ and $g(z)$ coincide in some subportion $D' \subset D$ or on a curve Γ interior to D, then $f(z) = g(z)$ everywhere in D.*

Proof   Corresponding to any point z_0 in D' (or on Γ) consider the largest circle C contained entirely within D. Both $f(z)$ and $g(z)$ may be represented by a Taylor series inside C, and by the uniqueness of Taylor series, $f(z) = g(z)$ inside C. Next pick a new interior point of C but near its boundary, and repeat the above Taylor series argument to find $f(z) = g(z)$ in an extended domain. This procedure can be repeated so as to entirely fill up the common domain D. (This statement, while intuitively clear, requires some analysis to substantiate – we shall omit it.)   □

Consequently, a function $f(z)$ which vanishes everywhere in a subdomain $D' \subset D$ or on a curve Γ entirely contained within D must vanish everywhere inside D (i.e., $g(z) = 0$ in the Theorem). The discussion in Theorem 3.2.8 provides us with a way of "analytically continuing" a known function in some domain to a larger domain. We remark that one must be careful when continuing a multivalued function; this is discussed further in Section 3.5.

In fact, analytic continuation of a function $f(z)$ to a function $g(z)$, with which it shares a common boundary, is closely related to the above. It can be described via the following theorem, which is proved with the aid of Morera's Theorem.

Theorem 3.2.9   *Let D_1 and D_2 be two disjoint domains, whose boundaries share a common contour Γ. Let $f(z)$ be analytic in D_1 and continuous in $D_1 \cup \Gamma$ and $g(z)$ be analytic in D_2 and continuous in $D_2 \cup \Gamma$, and let $f(z) = g(z)$ on Γ. Then the function*

$$H(z) = \begin{cases} f(z), & z \in D_1; \\ f(z) = g(z), & z \in \Gamma; \\ g(z), & z \in D_2 \end{cases}$$

is analytic in $D = D_1 \cup \Gamma \cup D_2$. We say that $g(z)$ is the **analytic continuation of** *$f(z)$.*

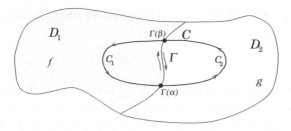

Figure 3.1 Analytic continuation

Proof Consider a closed contour C in D. If C does not intersect Γ, then $\oint_C H(z)\,dz = 0$ because C is entirely contained in D_1 or D_2. On the other hand, if C intersects Γ, then we have (referring to Figure 3.1)

$$\oint_C H(z)\,dz = \int_{C_1} f(z)\,dz + \int_{\Gamma(\alpha)}^{\Gamma(\beta)} f(z)\,dz$$

$$+ \int_{\Gamma(\beta)}^{\Gamma(\alpha)} g(z)\,dz + \int_{C_2} g(z)\,dz = 0, \qquad (3.2.27)$$

where we have divided the closed contour C into two other closed contours $C_1 + \Gamma$ and $C_2 - \Gamma$, and the endpoints of the contour Γ inside C are labeled $\Gamma(\alpha)$ and $\Gamma(\beta)$. Note that the two "intermediate" integrals in opposite directions along Γ mutually cancel because $f(z) = g(z)$ on Γ. Then, the fact that $f(z)$ and $g(z)$ are analytic in D_1 and D_2, respectively, ensures that $\oint_C H(z)\,dz = 0$, whereupon from Morera's Theorem, 2.6.5, we find that $H(z)$ is analytic in $D = D_1 \cup \Gamma \cup D_2$. (If f, g are analytic on Γ this statement follows immediately; otherwise some more analysis is needed.) $\square$

Example 3.2.10 Consider the function $f(z) = \sum_{n=0}^{\infty} z^n, |z| < 1$. We have seen that $\sum_{n=0}^{\infty} z^n = \frac{1}{1-z} = g(z)$ which is defined for all $z \neq 1$. We say that $g(z)$ is the analytic continuation of $f(z)$.

Example 3.2.11 Consider the function $f(z) = \sin^2 z + \cos^2 z - 1$. This is an analytic function which coincides with $g(x) = \sin^2 x + \cos^2 x - 1$ for $z = x$. Since $g(x) = 0$ we have from Theorem 3.2.8 that $f(z) = \sin^2 z + \cos^2 z - 1 = 0$ and hence $\sin^2 z + \cos^2 z = 1$ for all z. Indeed we can make similar statements about the well-known trigonometric identities.

Theorem 3.2.12 *If $f(z)$ is analytic and not identically zero in some domain D containing $z = z_0$, then its zeroes are isolated; that is, there is a neighborhood about $z = z_0$, $f(z_0) = 0$, in which $f(z)$ is nonzero.*

Proof   Because $f(z)$ is analytic at z_0 it has a Taylor series about $z = z_0$. If it has a zero of order n we write

$$f(z) = (z - z_0)^n g(z),$$

where $g(z)$ has a Taylor series about $z = z_0$, and $g(z_0) \neq 0$. There must exist a maximum integer n, otherwise $f(z)$ would be identically zero in a neighborhood of z_0, and, from Theorem 3.2.8, must vanish everywhere in D. Because $g(z)$ is analytic it follows that, for sufficiently small ϵ, $|g(z) - g(z_0)| < \epsilon$ whenever z is in the neighborhood of z_0; namely, $0 < |z - z_0| < \delta$. Hence $g(z)$ can be made as close to $g(z_0)$ as desired, hence $g(z) \neq 0$ and $f(z) \neq 0$ in this neighborhood.   □

3.2.2 *Analyticity of Functions Defined by Integrals*

Finally, to close this section we briefly discuss the behavior of functions that are represented by integrals. Such integrals arise frequently in applications. For example:

(a) the Fourier transform $F(z)$ of a function $f(t)$

$$F(z) = \int_{-\infty}^{\infty} f(t) e^{-izt} \, dt;$$

(b) the Cauchy-type integral $F(z)$ associated with a function $f(t)$ on a simple closed contour C with z inside C:

$$F(z) = \frac{1}{2\pi i} \oint_C \frac{f(t)}{t - z} \, dt.$$

These are two examples we will study in some detail in subsequent chapters, in which a given integral depends on another parameter, in this case z. Frequently one is interested in the question of when the function $F(z)$ is analytic, which we now address in some generality.

Consider integrals of the following form:

$$F(z) = \int_a^b g(z,t) \, dt \qquad a \leq t \leq b, \tag{3.2.28}$$

where
(a) for each t, $g(z,t)$ is an analytic function of z in a domain D; and
(b) for each z, $g(z,t)$ is a continuous function of t.

With these hypotheses it follows that $F(z)$ is analytic in D and

$$F'(z) = \int_a^b \frac{\partial g}{\partial z}(z,t) \, dt. \tag{3.2.29}$$

Because for each t, $g(z, t)$ is an analytic function of z, we find that

$$g(z, t) = \sum_{j=0}^{\infty} c_j(t)(z - z_0)^j, \qquad (3.2.30)$$

where z_0 is any point in D and from Eq. (3.2.9)

$$c_j(t) = \frac{1}{2\pi i} \oint_C \frac{g(\zeta, t)}{(\zeta - z_0)^{j+1}} \, d\zeta,$$

and C is a circle inside D centered at z_0. The continuity of $g(z, t)$ as a function of t implies the continuity of $c_j(t)$. The function $g(z, t)$ is bounded in D: $|g(z, t)| < M$ because it is analytic there. Hence

$$|c_j(t)| \le \frac{M}{2\pi} \oint_C \frac{|d\zeta|}{|\zeta - z_0|^{j+1}} = M/\rho^j,$$

where $\rho = |\zeta - z_0|$ is the radius of the circle C. Thus, the Taylor series given by Eq. (3.2.30) converges uniformly for $|z - z_0| < \rho$ and can therefore be integrated and/or differentiated termwise, from which Eq. (3.2.29) follows, using the series for g with (3.2.28).

We also note that a closed contour in t can be viewed (see Section 2.4) as a special case of a line integral of the form of Eq. (3.2.29) where $a \le t \le b$. Hence the above results apply when $F(z) = \oint_{\hat{C}} g(z, t) dt$ and $\hat{C}$ is a closed contour (such as a Cauchy-type integral).

On the other hand, if the contour becomes infinite (e.g. $a = -\infty$ or $b = \infty$), then, in addition to the hypotheses already stated, it is necessary to specify a uniformity restriction on $g(z, t)$ in order to have analyticity of $F(z)$ and then Eq. (3.2.28). Namely, in this case of infinite limits it is sufficient to add that $g(z, t)$ satisfies, for example, $|g(z, t)| \le G(t)$, where $\int_a^b G(t) \, dt < \infty$. We will not go into further details here.

Example 3.2.13 Using the above notation from Eq. (3.2.28), consider the function

$$F(z) = \int_{-\infty}^{\infty} \frac{dt}{1 + z^2 + t^2}, \qquad g(z, t) = \frac{1}{1 + z^2 + t^2}.$$

Then $g(z, t) = \frac{1}{1+z^2+t^2}$ is an analytic function of z for all t and is a continuous function of t for all z. Since

$$\frac{1}{1 + z^2 + t^2} \le \frac{1}{1 + t^2},$$

hence

$$\int_{-\infty}^{\infty} \frac{dt}{1 + t^2} = \pi < \infty,$$

we conclude that

$$F'(z) = -2z \int_{-\infty}^{\infty} \frac{dt}{(1 + z^2 + t^2)^2}.$$

In fact by methods to be discussed in Chapter 4 we can show that $F(z) = \frac{\pi}{\sqrt{1+z^2}}$ and $F'(z) = \frac{-\pi z}{(1+z^2)^{3/2}}$.

Example 3.2.14 Let

$$F(z) = \int_{-\infty}^{\infty} f(t) e^{-izt} \, dt,$$

where $f(t)$ is continuous and $f(t) < Ce^{-k|t|}$ where $C, k > 0$ are constants. We call $F(z)$ the Fourier transform of $f(t)$. We will study its properties later in Chapter 4. Using the above notation, we write $g(z, t) = f(t) e^{-izt}$; it is an analytic function of z for all t and is a continuous function of t for all z. Note that

$$|g(z, t)| = |f(t)||e^{-i(z_r + iz_I)t}| = |f(t)||e^{z_I t}| < Ce^{-k|t|+z_I t} \text{ where } z = z_R + iz_I.$$

Thus $|g(z, t)| < Ce^{-(k-z_I)t}$ for $t > 0$ and $|g(z, t)| < Ce^{(k+z_I)t}$ for $t < 0$. Consequently the integral

$$\int_{-\infty}^{\infty} |g(z, t)| \, dt < C \int_0^{\infty} e^{-(k-z_I)t} \, dt + C \int_{-\infty}^0 e^{(k+z_I)t} \, dt$$

converges when $k - z_I > 0$ and $k + z_I > 0$ or the strip $-k < z_I < k$. In this strip the function $F(z)$ is an analytic function of z.

3.2.3 Problems for Section 3.2

1. Obtain the radius of convergence of the series $\sum_{n=1}^{\infty} s_n(z)$, where $s_n(z)$ is given by

   (a) z^n;   (b) $\dfrac{z^n}{(n+1)!}$;   (c) $n^n z^n$;   (d) $\dfrac{z^{2n}}{(2n)!}$;   (e) $\dfrac{n!}{n^n} z^n$.

2. Find Taylor series expansions around $z = 0$ of the following functions in the given regions:

   (a) $\dfrac{1}{1 - z^2}$, $|z| < 1$;   (b) $\dfrac{z}{1 + z^2}$, $|z| < 1$;

   (c) $\cosh z$, $|z| < \infty$;   (d) $\dfrac{\sin z}{z}$, $0 < |z| < \infty$;

   (e) $\dfrac{\cos z - 1}{z^2}$, $0 < |z| < \infty$;   (f) $\dfrac{e^{z^2} - 1 - z^2}{z^3}$, $0 < |z| < \infty$.

3. Let the Euler numbers E_n be defined by the power series

$$\frac{1}{\cosh z} = \sum_{n=0}^{\infty} \frac{E_n}{n!} z^n.$$

   (a) Find the radius of convergence of this series.

   (b) Determine the first six Euler numbers.

4. Show that about any point $z = x_0$,

$$e^z = e^{x_0} \sum_{n=0}^{\infty} \frac{(z - x_0)^n}{n!}.$$

5. (a) Use the identity $\frac{1}{z} = 1/((z + 1) - 1)$ to establish

$$\frac{1}{z} = -\sum_{n=0}^{\infty} (z + 1)^n, \qquad |z + 1| < 1.$$

   (b) Use the above identity to establish in addition that

$$\frac{1}{z^2} = \sum_{n=0}^{\infty} (n + 1)(z + 1)^n, \qquad |z + 1| < 1.$$

   Verify that you get the same result by differentiation of the series in part (a).

6. Evaluate the integrals $\oint_C f(z) dz$ where C is the unit circle centered at the origin, and $f(z)$ is given by the following:

   (a) $\dfrac{\sin z}{z}$; (b) $\dfrac{\sin z}{z^2}$; (c) $\dfrac{\cosh z - 1}{z^4}$.

7. Use the Taylor series for $1/(1+z)$ about $z = 0$ to find the Taylor series expansion of $\log(1 + z)$ about $z = 0$ for $|z| < 1$.

8. Use the Taylor series representation of $1/(1 - z)$ around $z = 0$ for $|z| < 1$ to find a series representation of $1/(1 - z)$ for $|z| > 1$. (Hint: use $1/(1 - z) = -1/(z(1 - 1/z))$).

9. Use the Taylor series representation of $1/(1 - z)$ around $z = 0$, for $|z| < 1$, to deduce the series representation of $1/(1 - z)^2$, $1/(1 - z)^3$, ..., $1/(1 - z)^m$.

10. Use the binomial expansion and Cauchy's Integral Theorem to evaluate

$$\oint_C (z + 1/z)^{2n} \frac{dz}{z},$$

where C is the unit circle centered at the origin. Recall the binomial expansion

$$(a + b)^n = a^n + na^{n-1}b + \cdots = \sum_{k=0}^{n} \binom{n}{k} a^{n-k} b^k,$$

where

$$\binom{n}{k} = \frac{n!}{k!(n-k)!}.$$

Use this result to establish the following real integral formula:

$$\frac{1}{2\pi} \int_0^{2\pi} (\cos\theta)^{2n}\, d\theta = \frac{(2n)!}{4^n (n!)^2}.$$

11. Define $F(z) = \int_{-\infty}^{\infty} f(t)e^{izt}\, dt$ where $|f(t)| \le e^{-at^2}$, and $a > 0$. Discuss why $F(z)$ is analytic for all finite z.

12. Show that since:

(a) $\cosh^2 x - \sinh^2 x = 1$ on the real axis, this implies that $\cosh^2 z - \sinh^2 z = 1$ for all z;

(b) $\cos^2 x - \sin^2 x = \cos 2x$ on the real axis, this implies that $\cos^2 z - \sin^2 z = \cos 2z$ for all z.

3.3 Laurent Series

In many applications we encounter functions which are, in some sense, generalizations of analytic function. Typically they are not analytic at some point, points, or in some regions of the complex plane, and consequently, Taylor series cannot be employed in the neighborhood of such points. However, another series representation can frequently be found in which both positive and negative powers of $(z - z_0)$ exist. (Recall that Taylor series expansions contain only positive powers of $(z - z_0)$.) Such a series is valid for those functions which are analytic in and on a circular annulus, $R_1 \le |z - z_0| \le R_2$ (see Figure 3.2).

In the derivation of Laurent series it is convenient to work with the series about an arbitrary point $z = z_0$.

Theorem 3.3.1 (Laurent Series)   *A function $f(z)$ analytic in an annulus $R_1 \le |z - z_0| \le R_2$ can be represented by the series expansion*

$$f(z) = \sum_{n=-\infty}^{\infty} C_n (z - z_0)^n, \tag{3.3.1}$$

in the region $R_1 < R_a \le |z - z_0| \le R_b < R_2$, where

$$C_n = \frac{1}{2\pi i} \oint_C \frac{f(z)\, dz}{(z - z_0)^{n+1}} \tag{3.3.2}$$

and C is any simple closed contour in the region of analyticity enclosing the inner boundary $|z - z_0| = R_1$.

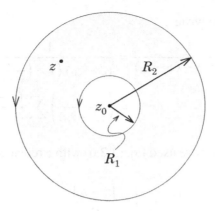

Figure 3.2 Circular annulus $R_1 \le |z - z_0| \le R_2$

(Note: if $f(z)$ is analytic in $R_1 < |z - z_0| < R_2$ then the same result holds.)

Proof We introduce the usual cross cut in the annulus (see Figure 3.3) where we denote by C_1 and C_2 the inside and outside contours surrounding the point $z = z_0$.

Contour C_1 lies on $|z - z_0| = R_1$, and contour C_2 lies on $|z - z_0| = R_2$. Application of Cauchy's formula to the crosscut region where $f(z)$ is analytic and the crosscut contributions cancel, lets us write $f(z)$ as follows:

$$f(z) = \frac{1}{2\pi i} \oint_{C_2} \frac{f(\zeta)}{\zeta - z} \, d\zeta - \frac{1}{2\pi i} \oint_{C_1} \frac{f(\zeta)}{\zeta - z} \, d\zeta. \tag{3.3.3}$$

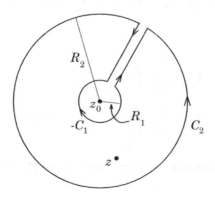

Figure 3.3 Inner contour C_1 and outer contour C_2

In the first integral we write

$$\frac{1}{\zeta - z} = \frac{1}{(\zeta - z_0) - (z - z_0)} = \frac{1}{(\zeta - z_0)\left(1 - \left(\frac{z - z_0}{\zeta - z_0}\right)\right)}$$

$$= \left(\frac{1}{(\zeta - z_0)}\right) \sum_{j=0}^{\infty} \frac{(z - z_0)^j}{(\zeta - z_0)^j}. \tag{3.3.4}$$

In the above equation we have used Eq. (3.2.6) with z replaced by $\frac{z - z_0}{\zeta - z_0}$, and we note that

$$\left|\frac{z - z_0}{\zeta - z_0}\right| = \frac{|z - z_0|}{R_2} < 1.$$

In the second integral we write

$$-\frac{1}{\zeta - z} = \frac{1}{z - z_0 - (\zeta - z_0)}$$

$$= \frac{1}{(z - z_0)\left(1 - \left(\frac{\zeta - z_0}{z - z_0}\right)\right)} = \frac{1}{z - z_0} \sum_{j=0}^{\infty} \left(\frac{\zeta - z_0}{z - z_0}\right)^j. \tag{3.3.5}$$

Again in Eq. (3.3.5) we have made use of Eq. (3.2.6), where z is now replaced by $\frac{\zeta - z_0}{z - z_0}$, and we note that

$$\left|\frac{\zeta - z_0}{z - z_0}\right| = \frac{R_1}{|z - z_0|} < 1.$$

Using Eq. (3.3.4) in the first integral of Eq. (3.3.3) and Eq. (3.3.5) in the second integral of Eq. (3.3.3) gives us the following representation for $f(z)$:

$$f(z) = \sum_{j=0}^{\infty} A_j (z - z_0)^j + \sum_{j=0}^{\infty} B_j (z - z_0)^{-(j+1)} \tag{3.3.6}$$

where

$$A_j = \frac{1}{2\pi i} \oint_{C_2} \frac{f(\zeta)}{(\zeta - z_0)^{j+1}} \, d\zeta, \tag{3.3.7a}$$

$$B_j = \frac{1}{2\pi i} \oint_{C_1} f(\zeta)(\zeta - z_0)^j \, d\zeta. \tag{3.3.7b}$$

We make the substitutions $n = j$ in the first sum of Eq. (3.3.6) and $n = -(j + 1)$ in the second to obtain

$$f(z) = \sum_{n=0}^{\infty} A_n (z - z_0)^n + \sum_{n=-\infty}^{-1} B_{-n-1} (z - z_0)^n, \tag{3.3.8}$$

where

$$A_n = \frac{1}{2\pi i} \oint_{C_2} \frac{f(\zeta)}{(\zeta - z_0)^{n+1}} \, d\zeta \tag{3.3.9}$$

and

$$B_{-n-1} = \frac{1}{2\pi i} \oint_{C_1} \frac{f(\zeta)}{(\zeta - z_0)^{n+1}} \, d\zeta. \tag{3.3.10}$$

Because $f(z)$ is analytic in the annulus, it follows from Cauchy's Theorem that each of the integrals $\oint_{C_1}$ and $\oint_{C_2}$ in Eqs. (3.3.9) and (3.3.10) can be deformed into any simple closed contour enclosing C_1. This yields

$$C_n = \frac{1}{2\pi i} \oint_C \frac{f(\zeta)}{(\zeta - z_0)^{n+1}} \, d\zeta, \tag{3.3.11}$$

where $C_n = A_n$ for $n \geq 0$ and $C_n = B_{-n-1}$ for $n \leq -1$. Thus, Eq. (3.3.8) becomes

$$f(z) = \sum_{n=-\infty}^{\infty} C_n(z - z_0)^n, \tag{3.3.12}$$

which proves the theorem. □

The coefficient of the term $1/(z - z_0)$, which is C_{-1} in Eq. (3.3.12), turns out to play a very special role in complex analysis. It is given a special name: the **residue** of the function $f(z)$ (see Chapter 4). The negative powers of the Laurent series are referred to as the **principal part** of $f(z)$.

We note two important special cases:

(a) Suppose $f(z)$ is analytic everywhere *inside* the circle $|z - z_0| = R_1$. Then by Cauchy's theorem, $C_n = 0$ for $n \leq -1$ because the integrand in Eq. (3.3.11) is analytic; in this case, Eq. (3.3.12) reduces to the Taylor series

$$f(z) = \sum_{n=0}^{\infty} C_n(z - z_0)^n, \tag{3.3.13}$$

where C_n is given by Eq. (3.3.11) for $n \geq 0$.

(b) Suppose $f(z)$ is analytic everywhere *outside* the circle $|z - z_0| = R_2$. Then the integral (3.3.11) yields $C_n = 0$ for $n \geq 1$. In particular for $n \geq 1$ we have that the contour C in Eq. (3.3.11) may be deformed to a large circle $|z| = R$. Letting $z - z_0 = Re^{i\theta}$, we find

$$C_n = \frac{1}{2\pi i} \int_0^{2\pi} \frac{f(z_0 + Re^{i\theta})}{(Re^{i\theta})^{n+1}} Re^{i\theta} i \, d\theta \tag{3.3.14}$$

where C_R denotes a circle of radius R. Because $f(z)$ is analytic at infinity, the function $f(z_0 + Re^{i\theta})$ is bounded; that is,

$$|f(z_0 + Re^{i\theta}| \le M \tag{3.3.15}$$

hence,

$$|C_n| \le \frac{1}{2\pi} \int_0^{2\pi} \frac{M}{R^{n+1}} R d\theta \le \frac{M}{R^n}. \tag{3.3.16}$$

It follows that:

$$|C_n| \to 0 \text{ as } R \to \infty, \text{ with } n \ge 1.$$

Thus, in this case $f(z)$ has the form:

$$f(z) = \sum_{n=0}^{\infty} \frac{c_n}{(z - z_0)^n}.$$

In practice, Laurent series frequently may be obtained from the Taylor series of a function by appropriate substitutions. (We shall consider a number of other examples later in this section.) For example, replacing z in the series expansion for e^z (see Eq. (3.2.10)) by $1/z$ yields a Laurent series for $e^{1/z}$:

$$e^{1/z} = \sum_{j=0}^{\infty} \frac{1}{j! z^j}, \tag{3.3.17}$$

which contains an infinite number of negative powers of z. This is an example of case (b) above; that is, $e^{1/z}$ is analytic for all $|z| > 0$.

Laurent series have properties very similar to those of Taylor series. For example the series converges uniformly.

Theorem 3.3.2   *The Laurent series, Eqs. (3.3.1) and (3.3.2), of a function $f(z)$ which is analytic in an annulus $R_1 \le |z - z_0| \le R_2$ converges uniformly to $f(z)$ for $\rho_1 < |z - z_0| < \rho_2$, where $R_1 \le \rho_1$ and $R_2 \ge \rho_2$.*

Proof   The derivation of Laurent series shows that $f(z)$ has two representative parts, given by the two sums in Eq. (3.3.6). We write $f(z) = f_1(z) + f_2(z)$. The first series in Eq. (3.3.6) is the Taylor series part and it converges uniformly to $f_1(z)$ by the proof given in Theorem 3.2.1. For the second sum,

$$f_2(z) = \sum_{j=0}^{\infty} B_j (z - z_0)^{-(j+1)}, \tag{3.3.18}$$

we can use the M test. For j large enough and for $z = z_1$ on $|z - z_0| = \rho_1$,

$$\left| B_j(z - z_0)^{-(j+1)} \right| = \frac{|B_j|}{|z_1 - z_0|^{j+1}} \left| \frac{z_1 - z_0}{z - z_0} \right|^{j+1}$$

$$< \left| \frac{z_1 - z_0}{z - z_0} \right|^{j+1} = M^{j+1}, \quad M < 1, \tag{3.3.19}$$

where $|B_j|/|z_1 - z_0|^{j+1} < 1$ is due to the convergence of the series (3.3.18) (see Theorem 3.3.1). $\qquad\square$

A corollary to this result is the fact that the Laurent series may be integrated termwise, a fact that follows from Theorem 3.1.2. Indeed we have

$$\int \sum_{-\infty}^{\infty} C_n(z - z_0)^n \, dz = \sum_{-\infty}^{\infty} \int C_n(z - z_0)^n \, dz$$

$$= \sum_{-\infty, n \neq -1}^{\infty} \frac{C_n(z - z_0)^{n+1}}{n + 1} + C_{-1} \log(z - z_0) + C,$$

where C is an arbitrary constant. But we also note that due to the $\log(z - z_0)$ term, the integral of a Laurent series is not a Laurent series unless $C_{-1} = 0$. Similarly, the Laurent series may be differentiated termwise; the proof is similar to Theorem 3.2.4 and is therefore omitted. It is also easily shown that the elementary operations such as addition, subtraction and multiplication for Laurent series behave just like Taylor series. Like Taylor series, Laurent series are unique; we now show this.

Theorem 3.3.3 *The Laurent expansion given by Eqs. (3.3.11) and (3.3.12) is unique.*

Proof We show that if there were two such uniformly convergent series inside the annulus $R_1 < |z - z_0| < R_2$ we would have

$$f(z) = \sum_{-\infty}^{\infty} a_n(z - z_0)^n = \sum_{-\infty}^{\infty} b_n(z - z_0)^n.$$

Then, multiplying the sums by $(z - z_0)^j$ and using

$$\frac{1}{2\pi i} \oint_C (\zeta - z_0)^{n+j} \, d\zeta = \begin{cases} 1 & \text{when } n + j = -1, \\ 0 & \text{when } n + j \neq -1, \end{cases} \tag{3.3.20}$$

we find $a_{-1-j} = b_{-1-j}$, which establishes that a Laurent series is unique. $\qquad\square$

This fact allows us to obtain Laurent expansions by elementary methods.

Before showing examples we note that if we have a uniformly convergent series

$$f(z) = \sum_{n=-\infty}^{\infty} b_n(z - z_0)^n \tag{3.3.21}$$

valid in the annulus $R_1 \le |z - z_0| \le R_2$, then $b_n = C_n$, with C_n given by Eq. (3.3.11).

We emphasize that in practice one does not use Eq. (3.3.2) to compute the coefficients of the Laurent expansion of a given function. Instead one often appeals to the above uniqueness result and uses well-known Taylor expansions and appropriate substitutions.

Example 3.3.4   Find the Laurent expansion of $f(z) = 1/(1 + z)$ for $|z| > 1$.
   The Taylor series expansion (3.2.6) of $(1 - z)^{-1}$ is

$$\frac{1}{1 - z} = \sum_{n=0}^{\infty} z^n \qquad \text{for } |z| < 1. \tag{3.3.22}$$

We write

$$\frac{1}{1 + z} = \frac{1}{z(1 + 1/z)} \tag{3.3.23}$$

and use Eq. (3.3.22) with z replaced by $-1/z$, noting that if $|z| > 1$ then $|-1/z| < 1$. We find

$$\frac{1}{1 + z} = \frac{1}{z} \sum_{n=0}^{\infty} \frac{(-1)^n}{z^n} = \sum_{n=0}^{\infty} \frac{(-1)^n}{z^{n+1}} = \frac{1}{z} - \frac{1}{z^2} + \frac{1}{z^3} - \cdots.$$

We note that for $|z| < 1$, $f(z) = 1/(1 + z) = \sum_{n=0}^{\infty}(-1)^n z^n$. Thus, there are different series expansions in different regions of the complex plane. In summary,

$$\frac{1}{1 + z} = \begin{cases} \sum_{n=0}^{\infty}(-1)^n z^n & |z| < 1, \\ \sum_{n=0}^{\infty} \frac{(-1)^n}{z^{n+1}} & |z| > 1. \end{cases}$$

Example 3.3.5   Find the Laurent expansion of

$$f(z) = \frac{1}{(z - 1)(z - 2)} \qquad \text{for } 1 < |z| < 2.$$

We use partial fraction decomposition to rewrite $f(z)$ as

$$f(z) = -\frac{1}{z - 1} + \frac{1}{z - 2}. \tag{3.3.24}$$

Anticipating the fact that we will use Eq. (3.3.22), we rewrite Eq. (3.3.24) as

$$f(z) = -\frac{1}{z}\left(\frac{1}{1 - 1/z}\right) - \frac{1}{2}\left(\frac{1}{1 - z/2}\right). \tag{3.3.25}$$

Because $1 < |z| < 2$, $|1/z| < 1$, and $|z/2| < 1$, we can use Eq. (3.2.27) to obtain

$$f(z) = -\frac{1}{z}\sum_{n=0}^{\infty}\frac{1}{z^n} - \frac{1}{2}\sum_{n=0}^{\infty}\left(\frac{z}{2}\right)^n$$

$$= -\left(\frac{1}{z} + \frac{1}{z^2} + \frac{1}{z^3} + \cdots\right) - \frac{1}{2}\left(1 + \frac{z}{2} + \left(\frac{z}{2}\right)^2 + \cdots\right). \tag{3.3.26}$$

Thus

$$f(z) = \sum_{n=-\infty}^{\infty} C_n z^n$$

where

$$C_n = \begin{cases} -1 & n \le -1; \\ \frac{-1}{2^{n+1}} & n \ge 0. \end{cases}$$

As with Example 3.3.4, there exist different Laurent series expansions for $|z| < 1$ and for $|z| > 2$.

A somewhat more complicated example follows.

Example 3.3.6 Find the first two nonzero terms of the Laurent expansion of the function $f(z) = \tan z$ about $z = \pi/2$.

Let us call $z = \pi/2 + u$, so

$$f(z) = \frac{\sin\left(\frac{\pi}{2} + u\right)}{\cos\left(\frac{\pi}{2} + u\right)} = -\frac{\cos u}{\sin u}. \tag{3.3.27}$$

This can be expanded using the Taylor series for $\sin u$ and $\cos u$:

$$f(z) = -\frac{\left(1 - \frac{u^2}{2!} + \cdots\right)}{\left(u - \frac{u^3}{3!} + \cdots\right)} = -\frac{1}{u}\frac{\left(1 - \frac{u^2}{2!} + \cdots\right)}{\left(1 - \frac{u^2}{3!} + \cdots\right)}.$$

The denominator can be expanded via Eq. (3.3.22) to obtain for the first two nonzero terms

$$f(z) = -\frac{1}{u}\left(1 - \frac{u^2}{2!} + \cdots\right)\left(1 + \frac{u^2}{3!} + \cdots\right) = -\frac{1}{u}\left(1 - \frac{u^2}{3} + \cdots\right)$$

$$= -\frac{1}{\left(z - \frac{\pi}{2}\right)} + \frac{\left(z - \frac{\pi}{2}\right)}{3} + \cdots.$$

We also note that this Laurent series converges for $|z - \pi/2| < \pi$ or $-\pi/2 < z < 3\pi/2$. Since $\cos z$ is analytic in this region it vanishes for $z = -\pi/2, 3\pi/2$.

3.3.1 Problems for Section 3.3

1. Expand the function $f(z) = 1/(1 + z^2)$ in a:

   (a) Taylor series for $|z| < 1$;

   (b) Laurent series for $|z| > 1$.

2. Given the function $f(z) = z/(a^2 - z^2)$, with $a > 0$, expand $f(z)$ in a Laurent series in powers of z in the regions:

$$\text{(a) } |z| < a; \qquad \text{(b) } |z| > a.$$

3. Given the function

$$f(z) = \frac{z}{(z - 2)(z + i)},$$

   expand $f(z)$ in a Laurent series in powers of z in the regions:

$$\text{(a) } |z| < 1; \qquad \text{(b) } 1 < |z| < 2; \qquad \text{(c) } |z| > 2.$$

4. Evaluate the integral $\oint_C f(z)\, dz$ where C is the unit circle centered at the origin and $f(z)$ is given as follows:

$$\text{(a) } \frac{e^z}{z^3}; \quad \text{(b) } \frac{1}{z^2 \sin z}; \quad \text{(c) } \tanh z; \quad \text{(d) } \frac{1}{\cos 2z}; \quad \text{(e) } e^{1/z}.$$

5. Let

$$e^{\frac{t}{2}(z - 1/z)} = \sum_{n=-\infty}^{\infty} J_n(t) z^n.$$

   Show from the definition of Laurent series and using properties of integration that

$$J_n(t) = \frac{1}{2\pi} \int_{-\pi}^{\pi} e^{-i(n\theta - t \sin \theta)}\, d\theta$$

$$= \frac{1}{\pi} \int_0^{\pi} \cos(n\theta - t \sin \theta)\, d\theta.$$

   The functions $J_n(t)$ are called Bessel functions, well known from mathematics and physics.

6. Given the function

$$A(z) = \int_z^{\infty} \frac{e^{-1/t}}{t^2}\, dt$$

   find a Laurent expansion in powers of z for $|z| > R$, $R > 0$. The same procedure fails if we consider

$$E(z) = \int_z^{\infty} \frac{e^{-t}}{t}\, dt;$$

   why? See also Problem 7.

7. Suppose we are given

$$E(z) = \int_z^\infty \frac{e^{-t}}{t}\, dt, \qquad \mathrm{Re}\, z > 0.$$

A formal series may be obtained by repeated integration by parts; that is,

$$E(z) = \frac{e^{-z}}{z} - \int_z^\infty \frac{e^{-t}}{t^2}\, dt$$

$$= \frac{e^{-z}}{z} + \frac{e^{-z}}{z^2} - \int_z^\infty \frac{2e^{-t}}{t^3}\, dt = \cdots .$$

If this procedure is continued, show that the series is given by

$$E(z) = \frac{e^{-z}}{z}\left(1 - \frac{1}{z} + \cdots + \frac{(-1)^n n!}{z^n}\right) + R_n(z),$$

$$R_n(z) = (-1)^{n+1}(n+1)! \int_z^\infty \frac{e^{-t}}{t^{n+2}}\, dt.$$

Explain why the series does not converge. See also Problem 8.

8. In Problem 7, consider $z = x$ real, $x > 0$. Show that

$$|R_n(z)| \le (n+1)! \frac{e^{-x}}{x^{n+2}}.$$

Explain how to approximate the integral $E(x)$ for large x, given some n. Find suitable values of x for $n = 1, 2, 3$ in order to approximate $E(x)$ to within 0.01, using the above inequality for $|R_n(x)|$. Why does the approximation fail as $n \to \infty$?

9. Find the first three nonzero terms of the Laurent series for the function $f(z) = [z(z-1)]^{1/2}$ for $|z| > 1$ on its principal branch.

10. Show that the first three terms of the Laurent series of $F(z) = \frac{1}{z^2 \sin z}$ for $0 < |z| < \pi$ is given by

$$F(z) = \frac{1}{z^3} + \frac{1}{6z} + \frac{7z}{360} + \cdots .$$

Use this result to deduce that $\oint_C F(z)\, dz = \frac{\pi i}{3}$ where C is the unit circle.

*3.4 Theoretical Results for Sequences and Series

Earlier in this chapter we introduced the notions of sequences, series and uniform convergence. Although the Weierstrass M test was stated, a proof was deferred to this section for those interested readers. We begin this section by discussing the notion of a Cauchy sequence.

Definition 3.4.1   A sequence of complex numbers $\{f_n\}$ forms a **Cauchy sequence** if, for every $\epsilon > 0$, there is an $N = N(\epsilon)$, such that whenever $n \geq N$ and $m \geq N$ we have $|f_n - f_m| < \epsilon$.

The same definition as the above applies to sequences of complex functions $\{f_n(z)\}$, where it is understood that $f_n(z)$ exists in some region $\mathcal{R}$, $z \in \mathcal{R}$. Here, in general, $N = N(\epsilon, z)$. Whenever $N = N(\epsilon)$ only, the sequence $\{f_n(z)\}$ is said to be a uniform Cauchy sequence. The following result is immediate.

Theorem 3.4.2   *If a sequence converges, then it is a Cauchy sequence.*

Proof   If $\{f_n(z)\}$ converges to $f(z)$, then for any $\epsilon > 0$ there is an $N = N(\epsilon, z)$ such that whenever $n > N$ and $m > N$,

$$|f_n(z) - f(z)| < \frac{\epsilon}{2} \quad \text{and} \quad |f_m(z) - f(z)| < \frac{\epsilon}{2}.$$

Hence

$$
\begin{aligned}
|f_n(z) - f_m(z)| &= |f_n(z) - f(z) - (f_m(z) - f(z))| \\
&\leq |f_n(z) - f(z)| + |f_m(z) - f(z)| \\
&< \epsilon,
\end{aligned}
$$

and so $\{f_n(z)\}$ is a Cauchy sequence.

We note that if $\{f(z)\}$ converges uniformly to $f(z)$ then $N = N(\epsilon)$ only, and the Cauchy sequence is uniform. $\square$

We shall next prove the converse, namely that every Cauchy sequence converges. We shall employ the following result of real analysis, namely *every real Cauchy sequence has a limit*.

Theorem 3.4.3   *If $\{f_n(z)\}$ is a Cauchy sequence then there is a function $f(z)$ such that $\{f_n(z)\}$ converges to $f(z)$.*

Proof   Let us call $f_n(z) = u_n(x, y) + iv_n(x, y)$. Since

$$
\begin{aligned}
|u_n(x, y) - u_m(x, y)| &\leq |f_n(z) - f_m(z)|, \\
|v_n(x, y) - v_m(x, y)| &\leq |f_n(z) - f_m(z)|,
\end{aligned}
$$

and $\{f_n(z)\}$ is a Cauchy sequence, we find that $\{u_n(x, y)\}$ and $\{v_n(x, y)\}$ are real Cauchy sequences and hence have limits $u(x, y)$ and $v(x, y)$, respectively. Thus, the function $f(z) = u(x, y) + iv(x, y)$ exists and is the limit of $\{f_n(z)\}$.

Convergence will be uniform if the number N for the Cauchy sequence $\{f_n(z)\}$ depends only on ϵ (i.e., $N = N(\epsilon)$) and not on both ϵ and z (i.e., $N = N(\epsilon, z)$). $\square$

The above theorem allows us to prove the Weierstrass M test given in Section 3.1. We repeat the statement of this theorem now for the convenience of the reader.

Theorem 3.4.4 *Let* $|b_j(z)| \leq M_j$, *with each* $M_j, j = 1, 2, \ldots$ *constant, in some region* $\mathcal{R}$. *If* $\sum_{j=1}^{\infty} M_j$ *converges, then the series* $f(z) = \sum_{j=1}^{\infty} b_j(z)$ *converges uniformly in* $\mathcal{R}$.

Proof Let $n > m$ and $f_n(z) = \sum_{j=1}^{n} b_j(z)$. Then

$$|f_n(z) - f_m(z)| = \left| \sum_{j=m+1}^{n} b_j(z) \right| \leq \sum_{j=m+1}^{n} |b_j(z)| \leq \sum_{j=m+1}^{n} M_j \leq \sum_{j=m+1}^{\infty} M_j.$$

Because $\sum_{j=1}^{\infty} M_j$ converges, we know that there is an $N = N(\epsilon)$ (the M_j are only constants) such that when $m > N$, $\sum_{j=m+1}^{\infty} M_j < \epsilon$. Thus $\{f_n(z)\}$ is a uniformly convergent Cauchy sequence in $\mathcal{R}$, and Theorem 3.4.4 follows. $\qquad\square$

Early in this chapter we proved Theorem 3.1.2, which allowed us to interchange the operation of integration with a limit of a uniformly convergent sequence of functions. A corollary of this result is Eq. (3.1.9), which allows the interchange of sum and an integral for a uniformly convergent series. A similar theorem holds for the operation of differentiation.

Theorem 3.4.5 *Let* $f_n(z)$ *be analytic in the disc* $|z - z_0| < R$, *and let* $\{f_n(z)\}$ *converge uniformly to* $f(z)$ *in* $|z - z_0| \leq R - \delta$, *for* $\delta > 0$. *Then:*

(a) $f(z)$ *is analytic for* $|z - z_0| < R$; *and*

(b) $\{f_n'(z)\}$, $\{f_n''(z)\}$, $\ldots$, *converge uniformly in* $|z - z_0| \leq R - \delta$, *to* $f'(z)$, $f''(z)$, $\ldots$.

Proof (a) Let C be any simple closed contour lying inside $|z - z_0| \leq R - \delta$ (see Figure 3.4) for all $R > \delta > 0$. Because $\{f_n(z)\}$ is uniformly convergent, we have, from Theorem 3.1.2,

$$\oint_C f(z)\, dz = \lim_{n \to \infty} \oint_C f_n(z)\, dz. \qquad (3.4.1)$$

Because $f_n(z)$ is analytic we conclude from Cauchy's theorem that $\oint_C f_n(z)\, dz = 0$, hence $\oint_C f(z)\, dz = 0$. Now from Morera's Theorem, 2.6.5 of Section 2.6, we find that $f(z)$ is analytic in $|z - z_0| < R$ (because δ may be made arbitrarily small). This proves part (a).

(b) Let C_1 be the circle $|z - z_0| = R - \nu$ for all $0 < \nu < \frac{R}{2}$ (see Figure 3.5). We next use Cauchy's theorem for $f'(z) - f_n'(z)$ (Theorem 2.6.3, Eq. (2.6.5)), which gives

$$(f'(z) - f_n'(z)) = \frac{1}{2\pi i} \oint_{C_1} \frac{(f(\zeta) - f_n(\zeta))}{(\zeta - z)^2}\, d\zeta. \qquad (3.4.2)$$

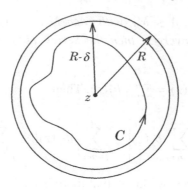

Figure 3.4 Region of analyticity in Theorem 3.4.5(a)

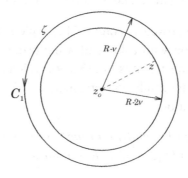

Figure 3.5 For Theorem 3.4.5(b); $|z - \zeta| > v$

Because $\{f_n(z)\}$ is a uniformly convergent sequence, we find that for $n > N$ and any z in C_1

$$|f(z) - f_n(z)| < \epsilon_1.$$

Thus

$$|f'(z) - f'_n(z)| < \frac{\epsilon_1}{2\pi} \oint_{C_1} \frac{|d\zeta|}{|\zeta - z|^2}.$$

If z lies inside C_1 say, $|z - z_0| = R - 2v$, then $|\zeta - z| \geq v$, $\oint_{C_1} |d\zeta| < 2\pi R$, hence

$$|f'(z) - f'_n(z)| < \frac{\epsilon_1 R}{v^2}. \tag{3.4.3}$$

Taking ϵ_1 as small as necessary; that is, $\epsilon_1 = \epsilon v^2/R$, ensures that $|f'(z) - f'_n(z)|$ is arbitrarily small; that is, $|f'(z) - f'_n(z)| < \epsilon$ and hence $\{f'_n(z)\}$ converges uniformly to $f'(z)$ inside $|z - z_0| \leq R - 2v$. Taking $\delta = 2v$ and then employing part (a) above establishes the theorem for the sequence $\{f'_n(z)\}$. (The values of δ and v can be taken arbitrarily small.) Because the sequence $\{f'_n(z)\}$ converges uniformly inside

$|z - z_0| \leq R - \delta$, for $R > \delta > 0$, we can repeat the above procedure in order to establish the Theorem for the sequence $\{f_n''(z)\}$; that is, the sequence $\{f_n''(z)\}$ converges uniformly to $f''(z)$ and so on for $\{f_n'''(z)\}$ to $f_n'''(z)$, etc.   □

An immediate consequence of this theorem is the result for series. To see this, write

$$S_n(z) = \sum_{j=1}^{n} f_j(z). \qquad (3.4.4)$$

If $S_n(z)$ satisfies the hypothesis of Theorem 3.4.5, then

$$\lim_{n \to \infty} S_n'(z) = \lim_{n \to \infty} \sum_{j=1}^{n} f_j'(z) = \sum_{j=1}^{\infty} f_j'(z) = S'(z). \qquad (3.4.5)$$

Similarly, another corollary of Theorem 3.4.5 is that power series

$$f(z) = \sum_{n=0}^{\infty} a_n(z - z_0)^n$$

may be differentiated termwise inside their radius of convergence. Indeed, we have already shown in Section 3.2 that any power series is really the Taylor series expansion of the represented function. Hence Theorem 3.4.5 could have alternatively been used to establish the validity of differentiating Taylor series inside their radius of convergence.

We remark that $\{f_n(z)\}$ being a uniformly convergent sequence of *analytic* functions gives us a much stronger result than we have for uniformly convergent sequences of only real functions. Namely sequences of derivatives of any order of $f_n(z)$ are uniformly convergent. For example consider the real sequence $\{u_n(x)\}$, where

$$u_n(x) = \frac{\cos n^2 x}{n}, \qquad |x| < \infty.$$

This sequence is uniformly convergent to zero because $|u_n(x)| \leq 1/n$ (independent of x), which converges to *zero*. However, the sequence of functions $\{u'_n(x)\}$

$$u_n'(x) = 2n \sin n^2 x, \qquad |x| < \infty, \qquad (3.4.6)$$

for $x \neq m\pi, m$ integer, has no limit whatsoever! The sequence $\{u_n'(x)\}$ is not uniformly convergent. We note also the above sequence $u_n(z)$ for $z = x + iy$ is not uniformly convergent for $|z| < \infty$ because $\cos n^2 z = \cos n^2 x \cosh n^2 y - i \sin n^2 x \sinh n^2 y$; and both $\cosh n^2 y$ and $\sinh n^2 y$ diverge as $n \to \infty$ for $y \neq 0$.

We conclude with an example.

Example 3.4.6 We are given

$$\zeta(z) = \sum_{n=1}^{\infty} \frac{1}{n^z}. \tag{3.4.7}$$

(The function $\zeta(z)$ is often called the **Riemann zeta function**; it appears in many branches of mathematics and physics.) Show that $\zeta(z)$ is analytic for all $x > 1$, where $z = x + iy$.

By definition, $n^z = e^{z \log n}$, where we take $\log n$ to be the principal branch of the log. Hence

$$n^z = e^{z \log n} = e^{(x+iy) \log n}$$

is analytic for all z because e^{kz} is analytic, and

$$|n^z| = e^{x \log n} = n^x.$$

Thus, from the Weierstrass M test (Theorem 3.1.4 or 3.4.4, proven in this section), we find that the series representing $\zeta(z)$ converges uniformly because the series $\sum_{n=1}^{\infty}(1/n^x)$ (for $x > 1$) is a convergent series of real numbers. That is, we may use the integral theorem for a series of real numbers. In other words, we may use the integral theorem for a series of real numbers as our upper bound to establish this. Note that

$$\int_{1}^{\infty} \frac{1}{n^x} \, dn = \frac{1}{x - 1}.$$

Thus, from Theorem 3.4.5 we have that because $\{\zeta_m(z)\} = \{\sum_{n=1}^{m} n^{-z}\}$ is a uniformly convergent sequence of analytic functions for all $x > 1$, the sum $\zeta(z)$ is analytic.

3.4.1 Problems for Section 3.4

1. Demonstrate whether or not the following sequences are Cauchy sequences:

   (a) $\{z^n\}_{n=1}^{\infty}$, $\quad |z| < 1$; $\qquad$ (b) $\left\{1 + \dfrac{z}{n}\right\}_{n=1}^{\infty}$, $\quad |z| < \infty$;

   (c) $\{\cos nz\}_{n=1}^{\infty}$, $\quad |z| < \infty$; $\qquad$ (d) $\left\{e^{-n/z}\right\}_{n=1}^{\infty}$, $\quad |z| < 1$.

2. Discuss whether the following series converge uniformly in the given domains.

   (a) $\displaystyle\sum_{j=1}^{n} z^j$, $\quad |z| \le R, \ R < 1$; $\qquad$ (b) $\displaystyle\sum_{j=0}^{n} e^{-jz}$, $\quad \dfrac{1}{2} < |z| \le R, \ R < 1$;

   (c) $\displaystyle\sum_{j=1}^{n} j! z^{2j}$, $\quad |z| < a, \ a > 0$.

3. Establish that the function $\sum_{n=1}^{\infty} \frac{1}{e^n n^z}$ is an analytic function of z for all z, i.e., it is an entire function.

4. Show that the following functions are analytic functions of z for all z; i.e., they are entire:

$$\text{(a)} \ \sum_{n=1}^{\infty} \frac{z^n}{(n!)^2}; \qquad \text{(b)} \ \sum_{n=1}^{\infty} \frac{\cosh nz}{n!}; \qquad \text{(c)} \ \sum_{n=1}^{\infty} \frac{z^{2n+1}}{[(2n+1)!]^{1/2}}.$$

5. Consider the function $f(z) = \sum_{n=1}^{\infty}(1/(z^2 + n^2))$. Break the function $f(z)$ into two parts, $f(z) = f_1(z) + f_2(z)$, where

$$f_1(z) = \sum_{n=1}^{N} \left(\frac{1}{z^2 + n^2} \right)$$

and

$$f_2(z) = \sum_{n=N+1}^{\infty} \left(\frac{1}{z^2 + n^2} \right).$$

For $|z| < R$, $N > 2R$, show that in the second sum

$$\left| \frac{1}{z^2 + n^2} \right| \leq \frac{1}{n^2 - R^2} \leq \frac{4}{3n^2},$$

whereupon explain why $f_2(z)$ converges uniformly and consequently, why $f(z)$ is analytic everywhere except at the distinct points $z = \pm in$.

6. Use the method of Problem 5 to investigate the analytic properties of $f(z) = \sum_{n=1}^{\infty} \frac{1}{(z+n)^2}$.

3.5 Singularities of Complex Functions

3.5.1 Singular Points

We begin this section by introducing the notion of an **isolated singular point**. The concept of a singular point was introduced in Section 2.1 as being a point where a given (single-valued) function is not analytic. Namely, $z = z_0$ is a singular point of $f(z)$ if $f'(z_0)$ does not exist. Suppose $f(z)$ (or any single-valued branch of $f(z)$, if $f(z)$ is multivalued) is analytic in the region $0 < |z - z_0| < R$ (i.e., in a neighborhood of $z = z_0$), and *not* at the point z_0. Then the point $z = z_0$ is called an **isolated singular point** of $f(z)$. In the neighborhood of an isolated singular point, the results of Section 3.3 show that $f(z)$ may be represented by a Laurent expansion:

$$f(z) = \sum_{n=-\infty}^{\infty} C_n(z - z_0)^n. \tag{3.5.1}$$

Suppose $f(z)$ has an isolated singular point and in addition it is bounded; that is, $|f(z)| \leq M$ where M is a constant. In order for $f(z)$ to be bounded it is clear that all coefficients $C_n = 0$ for $n < 0$. Thus such a function $f(z)$ is given by a power series expansion, $f(z) = \sum_{n=0}^{\infty} C_n (z - z_0)^n$, valid for $|z - z_0| < R$ *except possibly at* $z = z_0$. However, because a power series expansion converges at $z = z_0$, it follows that $f(z)$ would be analytic if $C_0 = f(z_0)$ (the $n = 0$ term is the only nonzero contribution), in which case $\sum_{n=0}^{\infty} C_n (z - z_0)^n$ is the Taylor series expansion of $f(z)$. If $C_0 \neq f(z_0)$, we call such a point a **removable singularity**, because by a slight redefinition of $f(z_0)$, the function $f(z)$ is analytic. For example, consider the function $f(z) = (\sin z)/z$ which, strictly speaking, is undefined at $z = 0$. If it were the case that $f(0) \neq 1$, then $z = 0$ is a removable singularity. Namely, by simply redefining $f(0) = 1$, then $f(z)$ is analytic for all z including $z = 0$ and is represented by the power series

$$f(z) = 1 - \frac{z^2}{3!} + \frac{z^4}{5!} - \frac{z^6}{7!} + \cdots = \sum_{n=0}^{\infty} \frac{(-1)^n z^{2n}}{(2n + 1)!}.$$

Stated differently, if $f(z)$ is analytic in the region $0 < |z - z_0| < R$, and if $f(z)$ can be made analytic at $z = z_0$ by assigning an appropriate value for $f(z_0)$, then $z = z_0$ is a removable singularity.

An isolated singularity at z_0 of $f(z)$ is said to be a pole if $f(z)$ has the following representation:

$$f(z) = \frac{\phi(z)}{(z - z_0)^N}, \tag{3.5.2}$$

where N is a positive integer, $N \geq 1$, $\phi(z)$ is analytic in a neighborhood of z_0, and $\phi(z_0) \neq 0$. We generally say $f(z)$ has an Nth-**order pole** if $N \geq 2$ and has a **simple pole** if $N = 1$. Equation (3.5.2) implies that the Laurent expansion of $f(z)$ takes the form $f(z) = \sum_{n=-N}^{\infty} C_n (z - z_0)^n$; that is, the first coefficient is $C_{-N} = \phi(z_0)$. The coefficient C_{-N} is often called the **strength of the pole**. Moreover it is also clear that in the neighborhood of $z = z_0$, the function $f(z)$ takes on arbitrarily large values, or $\lim_{z \to z_0} f(z) = \infty$.

Example 3.5.1   Describe the singularities of the function

$$f(z) = \frac{z^2 - 2z + 1}{z(z + 1)^3} = \frac{(z - 1)^2}{z(z + 1)^3}.$$

The function $f(z)$ has a simple pole at $z = 0$, and a third-order (or **triple**) pole at $z = -1$. The strength of the pole at $z = 0$ is 1, because the expansion of $f(z)$ near $z = 0$ has the form

$$f(z) = \frac{1}{z}(1 - 2z + \cdots)(1 - 3z + \cdots) = \frac{1}{z} - 5 + \cdots.$$

Similarly, the strength of the third-order pole at $z = -1$ is -4, since the leading term of the Laurent series near $z = -1$ is $f(z) = -4/(z + 1)^3$. A three-dimensional plot of the magnitude of this function $|f(z)| = |f(x, y)|$ is given in Figure 3.6. It is clear from this figure that the triple pole at $z = 0$ dominates the pole at $z = 0$.

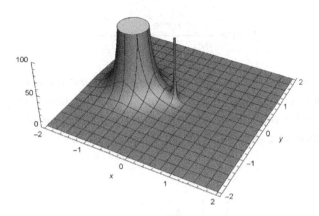

Figure 3.6 The magnitude of the function $|f(z) = \frac{(z-1)^2}{z(z+1)^3}|$ from Example 3.5.1

Example 3.5.2 Discuss the singularities of the function

$$f(z) = \frac{z + 1}{z \sin z}.$$

This function has poles at $z = 0$, and $z = n\pi$, $n \neq 0$. First, we find the Laurent series near $z = 0$. Using the Taylor series for $\sin z$ near $z = 0$,

$$f(z) = \frac{z + 1}{z \left(z - \frac{z^3}{3!} + \frac{z^5}{5!} - \cdots \right)} = \frac{z + 1}{z^2 \left(1 - \frac{z^2}{3!} + \frac{z^4}{5!} - \cdots \right)}$$

$$= \frac{z + 1}{z^2} \left[1 + \left(\frac{z^2}{3!} - \frac{z^4}{5!} + \cdots \right) + \left(\frac{z^2}{3!} - \frac{z^4}{5!} + \cdots \right)^2 + \cdots \right]$$

$$= \left(\frac{1}{z^2} + \frac{1}{z} \right) \left(1 + \frac{z^2}{3!} + \cdots \right) = \frac{1}{z^2} + \frac{1}{z} + \cdots,$$

we find that the function $f(z)$ has a second-order (**double**) pole at $z = 0$ with strength 1.

To find the Laurent series near $z = n\pi, n \neq 0$ it is convenient to transform variables $z = n\pi + u$ so that

$$f(u) = \frac{n\pi + 1 + u}{(n\pi + u) \cos n\pi \sin u} = \frac{n\pi + 1 + u}{(n\pi + u)(-1)^n(u - u^3/6)} = \frac{n\pi + 1}{n\pi(-1)^n u} + \cdots$$

or

$$f(z) = \frac{n\pi + 1}{n\pi(-1)^n(z - n\pi)} + \cdots .$$

Hence $f(z)$ has a simple pole in the neighborhood of $z = n\pi, n \neq 0$.

A three-dimensional plot of the magnitude of the function $|f(z)| = |f(x, y)|$ is given in Figure 3.7, from which we can see that the double pole at $z = 0$ has a larger effect than the pole at $z = n\pi, n = \pm 1, \pm 2, \pm 3, \ldots$

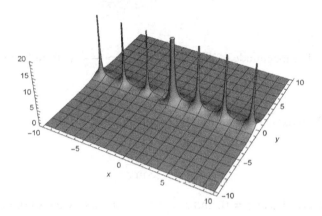

Figure 3.7 The magnitude of the function $|f(z) = \frac{z+1}{z \sin z}|$ from Example 3.5.2

Example 3.5.3 Describe the singularities of the function

$$f(z) = \tan z = \frac{\sin z}{\cos z}.$$

Here we will show the function $f(z)$ has simple poles with strength -1 at $z = \pi/2 + m\pi$ for $m = 0, \pm 1, \pm 2, \ldots$ As in the example above, it is useful to make a transformation of variables to transform the location of the poles to the origin: that is, $z = z_0 + z'$, where $z_0 = \pi/2 + m\pi$, so that

$$f(z) = \frac{\sin(\pi/2 + m\pi + z')}{\cos(\pi/2 + m\pi + z')}$$

$$= \frac{\sin(\pi/2 + m\pi)\cos z' + \cos(\pi/2 + m\pi)\sin z'}{\cos(\pi/2 + m\pi)\cos z' - \sin(\pi/2 + m\pi)\sin z'}$$

$$= \frac{(-1)^m \cos z'}{(-1)^{m+1}\sin z'}$$

$$= -\frac{\left(1 - (z')^2/2! + \cdots\right)}{(z' - (z')^3/3! + \cdots)} = -\frac{1}{z'}\frac{\left(1 - (z')^2/2! + \cdots\right)}{(1 - (z')^2/3! + \cdots)}$$

$$= -\frac{1}{z'}\left(1 - \left(\frac{1}{2!} - \frac{1}{3!}\right)z'^2 + \cdots\right) = -\frac{1}{z'}\left(1 - \frac{1}{3}z'^2 + \cdots\right)$$

$$= -\frac{1}{z'} + \frac{1}{3}z' + \cdots$$

$$= -\frac{1}{z - (\pi/2 + m\pi)} + \frac{1}{3}(z - (\pi/2 + m\pi)) + \cdots .$$

Hence $f(z) = \tan z$ always has a simple pole of strength -1 at $z = \frac{\pi}{2} + m\pi$.

A three-dimensional plot of the magnitude of the function: $|f(z)| = |f(x, y)|$ is given in Figure 3.8. You can see that the poles at $z = n\pi, n = 0, \pm1, \pm2, \pm3, \ldots$ are all the same size.

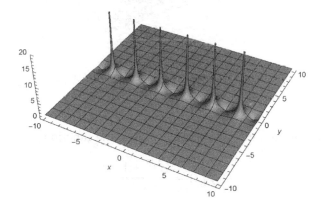

Figure 3.8 The magnitude of the function $|f(z) = \tan z = \frac{\sin z}{\cos z}|$ from Example 3.5.3

Example 3.5.4 Discuss the pole singularities of the function

$$f(z) = \frac{\log(z + 1)}{(z - 1)}.$$

The function $f(z)$ is multivalued with a branch point at $z = -1$, hence following the procedure in Section 2.2 we can make $f(z)$ single valued by introducing a branch cut. We take the cut from the branch point at $z = -1$ to $z = \infty$ along the negative real axis with $z = re^{i\theta}$ for $-\pi \le \theta < \pi$; this branch fixes $\log(1) = 0$. With this choice of branch, $f(z)$ has a simple pole at $z = 1$ with strength $\log 2$. We shall discuss the nature of branch point singularities later in this section.

Sometimes we might have different types of singularities depending on which branch of a multivalued function we select.

A three-dimensional plot of the magnitude of the function: $|f(z)| = |f(x, y)|$ is given in Figure 3.9; note that the direction of the positive x-axis is reversed. It is clear from the figure that the pole at $z = 1$ is much larger than the branch point singularity at $z = -1$. It is also clear that the function has a discontinuity in its derivative along the x-axis: $x \le -1$.

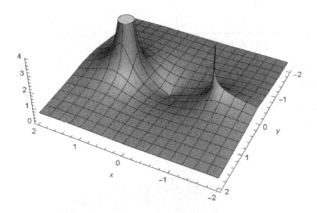

Figure 3.9 The magnitude of the function $|f(z) = \frac{\log(z+1)}{(z-1)}|$ from Example 3.5.4

Example 3.5.5 Discuss the pole singularities of the function

$$f(z) = \frac{z^{1/2} - 1}{z - 1}.$$

We let $z = 1 + t$, so that

$$f(z) = \frac{(1 + t)^{1/2} - 1}{t} = \frac{\pm\sqrt{1 + t} - 1}{t},$$

where $\pm$ denotes the two branches of the square root function with $\sqrt{x} \ge 0$ for $x \ge 0$. (The point $z = 0$ is a square root branch point.)

The Taylor series of $\sqrt{1 + t}$ is

$$\sqrt{1 + t} = 1 + \frac{1}{2}t - \frac{1}{8}t^2 + \cdots .$$

Thus for the "+" branch,

$$f(z) = \frac{\frac{t}{2} - \frac{1}{8}t^2 + \cdots}{t} = \frac{1}{2} - \frac{1}{8}t + \cdots,$$

whereas for the "−" branch,

$$f(z) = \frac{-2 - \frac{t}{2} + \frac{1}{8}t^2 - \cdots}{t} = \frac{-2}{t} - \frac{1}{2} + \frac{1}{8}t - \cdots.$$

On the + (principal) branch, $f(z)$ is analytic in the neighborhood of $t = 0$; that is, $t = 0$ is a removable singularity. For the − branch, $t = 0$ is a simple pole with strength −2.

An isolated singular point that is neither removable nor a pole is called an **essential singular point**. An essential singular point has a "full" Laurent series in the sense that given $f(z) = \sum_{n=-\infty}^{\infty} C_n(z - z_0)^n$, then for any $N > 0$ there is an $n < -N$ such that $C_n \neq 0$; that is, the series for negative n does not terminate. If this were not the case then $f(z)$ would have a pole (if $C_n = 0$ for $n < -N$ and $C_{-N} \neq 0$, then $f(z)$ would have a pole of order N with strength C_{-N}).

The prototypical example of an essential singular point is given by the function

$$f(z) = e^{1/z}, \tag{3.5.3}$$

which has the following Laurent series (Eq. (3.3.17)) about the essential singular point at $z = 0$:

$$f(z) = \sum_{n=0}^{\infty} \frac{1}{n!z^n}. \tag{3.5.4}$$

Because $f'(z) = -e^{1/z}/z^2$ exists for all points $z \neq 0$, it is clear that $f(z)$ is analytic in the neighborhood of $z = 0$; hence it is isolated (as it must be for $z = 0$ to be an essential singular point).

If we use polar coordinates $z = re^{i\theta}$, then Eq. (3.5.3) yields

$$f(z) = e^{\frac{1}{r}e^{-i\theta}} = e^{\frac{1}{r}(\cos\theta - i\sin\theta)} = e^{\frac{1}{r}\cos\theta}\left[\cos\left(\frac{\sin\theta}{r}\right) - i\sin\left(\frac{\sin\theta}{r}\right)\right],$$

whereupon the modulus of $f(z)$ is given by

$$|f(z)| = e^{\frac{1}{r}\cos\theta}.$$

Clearly for values of θ such that $\cos\theta > 0$, $f(z) \to \infty$ as $r \to 0$, and for $\cos\theta < 0$, $f(z) \to 0$ as $r \to 0$. Also for $\theta = \pm\pi/2$, $|f(z)| = 1$. Indeed, if we let r take values on a suitable curve, namely, $r = (1/R)\cos\theta$, $R \neq 0$ (i.e., the points (r, θ) lie on a circle of diameter $1/R$ tangent to the imaginary axis), then

$$f(z) = e^R[\cos(R\tan\theta) - i\sin(R\tan\theta)] \tag{3.5.5}$$

and

$$|f(z)| = e^R. \tag{3.5.6}$$

Thus $|f(z)|$ may take on any positive value in the neighborhood of $z = 0$. As $z \to 0$ on this circle, $\theta \to \pi/2$ (and $\tan\theta \to \infty$) with R fixed, then by considering a family of circles with different radii R the coefficient in brackets in Eq. (3.5.5) takes on *all values* on the unit circle *infinitely often*. Hence we see that $f(z)$ takes on *all* nonzero complex values with modulus Eq. (3.5.6) infinitely often.

In fact this example describes a general feature of essential singular points discovered by Picard (Picard's Theorem). He showed that in any neighborhood of an essential singularity of function, $f(z)$ assumes all values, except possibly one of them, an infinite number of times. The following result owing to Weierstrass is similar, and more easily shown.

Theorem 3.5.6 *If $f(z)$ has an essential singularity at $z = z_0$, then for any complex number w, $f(z)$ becomes arbitrarily close to w in a neighborhood of z_0. That is, given w, and any $\epsilon > 0$, $\delta > 0$, there is a z such that*

$$|f(z) - w| < \epsilon, \tag{3.5.7}$$

whenever $0 < |z - z_0| < \delta$.

Proof We prove this by contradiction. Suppose $|f(z) - w| > \epsilon$ whenever $|z - z_0| < \delta$, where δ is small enough such that $f(z)$ is analytic in the region $0 < |z - z_0| < \delta$. Thus in this region,

$$h(z) = \frac{1}{f(z) - w}$$

is analytic, and hence bounded; specifically, $|h(z)| < 1/\epsilon$. The function $f(z)$ is not identically constant, otherwise $f(z)$ would be analytic and hence would not possess an essential singular point. Because $h(z)$ is analytic and bounded, it is representable by a power series $h(z) = \sum_{n=0}^{\infty} C_n(z - z_0)^n$: thus, its only possible singularity is removable. By choosing $C_0 = h(z_0)$ it follows that $h(z)$ is analytic for $|z - z_0| < \delta$. Consequently

$$f(z) = w + \frac{1}{h(z)},$$

and $f(z)$ is either analytic with $h(z) \neq 0$ or else $f(z)$ has a pole of order N, strength C_N, where C_N is the first nonzero coefficient of the term $(z - z_0)^N$ in the Taylor series representation of $h(z)$. In either case, this contradicts the hypothesis that $f(z)$ has an essential singular point in the neighborhood of $z = z_0$. □

Functions that have only isolated singularities, while very special, turn out to be important in applications. An **entire** function is one which is analytic everywhere in the finite z-plane. As proved in Chapter 2, the only function analytic everywhere, including the point at infinity, is a constant (Liouville's Theorem, Section 2.6). Entire functions are either constant functions, or at infinity they have isolated poles or essential singularities. Some of the common entire functions include:

(a) polynomials;

(b) exponential functions;

(c) sine/cosine functions.

That is, $f(z) = z$, $f(z) = e^z$, $f(z) = \sin z$ are all entire functions.

As mentioned earlier, one can easily ascertain the nature of the singularity at $z = \infty$ by making the transformation $z = 1/t$ and investigating the behavior of the function near $t = 0$. Polynomials have poles at $z = \infty$, the order of which corresponds to the order of the polynomial. For example, $f(z) = z$ has a simple pole at infinity (of strength unity) because $f(t) = 1/t$. Similarly, $f(z) = z^2$ has a double pole at $z = \infty$, etc. The entire functions e^z and $\sin z$ have essential singular points at $z = \infty$. Indeed, the Taylor series for $\sin z$ shows that the Laurent series around $t = 0$ does not terminate in any finite negative power:

$$\sin \frac{1}{t} = \sum_{n=0}^{\infty} \frac{(-1)^n}{t^{2n+1}(2n+1)!},$$

hence it follows that $t = 0$ or $z = \infty$ is an essential singular point.

The next level of complication after an entire function is a function which has only poles in the finite z-plane. Such a function is called a **meromorphic function**. As with entire functions, meromorphic functions may have essential singular points at infinity. A meromorphic function is a ratio of entire functions. For example, a **rational function** (i.e., a ratio of polynomials),

$$R(z) = \frac{A_N z^N + A_{N-1} z^{N-1} + \cdots + A_1 z + A_0}{B_M z^M + B_{M-1} z^{M-1} + \cdots + B_1 z + B_0} \tag{3.5.8}$$

is meromorphic. It has only poles as its singular points. The denominator is a polynomial, whose zeroes correspond to the poles of $R(z)$. For example, the function

$$R(z) = \frac{z^2 - 1}{z^5 + 2z^3 + z} = \frac{(z+1)(z-1)}{z(z^4 + 2z^2 + 1)} = \frac{(z+1)(z-1)}{z(z^2+1)^2}$$
$$= \frac{(z+1)(z-1)}{z(z+i)^2(z-i)^2}$$

has poles at $z = 0$ (simple), at $z = \pm i$ (both double), and zeroes (simple) at $z = \pm 1$.

The function $f(z) = (\sin z)/(1 + z)$ is meromorphic. It has a pole at $z = -1$, due to the vanishing of $(1 + z)$, and an essential singular point at $z = \infty$ due to the behavior of $\sin z$ near infinity (as discussed earlier).

There are other types of singularities of a complex function which are nonisolated. In Sections 2.2–2.3, we discussed at length the various aspects of multivalued
functions. Multivalued functions have *branch points*. We recall that their character-
istic property is the following.

*If a circuit is made around a sufficiently small simple closed contour enclosing
the branch point, then the value assumed by the function at the end of the circuit
differs from its initial value.*

A branch point is an example of a nonisolated singular point, because a circuit (no
matter how small) around the branch point results in a discontinuity. We also recall
that in order to make a multivalued function $f(z)$ single-valued, we must introduce
a branch cut. Since $f(z)$ has a discontinuity across the cut, we shall consider the
branch cut as a singular curve (it is not simply a point). However, it is important
to recognize that a branch cut may be moved, as opposed to a branch point, and
therefore the nature of its singularity is somewhat artificial. Nevertheless, once a
concrete single-valued branch is defined, we must have an associated branch cut.
For example, the function

$$f(z) = \frac{\log z}{z}$$

has branch points at $z = 0$ and $z = \infty$. We may introduce a branch cut along the
positive real axis: $z = re^{i\theta}, 0 \leq \theta < 2\pi$. We note that $z = 0$ is a branch point and
not a pole because $\log z$ has a jump discontinuity as we encircle $z = 0$. It is not
analytic in a neighborhood of $z = 0$; hence $z = 0$ is not an isolated singular point.
(We note the difference between this example and Example 3.5.4 above.)

Another type of singular point is a **cluster point**. A cluster point is one in which
an infinite sequence of isolated singular points of a single-valued function $f(z)$
cluster about a point, or have a limit to a point, say $z = z_0$, in such a way that
there are an infinite number of isolated singular points in any arbitrarily small circle
about $z = z_0$. The standard example is given by the function $f(z) = \tan(1/z)$. As
$z \to 0$ along the real axis, $\tan(1/z)$ has poles at the locations $z_n = 1/(\pi/2 + n\pi)$,
with n integer, which cluster because any small neighborhood of the origin contains
an infinite number of them. There is no Laurent series representation valid in the
neighborhood of a cluster point.

Another singularity which arises in applications is associated with the case that
two analytic functions are separated by a closed curve or an infinite line. For
example, if C is a suitable closed contour and if $f(z)$ is defined as

$$f(z) = \begin{cases} f_i(z) & z \text{ inside } C; \\ f_o(z) & z \text{ outside } C, \end{cases} \tag{3.5.9}$$

where $f_i(z)$ and $f_o(z)$ are analytic in their respective regions have continuous limits
to the boundary C and are not equal on C, then the boundary C is a singular curve

across which the function has a jump discontinuity. We shall refer to this as a **boundary jump discontinuity**.

An example of such a situation is given by

$$f(z) = \frac{1}{2\pi i} \oint_C \frac{1}{\zeta - z} \, d\zeta = \begin{cases} f_i(z) = 1 & z \text{ inside } C; \\ f_o(z) = 0 & z \text{ outside } C. \end{cases} \tag{3.5.10}$$

The discontinuity depends entirely on the location of C, which is provided in the definition of the function $f(z)$ via the integral representation. We note that the functions $f_i(z) = 1$ and $f_o(z) = 0$ are analytic. Both of these functions can be continued beyond the boundary C in a natural way; just take $f_i(z) = 1$ and $f_o(z) = 0$, respectively. Indeed, functions obtained through integral representations such as Eq. (3.5.10) have a property by which the function $f(z)$ is comprised of functions such as $f_i(z)$ and $f_o(z)$, which are analytic inside and outside the original contour C.

In Ablowitz and Fokas (2003), Chapter 7, questions and applications very similar to Eq. (3.5.9) are studied. Such equations are called Riemann–Hilbert factorization problems.

3.5.2 Analytic Continuation and Natural Barriers

Frequently, one is given formulae that are valid in a limited region of space, and the goal is to find a representation, either in closed series form, integral representation, or otherwise, that is valid in a larger domain. The process of extending the range of validity of a representation or more generally extending the region of definition of an analytic function is called **analytic continuation**. This was briefly discussed at the end of Section 3.2 in Theorems 3.2.8 and 3.2.9. We shall elaborate further on this important issue in this section.

A typical example is the following. Consider the function defined by the series

$$f(z) = \sum_{n=0}^{\infty} z^n \tag{3.5.11}$$

when $|z| < 1$. When $|z| \to 1$, the series clearly diverges because z^n does not approach zero as $n \to \infty$. On the other hand, the function defined by

$$g(z) = \frac{1}{1 - z}, \tag{3.5.12}$$

which is defined *for all z except the point $z = 1$*, is such that $g(z) = f(z)$ for $|z| < 1$ because the Taylor series representation of Eq. (3.5.12) about $z = 0$ is Eq. (3.5.11) inside the unit circle. In fact, we claim that $g(z)$ is the unique analytic continuation of $f(z)$ outside the unit circle. The function $g(z)$ has a pole at $z = 1$. This example is representative of a far more general situation.

The relevant theorem was given earlier as Theorem 3.2.8, which implies the following.

Theorem 3.5.7   *A function that is analytic in a domain D is uniquely determined either by values in some interior domain of D or along an arc interior to D.*

The fact that a "global" analytic function can be deduced from such a relatively small amount of information illustrates just how powerful the notion of analyticity really is.

The example above – Eqs. (3.5.11) and (3.5.12) – shows that the function $f(z)$, which is represented by Eq. (3.5.11) inside the unit circle, uniquely determines the function $g(z)$ which is represented by Eq. (3.5.12) that is valid everywhere.

We remark that the function $1/(1 - z)$ is the only analytic function (analytic apart from a pole at $z = 1$) which can assume the values $f(x) = 1/(1 - x)$ along the real x-axis. This also shows how prescribing values along a curve fixes the analytic extension. Similarly, the function $f(z) = e^{kz}$ (for constant k) is the only analytic function which can be extended from $f(x) = e^{kx}$ on the real x-axis.

Chains of analytic continuations are sometimes required, and care may be necessary. For example consider the domains A, B, and C, and the associated analytic functions f, g, and h respectively (see Figure 3.10), and let $A \cap B$ denote the usual intersection of two sets.

Referring to Figure 3.10, Theorem 3.5.7 (or Theorem 3.2.8) implies that if $g(z)$ and $f(z)$ are analytic and have a domain $A \cap B$ in common, where $f(z) = g(z)$, then $g(z)$ is the analytic continuation of $f(z)$. Similarly, if $h(z)$ and $g(z)$ are analytic and have a domain $B \cap C$ in common, where $h(z) = g(z)$, then $h(z)$ is the analytic continuation of $g(z)$. However, we cannot conclude that $h(z) = f(z)$ because the intersecting domains A, B, C might enclose a branch point of a multivalued function.

The method of proof (of Theorems 3.2.8 or 3.5.7) extends the function locally by Taylor series arguments. We note that if we enclose a branch point, we move onto the next sheet of the corresponding Riemann surface.

For example consider the multivalued function

$$f(z) = \log z = \log r + i\theta, \tag{3.5.13}$$

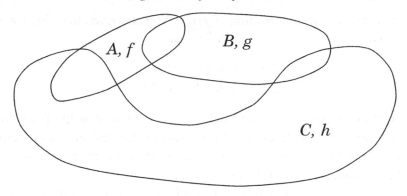

Figure 3.10 Analytic continuation in domains A, B, C

with three domains defined (see Figure 3.11) in the sectors

$$R_1 : 0 < \theta < \pi;$$

$$R_2 : \frac{3\pi}{4} < \theta < \frac{7\pi}{4};$$

$$R_3 : \frac{3\pi}{2} < \theta < \frac{5\pi}{2}.$$

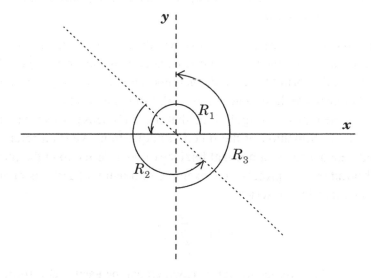

Figure 3.11 Overlapping domains R_1, R_2, R_3

The branches of $\log z$ defined by Eq. (3.5.13) in their respective domains R_1, R_2, and R_3 are related by analytic continuations. Namely, if we call $f_i(z)$ the function Eq. (3.5.13) defined in domain R_i, then $f_2(z)$ is the analytic continuation of $f_1(z)$, and $f_3(z)$ is the analytic continuation of $f_2(z)$. Note, however, that $f_3(z) \neq f_1(z)$; that is, the same point $z_0 = Re^{i\pi/4} = Re^{9i\pi/4}$ has

$$f_1(z_0) = \log R + i\pi/4,$$
$$f_3(z_0) = \log R + 9i\pi/4.$$

This example clearly shows that after analytic continuation the function does not return, upon a complete circuit, to the same value. Indeed, in this example we progress onto the adjacent sheet of the multivalued function because we have enclosed the branch point $z = 0$ of $f(z) = \log z$.

On the other hand, if in a simply-connected region there are no singular points enclosed between any two distinct paths of analytic continuation which together form a closed path, then we could cover the enclosed domain with small overlapping subregions and use Taylor series to analytically continue our function and obtain a single-valued function. This is frequently called the Monodromy Theorem which we now state without proof.

Theorem 3.5.8 (Monodromy Theorem)   *Let D be a simply-connected domain and $f(z)$ be analytic in D. If the function can be analytically continued along any two distinct contours C_1 and C_2 to a point in D, and if there are no singular points enclosed within C_1 and C_2, then the result of each analytic continuation is the same and the function is single valued.*

In fact, the theorem can be extended to cover the case where the region enclosed by contours C_1 and C_2 contains, at most, isolated singular points, $f(z)$ having a Laurent series of the form (3.5.1) in the neighborhood of any singular point. Thus, the enclosed domain can have poles or essential singular points.

There are some types of singularities that are, in a sense, so serious that they prevent analytic continuation of the function in question. We shall refer to such a (nonisolated) singularity as a **natural barrier** (often referred to in the literature as a **natural boundary**). A prototypical example of a natural barrier is contained in the function defined by the series

$$f(z) = \sum_{n=0}^{\infty} z^{2^n}. \tag{3.5.14}$$

The series (3.5.14) converges for $|z| < 1$, which can be easily seen from the ratio test. We shall sketch an argument which shows that analytic continuation to $|z| > 1$ is impossible.

Because

$$f(z^2) = \sum_{n=0}^{\infty} \left(z^2\right)^{2^n} = \sum_{n=0}^{\infty} z^{2^{n+1}} = \sum_{n=1}^{\infty} z^{2^n}, \tag{3.5.15}$$

it follows that f satisfies the functional equation

$$f(z^2) = f(z) - z. \tag{3.5.16}$$

From Eq. (3.5.14) it is clear that $z_0 = 1$ is a singular point because $f(1) = \infty$. It then follows from Eq. (3.5.16) that $f(z_1) = \infty$, where $z_1^2 = 1$ (i.e., $z_1 = \pm 1$). Similarly, $f(z_2) = \infty$, where $z_2^4 = 1$, because Eq. (3.5.14) implies

$$f(z^4) = f(z^2) - z^2 = f(z) - z - z^2. \tag{3.5.17}$$

Mathematical induction then yields

$$f(z^{2^m}) = f(z) - \sum_{j=0}^{m-1} z^{2^j}. \tag{3.5.18}$$

Hence the value of the function $f(z)$ at all points z_m on the unit circle satisfying $z^{2^m} = 1$ (i.e., all roots of unity) is infinite: $f(z_m) = \infty$, Therefore all these points are singular points. In order for the function (3.5.14) to be analytically continuable to $|z| \geq 1$, then at the very least we need $f(z)$ to be analytic on some small arc of the unit circle $|z| = 1$. However, no matter how small an arc we take on this circle, the above argument shows that there exist points z_m (roots of unity, satisfying $z^{2^m} = 1$) on any such arc such that $f(z_m) = \infty$. Because an analytic function must be bounded, analytic continuation is impossible.

Exotic singularities such as natural barriers are found in solutions of certain nonlinear differential equations arising in physical applications: see, for example, Eqs. (3.7.52) and (3.7.53). Consequently, their study is not merely a mathematical artifact.

3.5.3 Problems for Section 3.5

1. Discuss the type of singularity (removable, pole and order, essential, branch, cluster, natural barrier, etc.); if the type is a pole give the strength of the pole, and give the nature (isolated or not) of all singular points associated with the following functions. Include the point at infinity.

   (a) $\dfrac{e^{z^2} - 1}{z^2}$;   (b) $\dfrac{e^{2z} - 1}{z^2}$;   (c) $e^{\tan z}$;   (d) $\dfrac{z^3}{z^2 + z + 1}$;

   (e) $\dfrac{z^{1/3} - 1}{z - 1}$;   (f) $\log(1 + z^{1/2})$;   (g) $f(z) = \begin{cases} z^2 & |z| \leq 1, \\ 1/z^2 & |z| > 1; \end{cases}$

   (h) $f(z) = z^{4^n}$;   (i) $\operatorname{sech} z$;   (j) $\coth 1/z$.

2. Evaluate the integral $\oint_C f(z)\,dz$ where C is a unit circle centered at the origin and where $f(z)$ is given below.

(a) $\dfrac{g(z)}{z-w}$, $g(z)$ entire;   (b) $\dfrac{z}{z^2-w^2}$;   (c) ze^{1/z^2};

(d) $\cot z$;   (e) $\dfrac{1}{8z^3+1}$.

3. Show that the functions below are meromorphic; that is, the only singularities in the finite z-plane are poles. Determine the location, order and strength of the poles.

(a) $\dfrac{z}{z^4+2}$;   (b) $\tan z$;   (c) $\dfrac{z}{\sin^2 z}$;

(d) $\dfrac{e^z-1-z}{z^4}$;   (e) $\dfrac{1}{2\pi i}\oint_C \dfrac{w\,dw}{(w^2-2)(w-z)}$,

where C is the unit circle centered at the origin. First find the function for $|z|<1$, then analytically continue the function to $|z|\ge 1$.

4. (a) Discuss the nature of all singular points of the function

$$f(z) = \frac{1}{z^3(z-2i)}.$$

Explain why this is a meromorphic function.

(b) Show that $\oint_{C_0} f(z)\,dz = 0$, where C_0 is the unit circle.

5. Discuss the analytic continuation of the following functions:

(a) $\displaystyle\sum_{n=0}^{\infty} z^{2n}$, $|z|<1$;

(b) $\displaystyle\sum_{n=0}^{\infty} \frac{z^{n+1}}{n+1}$, $|z|<1$;

(c) $\displaystyle\sum_{n=0}^{\infty} z^{4^n}$.

Hint: (b) is also represented by the integral

$$\int_0^z \left(\sum_{n=0}^{\infty} z'^n\right) dz'.$$

6. Suppose we know a function $f(z)$ is analytic in the finite z-plane apart from singularities at $z=i$ and $z=-i$. Moreover, let $f(z)$ be given by the Taylor series

$$f(z) = \sum_{j=0}^{\infty} a_j z^j,$$

where a_j is known. Suppose we calculate $f(z)$ and its derivatives at $z = 3/4$ and compute a Taylor series in the form

$$f(z) = \sum_{j=0}^{\infty} b_j \left(z - \frac{3}{4}\right)^j.$$

Where would this series converge? How could we use this to compute $f(z)$? Suppose we wish to compute $f(2.5)$; how could we do this by series methods?

*3.6 Infinite products and Mittag-Leffler expansions

In previous sections we have considered various kinds of infinite series representations (i.e., Taylor series, Laurent series) of functions that are analytic in suitable domains. Sometimes in applications it is useful to consider infinite products to represent our functions.

If $\{a_k\}$ is a sequence of complex numbers, then an infinite product is denoted by

$$P = \prod_{k=1}^{\infty}(1 + a_k). \tag{3.6.1}$$

We say that the infinite product (3.6.1) converges if:
(a) the sequence of partial products P_n

$$P_n = \prod_{k=1}^{n}(1 + a_k) \tag{3.6.2}$$

converge to a finite limit; and
(b) that for N_0 large enough

$$\lim_{N \to \infty} \prod_{k=N_0}^{N}(1 + a_k) \neq 0. \tag{3.6.3}$$

If Eq. (3.6.3) is violated – that is, $\lim_{N\to\infty} \prod_{k=N_0}^{N}(1 + a_k) = 0$ for all N_0 – then we will consider the product to diverge. The reason for this is that the following infinite sum turns out to be intimately connected to the infinite product (see Eq. (3.6.4) below):

$$S = \sum_{k=1}^{\infty} \log(1 + a_k),$$

and it would not make sense if $P = 0$.

Moreover, in analogy with infinite sums, if the infinite product $\prod_{k=1}^{\infty}(1+|a_k|)$ converges we say P **converges absolutely**. If $\prod_{k=1}^{\infty}(1 + |a_k|)$ diverges but P converges, we say that P converges conditionally. Clearly, if one of the $a_k = -1$, then

$P_n = P = 0$. For now we shall exclude this trivial case and assume $a_k \neq -1$, for all k.

Equation (3.6.2) implies that $P_n = (1 + a_n)P_{n-1}$, whereupon $a_n = (P_n - P_{n-1})/P_{n-1}$. Thus if $P_n \to P$, we find that $a_n \to 0$, which is a necessary but not sufficient condition for convergence (note this necessary condition would also imply Eq. (3.6.3) for N_0 large enough).

A useful test for convergence of an infinite product is the following:

If the sum

$$S = \sum_{k=1}^{\infty} \log(1 + a_k) \tag{3.6.4}$$

converges, then so does the infinite product (3.6.1). We shall restrict $\log z$ to its principal branch.

Calling $S_n = \sum_{k=1}^{n} \log(1 + a_k)$, the nth partial sum of S, then

$$e^{S_n} = e^{\sum_{k=1}^{n} \log(1 + a_k)} = P_n$$

and $e^{S_n} \to e^S = P$ as $n \to \infty$. Note again that if $P = 0$, then $S = -\infty$, which we shall not allow, excluding the case where individual factors vanish.

The above definition also applies to products of *functions* where, for example, a_k is replaced by $a_k(z)$ for z in a region $\mathcal{R}$. We say that if a product of functions converges for each z in a region $\mathcal{R}$, then it converges in $\mathcal{R}$. The convergence is said to be uniform in $\mathcal{R}$ if the partial sequence of products obey $P_n(z) \to P(z)$ uniformly in $\mathcal{R}$. Uniformity is the same concept as that discussed in Section 3.4; namely, the estimate involved is independent of z. Moreover, there is an extension of the Weierstrass M test for sums, discussed in Section 3.1, to products of functions, which we now give; see also Levinson and Redheffer (1970).

Theorem 3.6.1   *Let $a_k(z)$ be analytic in a domain D for all k. Suppose that for all $z \in D$ and $k \geq N$ either*

$$\text{(a) } |\log(1 + a_k(z))| \leq M_k$$

or

$$\text{(b) } |a_k(z)| \leq M_k,$$

where $\sum_{k=N}^{\infty} M_k < \infty$, and the M_k are constants. Then the product

$$P(z) = \prod_{k=1}^{\infty} (1 + a_k(k))$$

is uniformly convergent to an analytic function $P(z)$ in D. Furthermore $P(z)$ is zero only when a finite number of its factors $1 + a_k(z)$ are zero in D.

Proof For $n \geq N$, define

$$P_n(z) = \prod_{k=N}^{n} (1 + a_k(z)),$$

$$S_n(z) = \sum_{k=N}^{n} \log (1 + a_k(z)).$$

Using inequality (a) in the hypothesis of Theorem 3.6.1 for any $z \in D$ with $m > N$ yields,

$$|S_m(z)| \leq \sum_{k=N}^{m} M_k \leq \sum_{k=1}^{\infty} M_k = M < \infty.$$

Similarly, for any $z \in D$ with $n > m \geq N$, we have

$$|S_n(z) - S_m(z)| \leq \sum_{k=m+1}^{n} M_k \leq \sum_{k=m+1}^{\infty} M_k \leq \epsilon_m,$$

where $\epsilon_m \to 0$ as $m \to \infty$, and $S_n(z)$ is a uniformly convergent Cauchy sequence. Because $P_k(z) = \exp S_k(z)$ it follows that

$$(P_n(z) - P_m(z)) = e^{S_m(z)} \left(e^{S_n(z) - S_m(z)} - 1 \right).$$

From the Taylor series $e^w = \sum_{n=0}^{\infty} w^n / n!$, we have

$$|e^w| \leq e^{|w|}, \qquad |e^w - 1| \leq e^{|w|} - 1,$$

whereupon from the above estimates we have

$$|P_n(z) - P_m(z)| \leq e^M \left(e^{\epsilon_m} - 1 \right),$$

and hence $\{P_n(z)\}$ is a uniform Cauchy sequence. Thus, (see Section 3.4) $P_n(z) \to P(z)$ uniformly in D and $P(z)$ is analytic because $P_n(z)$ is a sequence of analytic functions. Moreover, we have

$$|P_m(z)| = \left| e^{\operatorname{Re} S_m(z) + i \operatorname{Im} S_m(z)} \right| = \left| e^{\operatorname{Re} S_m(z)} \right|$$
$$\geq e^{-|\operatorname{Re} S_m(z)|} \geq e^{-|\operatorname{Re} S_m(z) + i \operatorname{Im} S_m(z)|}$$
$$\geq e^{-M}.$$

Thus $P_m(z) \geq e^{-M}$. Because M is independent of m, $P(z) \neq 0$ in D.

Because we may write $P(z) = \prod_{k=1}^{N-1}(1 + a_k(z)) \tilde{P}(z)$, we see that $P(z) = 0$ only if any of the factors $(1 + a_k(z)) = 0$, for $k = 1, 2, \ldots, N - 1$. (The estimate (a) of

Theorem 3.6.1 is invalid for such a possibility.) It also follows directly from the analyticity of $a_k(z)$ that

$$P_n(z) = \prod_{k=1}^{N-1} (1 + a_k(z))\tilde{P}_n(z)$$

is a uniformly convergent sequence of analytic functions.

Finally we note that the first hypothesis, (a), follows from the second hypothesis, (b), as is shown next.

The Taylor series of $\log(1 + w)$, $|w| < 1$, is given by

$$\log(1 + w) = \left(w - \frac{w^2}{2} + \frac{w^3}{3} + \cdots + (-1)^{n-1}\frac{w^n}{n} + \cdots\right).$$

Hence

$$|\log(1 + w)| \le |w| + \frac{|w|^2}{2} + \frac{|w|^3}{3} + \cdots + \frac{|w|^n}{n} + \cdots$$

and, for $|w| \le 1/2$, we have

$$|\log(1 + w)| \le |w|\left(1 + \frac{1}{2} + \frac{1}{2^2} + \cdots + \frac{1}{2^n} + \cdots\right)$$

$$\le |w|\left(\frac{1}{1 - 1/2}\right)$$

$$= 2|w|.$$

Thus, for $|a_k(z)| < 1/2$,

$$|\log(1 + a_k(z))| \le 2|a_k(z)|.$$

If we assume that $|a_k(z)| < M_k$, with $\sum_{k=1}^{\infty} M_k < \infty$, it is clear that there is a $k > N$ such that $|a_k(z)| < 1/2$, and we have hypothesis (a). The theorem goes through as before simply with M_k replaced by $2M_k$. □

As an example, consider the product

$$F(z) = \prod_{k=1}^{\infty} \left(1 - \frac{z^2}{k^2}\right). \tag{3.6.5}$$

Theorem 3.6.1 implies that $F(z)$ represents an entire function with simple zeroes at $z = \pm 1, \pm 2, \ldots$. In this case $a_k(z) = -z^2/k^2$. Inside the circle $|z| < R$ we have $|a_k(z)| \le R^2/k^2$. Because

$$\sum_{k=1}^{\infty} \frac{R^2}{k^2} < \infty.$$

Theorem 3.6.1 shows that the function $F(z)$ is analytic for all finite z inside the circle $|z| < R$. Because R can be made arbitrarily large, $F(z)$ is entire, and the only zeroes of $F(z)$ correspond to the vanishing of $(1 - z^2/k^2)$, for $k = 1, 2, \ldots$

Next we construct a function with simple zeroes at $z = 1, 2, \ldots$, and no other zeroes. We shall show that

$$G(z) = \prod_{k=1}^{\infty} \left\{ \left(1 - \frac{z}{k}\right) e^{z/k} \right\} \tag{3.6.6}$$

is one such function. At first one may think that the "convergence factor" $e^{z/k}$ could be dropped. But we will show that without this term, the product would diverge. In fact, the $e^{z/k}$ term is such that the contributions of the $(1/k)$ term inside the product cancels exactly; that is,

$$\left(1 - \frac{z}{k}\right) e^{z/k} = \left(1 - \frac{z}{k}\right)\left(1 + \frac{z}{k} + \frac{z^2}{2!k^2} + \cdots\right) = 1 - \frac{z^2}{2!k^2} + \cdots .$$

We note the Taylor series

$$\log\left((1 - w)e^w\right) = \log(1 - w) + w$$
$$= -\left(\frac{w^2}{2} + \frac{w^3}{3} + \frac{w^4}{4} + \cdots\right)$$
$$= -\left(w^2\right)\left(\frac{1}{2} + \frac{w}{3} + \frac{w^2}{4} + \cdots\right) ;$$

hence, for $|w| < 1/2$,

$$|\log(1 - w)e^w| \le |w|^2 \left(\frac{1}{2} + \frac{1}{2^2} + \frac{1}{2^3} + \cdots\right)$$
$$= |w|^2 \left(\frac{1}{2}\right)\left(\frac{1}{1 - 1/2}\right)$$
$$= |w|^2 .$$

Thus, for $|z| < R$ and $k > 2R$, for any fixed value R,

$$\left|\log\left(1 - \frac{z}{k}\right) e^{z/k}\right| \le \left|\frac{z}{k}\right|^2 \le \left(\frac{R}{k}\right)^2 ,$$

hence, from Theorem 3.6.1, the product (3.6.6) converges uniformly to an entire function with simple zeroes at $(1 - z/k) = 0$ for $k = 1, 2, \ldots$; that is, for $z = 1, 2, \ldots$.

We now show that $\prod_{k=1}^{\infty}(1 - z/k)$ diverges for $z \neq 0$. We note that for any integer $n \geq 1$,

$$H_n = \prod_{k=1}^{n}\left(1 - \frac{z}{k}\right)$$

$$= \prod_{k=1}^{n}\left(1 - \frac{z}{k}\right)e^{z/k} \cdot e^{-z/k}$$

$$= e^{-zS(n)}\prod_{k=1}^{n}\left\{\left(1 - \frac{z}{k}\right)e^{z/k}\right\},$$

where $S(n) = 1 + 1/2 + 1/3 + \cdots + 1/n$. Thus using the above result that Eq. (3.6.6) converges and because $S(n) \to \infty$, as $n \to \infty$ we have that for Re $z < 0$, $H_n \to \infty$; for Re $z > 0$, $H_n \to 0$. Also for Re $z = 0$ and Im $z \neq 0$, H_n does not have a limit as $n \to \infty$. By our definition of convergence of an infinite product (Eq. (3.6.1) below) we conclude that H is a divergent product.

Often the following observation is useful. If $F(z)$ and $G(z)$ are two entire functions which have the same zeroes and multiplicities, then there is an entire function $h(z)$ satisfying

$$F(z) = e^{h(z)}G(z) . \tag{3.6.7}$$

This follows from the fact that the function $F(z)/G(z)$ is entire with no zeroes; the ratio making all other zeroes of F and G removable singularities. Because F/G is analytic without zeroes, it has a logarithm which is everywhere analytic: $\log(F/G) = h(z)$.

It is natural to ask whether an entire function can be constructed that has zeroes of specified orders at assigned points with no other zeroes, or similarly, whether a meromorphic function can be constructed which has poles of specified orders at assigned points with no other poles. These questions lead to certain infinite products (the so-called Weierstrass products for entire functions) and infinite series (Mittag-Leffler expansions, for meromorphic functions). These notions extend our ability to represent functions of a certain specified character. Earlier we only had Taylor or Laurent series representations available.

First we shall discuss representations of meromorphic functions. In what follows we shall use certain portions of the Laurent series of a given meromorphic function. Namely, near any pole (of order N_j at $z = z_j$) of a meromorphic function we have the Laurent expansion

$$f(z) = \sum_{n=1}^{N_j}\frac{a_{n,j}}{(z - z_j)^n} + \sum_{n=0}^{\infty}b_{n,j}(z - z_j)^n .$$

The first part contains the pole contribution and is called the **principal part** at $z = z_j$, $p_j(z)$:

$$p_j(z) = \sum_{n=1}^{N_j} \frac{a_{n,j}}{(z - z_j)^n} . \tag{3.6.8}$$

We shall order points as follows: $|z_r| \leq |z_s|$ if $r < s$, with $z_0 = 0$ if the origin is one of the points to be included.

If the number of poles of the meromorphic function is finite, then the representation

$$f(z) = \sum_{j=1}^{m} p_j(z) \tag{3.6.9}$$

is nothing more than the partial fraction decomposition of a rational function vanishing at infinity, where the right-hand side of Eq. (3.6.9) has poles of specified character at the points $z = z_j$. A more general formula representing a meromorphic function with a finite number of poles is obtained by adding to the right side of Eq. (3.6.9) a function $h(z)$ which is entire. On the other hand, if the number of points z_j is infinite, the sum in Eq. (3.6.9) might or might not converge; for example, the partial sum

$$\sum_{\substack{k=-n \\ k \neq 0}}^{n} \frac{1}{z - k} = 2z \left(\frac{1}{z^2 - 1^2} + \frac{1}{z^2 - 2^2} + \cdots + \frac{1}{z^2 - n^2} \right)$$

converges uniformly for finite z, whereas the partial sum $\sum_{k=1}^{n} 1/(z - k)$ diverges, as can be verified from the elementary convergence criteria of infinite series. In general, we will need a suitable modification of Eq. (3.6.9) with the addition of an entire function $h(z)$ in order to find a rather general formula for a meromorphic function with prescribed principal parts. In what follows we take the case $\{z_j\}$; $|z_j| \to \infty$ as $j \to \infty$, with $z_0 = 0$.

Mittag-Leffler expansions involve the following. One wishes to represent a given meromorphic function $f(z)$ with prescribed principal parts $\{p_j(z)\}_{j=0}^{\infty}$ in terms of suitable functions. The aim is to find polynomials $\{g_j(z)\}_{j=0}^{\infty}$, where $g_0(z) = 0$, such that

$$f(z) = p_0(z) + \sum_{j=1}^{\infty} \left(p_j(z) - g_j(z) \right) + h(z) = \tilde{f}(z) + h(z), \tag{3.6.10}$$

where $h(z)$ is an entire function. The part of Eq. (3.6.10) that is denoted $\tilde{f}(z)$ has the same principal part (i.e., the same number, strengths and locations of poles) as $f(z)$. The difference $h(z)$, between $f(z)$ and $\tilde{f}(z)$, is necessarily entire. In order to pin down the entire function $h(z)$, more information about the function $f(z)$ is required.

When the function $f(z)$ has only simple poles ($N_j = 1$), the situation is considerably simpler, and we now discuss this situation in detail.

In the case of simple poles,

$$p_j(z) = \frac{a_j}{z - z_j} = -\frac{a_j}{z_j}\left(\frac{1}{1 - z/z_j}\right). \qquad (3.6.11)$$

Then there is an m such that for $|z/z_j| < 1$, the finite series

$$g_j(z) = -\frac{a_j}{z_j}\left(1 + \left(\frac{z}{z_j}\right) + \cdots + \left(\frac{z}{z_j}\right)^{m-1}\right), \qquad (3.6.12)$$

for $m \geq 1$, integer (if $m = 0$ we can take $g_j(z) = 0$), approximates $p_j(z)$ arbitrarily closely; a_j is the residue of the pole $z = z_j$. If we call

$$L(w, m) = \frac{1}{w - 1} + 1 + w + w^2 + \cdots + w^{m-1}, \qquad (3.6.13)$$

then, assuming convergence of the infinite series, Eq. (3.6.10) takes the form

$$f(z) = p_0(z) + \sum_{j=1}^{\infty} \left(\frac{a_j}{z_j}\right) L\left(\frac{z}{z_j}, m\right) + h(z), \qquad (3.6.14)$$

where $h(z)$ is an entire function and the following theorem holds.

Theorem 3.6.2 (Mittag-Leffler – simple poles)   *Let $\{z_k\}$ and $\{a_k\}$ be sequences with z_k distinct, $|z_k| \to \infty$ as $k \to \infty$, and $m \geq 1$, an integer, such that*

$$\sum_{j=1}^{\infty} \frac{|a_j|}{|z_j|^{m+1}} < \infty. \qquad (3.6.15)$$

Then Eq. (3.6.14) with $L(w, m)$ defined in Eq. (3.6.13), $h(z)$ an entire function, and $p_0(z)$ defined in Eq. (3.6.11) with $j = 0$, represents a meromorphic function whose only singularities are simple poles at z_k with residue a_k for $k = 1, 2, \ldots$.

Proof   From the fact that

$$1 + w + w^2 + \cdots + w^{m-1} = \frac{1}{1 - w} - \frac{w^m}{1 - w}$$

we have

$$L(w, m) = -\frac{w^m}{1 - w}.$$

For $|w| < 1/2$ we have $|1 - w| \geq 1 - |w| \geq 1/2$, and hence

$$|L(w, m)| \leq 2|w|^m. \qquad (3.6.16)$$

Let $|z| < R$ and for J large enough take $j > J$, $|z_j| > 2R$, then $|z/z_j| < 1/2$, hence the estimate (3.6.16) holds for $w = z/z_j$, and

$$\left|\frac{a_j}{z_j}L\left(\frac{z}{z_j}, m\right)\right| \le \left|\frac{a_j}{z_j}\right| 2 \left|\frac{z}{z_j}\right|^m \le \frac{2|R|^m |a_j|}{|z_j|^{m+1}}.$$

Thus with Eq. (3.6.15) we have that the series in Eq. (3.6.14) converges uniformly for $|z| < R$ (for arbitrarily large R), and Eq. (3.6.14) therefore represents a meromorphic function with the desired properties.   □

Using Theorem 3.6.2, we may determine which value of m ensures the convergence of the sum in Eq. (3.6.15), and consequently we may determine the function $L(w, m)$ in Eqs. (3.6.13)–(3.6.14).

For example, let us consider the function

$$f(z) = \pi \cot \pi z.$$

This function has simple poles at $z_j = j$, $j = 0, \pm1, \pm2, \ldots$ The strength of any of these poles is $a_j = 1$, which can be ascertained from the Laurent series of $f(z)$ in the neighborhood of z_j; that is, calling $z' = z - j$,

$$f(z') = \pi \frac{\cos \pi z'}{\sin \pi z'} = \frac{\pi\left(1 - \frac{(\pi z')^2}{2!} + \cdots\right)}{\pi z'\left(1 - \frac{(\pi z')^2}{3!} + \cdots\right)} = \frac{1}{z'}\left(1 - \frac{1}{3}(\pi z')^2 + \cdots\right).$$

The principal part at each z_j is therefore given by $p_j(z) = \frac{1}{z-j}$. Then the series (3.6.15) in Theorem 3.6.2

$$\sum_{\substack{j=-\infty \\ j\neq 0}}^{\infty} \frac{1}{|j|^{m+1}}$$

converges for $m = 1$.   Consequently from Theorem 3.6.2 and Eq. (3.6.14) the general form of the function is fixed to be

$$\pi \cot \pi z = \frac{1}{z} + \sum_{j=-\infty}^{\infty}{}' \left(\frac{1}{z-j} + \frac{1}{j}\right) + h(z) \tag{3.6.17a}$$

$$= \frac{1}{z} + 2\sum_{j=1}^{\infty} \frac{z}{z^2 - j^2} + h(z) = \sum_{j=-\infty}^{\infty} \frac{z}{z^2 - j^2} + h(z), \tag{3.6.17b}$$

where the prime in the sum means that the term $j = 0$ is excluded and where $h(z)$ is an entire function. Note that the $(1/j)$ term in Eq. (3.6.17a) is a necessary condition for the series to converge.

Later, in Section 4.2, we will show that by considering the integral

$$I = \frac{1}{2\pi i} \oint_C \pi \cot \pi \zeta \left(\frac{1}{z - \zeta} + \frac{1}{\zeta} \right) d\zeta, \tag{3.6.18}$$

where C is an appropriate closed contour, that the representation (3.6.17a–3.6.17b) holds with $h(z) = 0$.

The general case in which the principal parts contain an arbitrary number of poles – Eq. (3.6.8) with finite N_j – is more complicated. Nevertheless, as long as the locations of the poles are distinct, polynomials $g_j(z)$ can be found that establish the following (see, e.g., Henrici, volume 1 1977).

Theorem 3.6.3 (Mittag-Leffler – general case) *Let $f(z)$ be a meromorphic function in the complex plane with poles $\{z_j\}$ and corresponding principal parts $\{p_j(z)\}$. Then there exist polynomials $\{g_j(z)\}_{j=1}^{\infty}$ such that Eq. (3.6.10) holds and the series $\sum_{j=1}^{\infty} \left(p_j(z) - g_j(z) \right)$ converges uniformly on every bounded set not containing the points $\{z_j\}_{j=0}^{\infty}$.*

Proof We only sketch the essential idea behind the proof; the details are cumbersome. Each of the principal parts $\{p_j(z)\}_{j=1}^{\infty}$ can be expanded in a convergent Taylor series (around $z = 0$) for $|z| < |z_j|$. It can be shown that enough terms can be taken in this Taylor series so that the polynomials $g_j(z)$ obtained by truncation of the Taylor series of $p_j(z)$ at order z^{K_j}, i.e.,

$$g_j(z) = \sum_{k=0}^{K_j} B_{k,j} \, z^k,$$

ensure that the difference $|p_j(z) - g_j(z)|$ is suitably small. It can be shown (e.g. Henrici, volume I, 1977) that, for any $|z| < R$, the polynomials $g_j(z)$ of order K_j ensure that the series

$$\sum_{j=1}^{\infty} \left| p_j(z) - g_j(z) \right|$$

converges uniformly.   □

It should also be noted that even when we only have simple poles for the $p_j(z)$ there may be cases where we need to use the more general polynomials described in Theorem 3.6.3; for example, if we have $p_j(z) = 1/(z - z_j)$ where $z_j = \log(1 + j)$, $(a_j = 1)$. Then we see that in this case Eq. (3.6.15) is not true for any integer m.

A similar question to the one we have been asking is how to represent an entire function with specified zeroes at location z_k. We use the same notation as before: $z_0 = 0$, $|z_1| \le |z_2| \le \cdots$, and $|z_k| \to \infty$ as $k \to \infty$. The aim is to generalize the notion of factoring of a polynomial to factoring an entire function. We specify the order of each zero by a_k. One method to derive such a representation is to use the fact that if $f(z)$ is entire, then $f'(z)/f(z)$ is meromorphic with simple poles. Note near any isolated zero z_k with order a_k of $f(z)$ we have $f(z) \approx b_k(z - z_k)^{a_k}$; hence $f'(z)/f(z) \approx a_k/(z - z_k)$. Thus the order of the zero plays the same role as the residue in the Mittag-Leffler Theorem.

From the proof of Eq. (3.6.10) in the case of simple poles, using Eqs. (3.6.11)–(3.6.15), we have the uniformly convergent series representation

$$\frac{f'(z)}{f(z)} = \frac{a_0}{z} + \sum_{j=1}^{\infty} \left(\frac{a_j}{z - z_j} + \frac{a_j}{z_j} \sum_{k=0}^{m-1} \left(\frac{z}{z_j} \right)^k \right) + h(z), \qquad (3.6.19)$$

where $h(z)$ is an arbitrary entire function. Integrating and taking the exponential yields (care must be taken with regard to the constants of integration, cf. Eq. (3.6.21) below)

$$f(z) = z^{a_0} \prod_{j=1}^{\infty} \left\{ (1 - z/z_j) \exp \left(\sum_{k=0}^{m-1} \frac{(\frac{z}{z_j})^{k+1}}{k+1} \right) \right\}^{a_j} g(z), \qquad (3.6.20)$$

where $g(z) = \exp(\int h(z)dz)$ is an entire function without zeroes. The function (3.6.20) is, in fact, the most general entire function with such specified behavior. Equation (3.6.20) could, of course, be proven independently without recourse to the series representations discussed earlier. This result is referred to as the Weierstrass factor theorem. When $a_j = 1$, $j = 0, 1, 2, \ldots$, Eq. (3.6.20) with (3.6.15) gives the representation of an entire function with simple zeroes.

Theorem 3.6.4 (Weierstrass factor theorem) *An entire function with isolated zeroes at $z_0 = 0$, $\{z_j\}_{j=1}^{\infty}$, for $|z_1| \leq |z_2| \leq \cdots$, where $|z_j| \to \infty$ as $j \to \infty$, and with orders a_j, is given by Eq. (3.6.20) with z_j, a_j satisfying the constraint (3.6.15).*

A more general result is obtained by replacing m by j in (3.6.20). This is sometimes used when (3.6.15) diverges.

We note that z_j cannot have a limit point other than ∞. If z_j has a limit point, say z_*, then z_j can be taken arbitrarily close to z_*; therefore $f(z)$ would not be entire, resulting in a contradiction. Recall that an analytic function must have its zeroes isolated (Theorem 3.2.12).

In practice it is usually easiest to employ the Mittag–Leffler expansion for $f'(z)/f(z)$, with $f(z)$ entire, as we have done above, in order to represent an entire function. Note that the expansion (3.6.17a,b), with $h(z) = 0$, can be integrated using the principal branch of the logarithm function to find

$$\log \sin \pi z = \log z + A_0 + \sum_{n=1}^{\infty} \left[\log(z^2 - n^2) - A_n \right],$$

where A_0 and A_n are constant. Using

$$\lim_{z \to 0} \frac{\sin \pi z}{z} = \pi$$

the constants can be evaluated: $A_0 = \log \pi$ and $A_n = \log(-n^2)$. Taking the exponential of both sides yields

$$\frac{\sin \pi z}{\pi} = z \prod_{n=1}^{\infty} \left(1 - \left(\frac{z}{n}\right)^2\right), \tag{3.6.21}$$

which provides a concrete example of a Weierstrass expansion.

3.6.1 Problems for Section 3.6

1. Discuss where the following infinite products converge as a function of z:

(a) $\displaystyle \prod_{n=0}^{\infty} (1 + z^n);$ (b) $\displaystyle \prod_{n=0}^{\infty} \left(1 + \frac{z^n}{n!}\right);$

(c) $\displaystyle \prod_{n=1}^{\infty} \left(1 + \frac{2z}{n}\right);$ (d) $\displaystyle \prod_{n=1}^{\infty} \left(1 + \left(\frac{2z}{n}\right)^2\right).$

2. Show that the product

$$\prod_{k=1}^{\infty} \left(1 - \frac{z^4}{k^4}\right)$$

represents an entire function with zeroes at $z = \pm k, \pm ik$ for $k = 1, 2, \ldots$.

3. Using the expansion

$$\frac{\sin \pi z}{\pi z} = \prod_{n=1}^{\infty} \left(1 - \left(\frac{z}{n}\right)^2\right)$$

show that we also have

$$\frac{\sin \pi z}{\pi z} = \prod_{n=-\infty}^{\infty}{}' \left(1 - \frac{z}{n}\right) e^{z/n},$$

where the prime means that the $n = 0$ term is omitted. (Also see Problem 4 below.)

4. Use the representation

$$\frac{\sin \pi z}{\pi z} = \prod_{n=-\infty}^{\infty}{}' \left(1 - \frac{z}{n}\right) e^{z/n}$$

to deduce, by differentiation, that

$$\pi \cot \pi z = \frac{1}{z} + \sum_{n=-\infty}^{\infty}{}' \left(\frac{1}{z-n} + \frac{1}{n}\right),$$

where the prime means that the $n = 0$ term is omitted. Repeat the process to find

$$\pi \csc^2 \pi z = \sum_{n=-\infty}^{\infty} \frac{1}{(z-n)^2}.$$

5. Show that if $f(z)$ is meromorphic in the finite z-plane, then $f(z)$ must be the ratio of two entire functions.

6. Let $\Gamma(z)$ be given by

$$\frac{1}{\Gamma(z)} = z e^{\gamma z} \prod_{n=1}^{\infty} \left(1 + \frac{z}{n}\right) e^{-z/n}$$

for $z \neq 0, -1, -2, \ldots$ and γ a constant.

(a) Show that

$$\frac{\Gamma'(z)}{\Gamma(z)} = -\frac{1}{z} - \gamma - \sum_{n=1}^{\infty} \left(\frac{1}{z+n} - \frac{1}{n}\right).$$

(b) Show that

$$\frac{\Gamma'(z+1)}{\Gamma(z+1)} - \frac{\Gamma'(z)}{\Gamma(z)} - \frac{1}{z} = 0$$

whereupon

$$\Gamma(z+1) = Cz\Gamma(z), \qquad C \text{ a constant.}$$

(c) Show that $\lim_{z \to 0} z\Gamma(z) = 1$, to find that $C = \Gamma(1)$.

(d) Determine the following representation for the constant γ so that $\Gamma(1) = 1$:

$$e^{-\gamma} = \prod_{n=1}^{\infty} \left(1 + \frac{1}{n}\right) e^{-1/n}.$$

(e) Show that

$$\prod_{n=1}^{\infty} \left(1 + \frac{1}{n}\right) e^{-1/n} = \lim_{n \to \infty} \frac{2}{1} \frac{3}{2} \frac{4}{3} \cdots \frac{n+1}{n} e^{-S(n)} = \lim_{n \to \infty} (n+1) e^{-S(n)},$$

where $S(n) = 1 + \frac{1}{2} + \frac{1}{3} + \cdots + \frac{1}{n}$. Consequently obtain the limit

$$\gamma = \lim_{n \to \infty} \left(\sum_{k=1}^{n} \frac{1}{k} - \log(n+1)\right).$$

The constant $\gamma = 0.5772157\ldots$ is referred to as Euler's constant.

7. In Section 3.6 we showed that

$$\pi \cot \pi z - \left(\frac{1}{z} + \sum_{j=-\infty}^{\infty}{}' \left(\frac{1}{z-j} + \frac{1}{j} \right) \right) = h(z),$$

where $\sum'$ indicates that the $j = 0$ term is omitted, and where $h(z)$ is entire. We now show how to establish that $h(z) = 0$.

(a) Show that $h(z)$ is periodic of period 1 by establishing that the left-hand side of the formula is periodic of period 1. (Show that the second term on the left side doesn't change when z is replaced by $z + 1$.)

(b) Because $h(z)$ is periodic and entire, we need only establish that $h(z)$ is bounded in the strip $0 \le \mathrm{Re}\, z \le 1$ to ensure, by Liouville's Theorem (Section 2.6.2) that it is a constant. For all finite values of $z = x + iy$ in the strip away from the poles, explain why both terms are bounded, and because the pole terms cancel, the difference is in fact bounded. Verify that as $y \to \pm\infty$ the term $\pi \cot \pi z$ is bounded.

To establish the boundedness of the second term on the left, rewrite it as follows:

$$S(z) = \frac{1}{z} + \sum_{n=1}^{\infty} \frac{2z}{z^2 - n^2}.$$

Use the fact that in the strip $0 < x < 1$, $y > 2$, $|z| \le \sqrt{2}y$, we have $|z^2 - n^2| \ge \frac{1}{\sqrt{2}}(y^2 + n^2)$ (note that these estimates are not sharp), and show that

$$|S(z)| \le \frac{1}{|z|} + 4y \sum_{n=1}^{\infty} \left(\frac{1}{y^2 + n^2} \right).$$

Explain why

$$\sum_{n=1}^{\infty} \frac{y}{y^2 + n^2} = \frac{1}{y} \sum_{n=1}^{\infty} \frac{1}{1 + (n/y)^2} \le \int_0^{\infty} \frac{1}{1 + u^2}\, du = \frac{\pi}{2},$$

and therefore conclude that $S(z)$ is bounded for $0 < x < 1$ and $y \to \infty$. The same argument works for $y \to -\infty$. Hence $h(z)$ is a constant.

(c) Because both terms on the left are odd in z, i.e., $f(z) = -f(-z)$, conclude that $h(z) = 0$.

8. Consider the function $f(z) = (\pi^2)/(\sin^2 \pi z)$.

(a) Establish that near every integer $z = j$ the function $f(z)$ has the singular part $p_j(z) = 1/(z - j)^2$.

(b) Explain why the series

$$S(z) = \sum_{j=-\infty}^{\infty} \frac{1}{(z-j)^2}$$

converges for all $z \neq j$.

(c) Because the series in part (b) converges, explain why the representation

$$\frac{\pi^2}{\sin^2 \pi z} = \sum_{j=-\infty}^{\infty} \frac{1}{(z-j)^2} + h(z),$$

where $h(z)$ is entire, is valid.

(d) Show that $h(z)$ is periodic of period 1 by showing that each of the terms $(\pi/\sin \pi z)^2$ and $S(z)$ are periodic of period 1. Explain why $(\pi/\sin \pi z)^2 - S(z)$ is a bounded function, and show that each term vanishes as $|y| \to \infty$. Hence conclude that $h(z) = 0$.

(e) Integrate termwise to find

$$\pi \cot \pi z = \frac{1}{z} + \sum_{n=-\infty}^{\infty}{}' \left(\frac{1}{z-n} + \frac{1}{n} \right),$$

where the prime indicates that the $n = 0$ term is omitted.

9. (a) Let $f(z)$ have simple poles at $z = z_n$, for $n = 1, 2, 3, \ldots, N$, with strengths a_n, and be analytic everywhere else. Show by contour integration (the reader may wish to consult Theorem 4.1.1) that

$$\frac{1}{2\pi i} \oint_{C_N} \frac{f(z')}{z'-z} \, dz' = f(z) + \sum_{n=1}^{N} \frac{a_n}{z_n - z}, \qquad (3.6.22)$$

where C_N is a large circle of radius R_N enclosing all the poles. Evaluate (3.6.22) at $z = 0$ to obtain

$$\frac{1}{2\pi i} \oint_{C_N} \frac{f(z')}{z'} \, dz' = f(0) + \sum_{n=1}^{N} \frac{a_n}{z_n}. \qquad (3.6.23)$$

(b) Subtract equation (3.6.23) from equation (3.6.22) to obtain

$$\frac{1}{2\pi i} \oint_{C_N} \frac{z f(z')}{z'(z'-z)} \, dz' = f(z) - f(0) + \sum_{n=1}^{N} a_n \left(\frac{1}{z_n - z} - \frac{1}{z_n} \right). \qquad (3.6.24)$$

(c) Assume that $f(z)$ is bounded for large z, to establish that the left-hand side of (3.6.24) vanishes as $R_N \to \infty$. Conclude that if the sum on the right hand side of (3.6.24) converges as $N \to \infty$, then

$$f(z) = f(0) + \sum_{n=1}^{\infty} a_n \left(\frac{1}{z - z_n} + \frac{1}{z_n} \right); \qquad (3.6.25)$$

this is a special case of the Mittag-Leffler Theorems 3.6.2 and 3.6.3.

(d) Let $f(z) = \pi \cot \pi z - 1/z$, and show that

$$\pi \cot \pi z - \frac{1}{z} = \sum_{n=-\infty}^{\infty}{}' \left(\frac{1}{z - n} + \frac{1}{n} \right), \qquad (3.6.26)$$

where the prime again indicates that the $n = 0$ term is omitted. (Equation (3.6.26) is another derivation of the result in this section.) We see that an infinite series of poles can represent the function $\cot \pi z$. Thus, in Section 3.6 we have established that Taylor and Laurent series are not the only ones that can be used for representations of functions.

*3.7 Differential Equations in the Complex Plane; Painlevé Equations

In this section we investigate various properties associated with solutions to ordinary differential equations in the complex plane.

In what follows we assume some basic familiarity with ordinary differential equations (ODEs) and their solutions. There are numerous texts on the subject; however, with regard to ODEs in the complex plane, the reader may wish to consult the treatises of Ince (1956) or Hille (1976) for an in-depth discussion, though these books contain much more advanced material. The purpose of this section is to outline some of the fundamental ideas underlying this topic and introduce the reader to concepts which appear frequently in the physics and applied mathematics literature.

We shall consider nth-order nonlinear ODEs in the complex plane, with the following structure:

$$\frac{d^n w}{dz^n} = F\left(w, \frac{dw}{dz}, \ldots, \frac{d^{n-1} w}{dz^{n-1}}; z \right), \qquad (3.7.1)$$

where F is assumed to be a locally analytic function of all its arguments, i.e., F has derivatives with respect to each argument in some domain D; thus F can have isolated singularities, branch points, etc. A system of such ODEs takes the form

$$\frac{dw_i}{dz} = F_i(w_1, \ldots, w_n; z), \qquad i = 1, \ldots, n, \qquad (3.7.2)$$

where again F_i is assumed to be a locally analytic function of its arguments. The scalar problem (3.7.1) is a special case of Eq. (3.7.2). To see this, we associate w_1 with w and take

$$\frac{dw_1}{dz} = w_2 \equiv F_1,$$

$$\frac{dw_2}{dz} = w_3 \equiv F_2$$

$$\vdots$$

$$\frac{dw_{n-1}}{dz} = w_n \equiv F_{n-1},$$

$$\frac{dw_n}{dz} = F(w_1, \ldots, w_n; z), \tag{3.7.3}$$

whereupon

$$w_{j+1} = \frac{d^j w_1}{dz^j}, \quad j = 1, \ldots, n-1 \tag{3.7.4a}$$

and

$$\frac{d^n w_1}{dz^n} = F\left(w_1, \frac{dw_1}{dz}, \ldots, \frac{d^{n-1} w_1}{dz^{n-1}}; z\right). \tag{3.7.4b}$$

A natural question one asks is the following. Is there an analytic solution to these ODEs? Given bounded initial values for Eq. (3.7.2) at $z = z_0$:

$$w_j(z_0) = w_{j,0} < \infty, \quad j = 1, 2, \ldots, n, \tag{3.7.5}$$

the answer is affirmative in a *small enough* region about $z = z_0$. We state this as a theorem.

Theorem 3.7.1 (Cauchy) *The system* (3.7.2) *with initial values* (3.7.5), *and with* $F_i(w_1, \ldots, w_n; z)$ *as an analytic function of each of its arguments in a domain* D *containing* $z = z_0$, *has a unique analytic solution in a neighborhood of* $z = z_0$.

Proof There are numerous ways of establishing this theorem, a common one being the method of **majorants**; that is, finding a convergent series which dominates the true series representation of the solution. The basic ideas are most easily illustrated by the scalar first-order nonlinear equation

$$\frac{dw}{dz} = f(w, z) \tag{3.7.6}$$

subject to the initial conditions $w(0) = 0$. Initial values $w(z_0) = w_0$ could be reduced to this case by translating variables, letting $z' = z - z_0$, $w' = w - w_0$, and writing the Eq. (3.7.6) in terms of w' and z'. The function $f(w, z)$ is assumed to be analytic and bounded when w and z lie inside the circles $|z| \leq a$ and $|w| \leq b$, with $|f| \leq M$, for some a, b and M. The series expansion of the solution to Eq. (3.7.6) may be computed by taking successive derivatives of Eq. (3.7.6); that is,

$$\frac{d^2 w}{dz^2} = \frac{\partial f}{\partial z} + \frac{\partial f}{\partial w} \frac{dw}{dz},$$

$$\frac{d^3 w}{dz^3} = \frac{\partial^2 f}{\partial z^2} + 2 \frac{\partial^2 f}{\partial z \partial w} \frac{dw}{dz} + \frac{\partial^2 f}{\partial w^2} \left(\frac{dw}{dz} \right)^2 + \frac{\partial f}{\partial w} \frac{d^2 w}{dz^2}$$

$$\vdots \tag{3.7.7}$$

This allows us to compute

$$w = \left(\frac{dw}{dz} \right)_0 z + \left(\frac{d^2 w}{dz^2} \right)_0 \frac{z^2}{2!} + \left(\frac{d^3 w}{dz^3} \right)_0 \frac{z^3}{3!} + \cdots . \tag{3.7.8}$$

The technique is to consider a comparison equation with the same initial condition

$$\frac{dW}{dz} = F(W, z), \qquad W(0) = 0, \tag{3.7.9}$$

in which each term in the series representation of $F(w, z)$ dominates that of $f(w, z)$. Specifically, the series representation for $f(w, z)$, which is assumed to be analytic in both variables w and z, is

$$f(w, z) = \sum_{j=0}^{\infty} \sum_{k=0}^{\infty} C_{jk} z^j w^k, \tag{3.7.10}$$

$$C_{jk} = \frac{1}{j!k!} \left(\frac{\partial^{j+k} f}{\partial z^j \partial w^k} \right)_0 . \tag{3.7.11}$$

At $w = b$ and $z = a$, we have assumed that f is bounded and we take the bound on f to be

$$|f(w, z)| \leq \sum_{j=0}^{\infty} \sum_{k=0}^{\infty} |C_{jk}| a^j b^k = M . \tag{3.7.12}$$

Each term of this series is bounded by M; hence

$$|C_{jk}| \leq M a^{-j} b^{-k} . \tag{3.7.13}$$

We take $F(w, z)$ to be

$$F(W, z) = \sum_{j=0}^{\infty} \sum_{k=0}^{\infty} \frac{M}{a^j b^k} z^j W^k . \tag{3.7.14}$$

So from Eqs. (3.7.13) and (3.7.10) the function $F(W, z)$ majorizes $f(w, z)$ termwise. Because the solution $W(z)$ is computed exactly the same way as for

Eq. (3.7.6) – that is, we only replace w and f by W and F in Eq. (3.7.7) – clearly the series solution (Eq. (3.7.8) with w replaced by W) for $W(z)$ would dominate that for w. Next we show that $W(z)$ has a solution in a neighborhood of $z = 0$. Summing the series (3.7.14) yields

$$F(w, z) = \frac{M}{\left(1 - \frac{z}{a}\right)\left(1 - \frac{W}{b}\right)}, \tag{3.7.15}$$

whereupon Eq. (3.7.9) yields

$$\left(1 - \frac{W}{b}\right)\frac{dW}{dz} = \frac{M}{1 - \frac{z}{a}}. \tag{3.7.16}$$

Hence by integration,

$$W(z) - \frac{1}{2b}(W(z))^2 = -Ma \, \log\left(1 - \frac{z}{a}\right) \tag{3.7.17}$$

and therefore

$$W(z) = b - b\left[1 + \frac{2aM}{b}\log\left(1 - \frac{z}{a}\right)\right]^{1/2}. \tag{3.7.18}$$

In Eq. (3.7.18) we take the positive value for the square root and the principal value for the log function so that $W(0) = 0$. The series representation (expanding the log, square root, etc.) of $W(z)$ dominates the series $w(z)$. The series for $W(z)$ converges up and until the nearest singularity: $z = a$ for the log function, or to $z = R$ where $[\cdot]^{1/2} = 0$, whichever is smaller.

Because R is given by

$$1 + \frac{2aM}{b}\log\left(1 - \frac{R}{a}\right) = 0, \tag{3.7.19a}$$

we have

$$R = a\left(1 - \exp\left(-\frac{b}{2Ma}\right)\right). \tag{3.7.19b}$$

Because $R < a$, the series representation of Eq. (3.7.9) converges absolutely for $|z| < R$. Hence a solution $w(z)$ satisfying Eq. (3.7.6) must exist for $|z| < R$, by comparison. Moreover, as long as we stay within the class of analytic functions, any series representation obtained this way will be unique because the Taylor series uniquely represents an analytic function.

The method described above can be readily extended to apply to the system of equations (3.7.2). Without loss of generality, taking initial values $w_j = 0$, for $j = 1, 2, \ldots, n$ at $z = 0$ and functions $F_i(w_1, \ldots, w_n, z)$ analytic inside $|z| \leq a$,

$|w_j| \leq b$, for $j = 1, \ldots, n$, then we can take $|F_i| \leq M$ in this domain. For the majorizing function, similar arguments as before yield

$$\frac{dW_1}{dz} = \frac{dW_2}{dz} = \cdots = \frac{dW_n}{dz}$$

$$= \frac{M}{\left(1 - \frac{z}{a}\right)\left(1 - \frac{W_1}{b}\right)\cdots\left(1 - \frac{W_n}{b}\right)}, \quad (3.7.20)$$

where

$$W_j(z) = 0 \quad \text{for } 1, \ldots, n.$$

Solving

$$\frac{dW_i}{dz} = \frac{dW_{i+1}}{dz}$$

with $W_i(0) = W_{i+1}(0) = 0$ for $i = 1, 2, \ldots, n - 1$, implies that

$$W_1 = W_2 = \cdots = W_n \equiv W,$$

whereupon Eq. (3.7.20) gives

$$\frac{dW}{dz} = \frac{M}{\left(1 - \frac{z}{a}\right)\left(1 - \frac{W}{b}\right)^n}, \quad W(0) = 0. \quad (3.7.21)$$

Solving Eq. (3.7.21) yields

$$W = b - b\left[1 + \frac{(n + 1)}{b}Ma\log\left(1 + \frac{z}{a}\right)\right]^{\frac{1}{n+1}} \quad (3.7.22)$$

with a radius of convergence given by $|z| \leq R$ where

$$R = a\left(1 - e^{-\frac{b}{(n+1)Ma}}\right). \quad (3.7.23)$$

Hence the series solution to the system (3.7.2), $w_j(0) = 0$ converges absolutely and uniformly inside the circle of radius R. $\qquad\square$

Thus, Theorem 3.7.1 establishes the fact that as long as $f_i(w_1, \ldots, w_n; z)$ in Eq. (3.7.2) is an analytic function of its arguments, then there is an analytic solution in a neighborhood (albeit small) of the initial values $z = z_0$. We may analytically continue our solution until we reach a singularity. This is due to the following.

Theorem 3.7.2 (Continuation Principle) *The function obtained by analytic continuation of the solution of Eq. (3.7.2), along any path in the complex plane, is a solution of the analytic continuation of the equation.*

Proof We note that because $g_i(z) = w'_i - F_i(w_1, \ldots, w_n; z)$, for $i = 1, 2, \ldots, n$, is zero inside the domain where we have established the existence of our solution, then any analytic continuation of $g_i(z)$ will necessarily be zero. Because the solution $w_i(z)$ satisfies $g_i(z) = 0$ inside the domain of its existence, and because the operations in $g_i(z)$ maintain analyticity, then analytically extending $w_i(z)$ gives the analytic extension of $g_i(z)$, which is identically zero.   □

Thus, we have that our solution may be analytically continued until we reach a singularity. A natural question to ask is *where* can we expect a singularity? There are two types: **fixed** and **movable**. A fixed singularity is one that is determined by the explicit singularities of the functions $f_i(., z)$. For example,

$$\frac{dw}{dz} = \frac{w}{z^2}$$

has a fixed singular point **(SP)** at $z = 0$. The solution reflects this fact:

$$w = Ae^{-1/z},$$

whereby we have an essential singularity at $z = 0$.

Movable SPs, on the other hand, depend on the initial conditions imposed. In a sense they are internal to the equation. For example, consider

$$\frac{dw}{dz} = w^2 . \tag{3.7.24a}$$

There are no fixed singular points, but the solution is given by

$$w = -\frac{1}{z - z_0} , \tag{3.7.24b}$$

where z_0 is arbitrary. The value of z_0 depends on the initial value; that is, if $w(z = 0) = w_0$, then $z_0 = 1/w_0$. Equation (3.7.24b) is an example of a **movable pole** (this is a simple pole). If we consider different equations, we could have different kinds of movable singularities, for example, movable branch points, movable essential singularities, etc. For example

$$\frac{dw}{dz} = w^p, \qquad p \geq 2, \tag{3.7.25a}$$

has the solution

$$w = ((p - 1)(z_0 - z))^{1/1-p} , \tag{3.7.25b}$$

which has a **movable branch point** for $p \geq 3$.

In what follows we shall, for the most part, quote some well-known results, regarding differential equations with fixed and movable singular points. We refer the reader to the monographs of Ince (1956) and Hille (1976), for the rigorous development, which would otherwise take us well outside the scope of the present text.

It is reasonable to start with the linear case. The linear homogeneous analog of Eq. (3.7.2) is

$$\frac{d\mathbf{w}}{dz} = A(z)\mathbf{w}, \qquad \mathbf{w}(z_0) = \mathbf{w}_0, \tag{3.7.26}$$

where $\mathbf{w}$ is an $(n \times 1)$ column vector, and $A(z)$ is an $(n \times n)$ matrix; that is,

$$A = \begin{pmatrix} a_{11} & \cdots & a_{1n} \\ \vdots & \ddots & \vdots \\ a_{n1} & \cdots & a_{nn} \end{pmatrix}, \qquad \mathbf{w} = \begin{pmatrix} w_1 \\ \vdots \\ w_n \end{pmatrix}.$$

The linear homogeneous scalar problem is obtained by specializing Eq. (3.7.26):

$$\frac{d^n w}{dz^n} = p_1(z)\frac{d^{n-1}w}{dz^{n-1}} + p_2(z)\frac{d^{n-2}w}{dz^{n-2}} + \cdots + p_n(z)w, \tag{3.7.27}$$

where we take, in Eq. (3.7.26),

$$A = \begin{pmatrix} 0 & 1 & 0 & \cdots & 0 & 0 \\ 0 & 0 & 1 & \cdots & 0 & 0 \\ \vdots & & & \ddots & \vdots & \vdots \\ 0 & 0 & 0 & \cdots & 0 & 1 \\ p_n(z) & p_{n-1}(z) & p_{n-2}(z) & \cdots & p_2(z) & p_1(z) \end{pmatrix} \tag{3.7.28a}$$

and

$$w_2 = \frac{dw_1}{dz}, \ldots, w_n = \frac{dw_{n-1}}{dz}, \qquad w_1 \equiv w. \tag{3.7.28b}$$

The relevant result is the following.

Theorem 3.7.3   *If $A(z)$ is analytic in a simply connected domain D containing z_0, then the linear initial value problem* (3.7.26) *has a unique analytic solution in D.*

A consequence of this theorem, insofar as singular points (SPs) are concerned, is that the general linear equation (3.7.26) has no movable SPs; its SPs are fixed purely by the singularities of the coefficient matrix $A(z)$, or in the scalar problem (3.7.27), by the singularities in the coefficients $\{p_j(z)\}_{j=1}^n$. One can prove Theorem 3.7.3 by an extension of what was done earlier. Namely, by looking for a series solution about a point of singularity, say, $z = 0$, $w(z) = \sum_{k=0}^{\infty} c_k z^k$, one can determine the coefficients c_k and show that the series converges until the nearest singularity of $A(z)$. Because this is fixed by the equation we have the result that *linear equations have only fixed singularities.*

For example, the scalar first-order equation

$$\frac{dw}{dz} = p(z)w, \qquad w(z_0) = w_0 \tag{3.7.29a}$$

has the explicit solution

$$w(z) = w_0 e^{\int_{z_0}^{z} p(\zeta)\, d\zeta}. \tag{3.7.29b}$$

Clearly, if $p(z)$ is analytic, then so is $w(z)$.

For linear differential equations there is great interest in a special class of differential equations which arise frequently in physical applications. These are so-called linear differential equations with **regular singular points**. Equation (3.7.26) is said to have a singular point in domain D if $A(z)$ has a singular point in D. We say $z = z_0$ is a regular singular point of Eq. (3.7.26) if the matrix $A(z)$ has a simple pole at $z = z_0$:

$$A(z) = \sum_{k=0}^{\infty} a_k (z - z_0)^{k-1},$$

where a_0 is not the zero matrix. The scalar equation (3.7.27) is said to have a regular singular point at $z = z_0$ if $p_k(z)$ has a kth-order pole, i.e., $(z - z_0)^k p_k(z)$, for $k = 1, \ldots, n$ is analytic at $z = z_0$. Otherwise, a singular point of a linear differential equation is said to be an irregular singular point. As mentioned earlier, Eq. (3.7.27) can be written as a matrix equation, Eq. (3.7.26), and the statements made here about scalar and matrix equations are easily seen to be consistent.

We may recast Eq. (3.7.27) by writing $Q_j(z) = -(z - z_0)^j p_j(z)$

$$(z - z_0)^n \frac{d^n w}{dz^n} + \sum_{j=1}^{n} Q_j(z)(z - z_0)^{n-j} \frac{d^{n-j} w}{dz^{n-j}} = 0, \tag{3.7.30}$$

where all the $Q_j(z)$ are analytic at $z = z_0$ for $j = 1, 2, \ldots$.

Fuchs and Frobenius showed that series methods may be applied to solve Eq. (3.7.30) and that, in general, the solution contains branch points at $z = z_0$. Indeed, if we expand $Q_j(z)$ about $z = z_0$ as

$$Q_j(z) = \sum_{k=0}^{\infty} c_{jk} (z - z_0)^k$$

then, the solution to Eq. (3.7.30) has the form

$$w(z) = \sum_{k=0}^{\infty} a_k (z - z_0)^{k+r}, \tag{3.7.31}$$

where r satisfies the so-called **indicial equation**

$$r(r-1)(r-2)\cdots(r-n+1)$$

$$+ \sum_{j=1}^{n-1} c_{j0}r(r-1)(r-2)\cdots(r-n+j+1) + c_{n0} = 0. \qquad (3.7.32)$$

There is always one solution of (3.7.30) of the form (3.7.31) with a root r obtained from (3.7.32). In fact there are n such linearly independent solutions so long as no two roots of this equation differ by an integer or zero (i.e., multiple root). In this special case the solution form (3.7.31) must in general be supplemented by appropriate terms containing powers of $\log(z - z_0)$. Equation (3.7.32) is obtained by inserting the expansion (3.7.31) into Eq. (3.7.30): then a recursion relation for the coefficients a_k is obtained by equating powers of $(z - z_0)$. Convergence of the series (3.7.31) is to the nearest singularity of the coefficients $Q_j(z)$, $j = 1, 2, \ldots, n$. If all the functions $Q_j(z)$ were indeed constant, c_{j0}, then Eq. (3.7.32) would lead to the roots associated with the solutions to Euler's equation.

The standard case is the second-order equation $n = 2$, which is covered in most elementary texts on differential equations:

$$(z - z_0)^2 \frac{d^2w}{dz^2} + (z - z_0)Q_1(z)\frac{dw}{dz} + Q_2(z)w = 0, \qquad (3.7.33)$$

where $Q_1(z)$ and $Q_2(z)$ are analytic in a neighborhood of $z = z_0$. The indicial equation (3.7.33) in this case satisfies

$$r(r-1) + c_{10}r + c_{20} = 0, \qquad (3.7.34)$$

where c_{10} and c_{20} are the first terms in the Taylor expansion of $Q_1(z)$ and $Q_2(z)$ about $z = z_0$; that is, $c_{10} = Q_1(z_0)$ and $c_{20} = Q_2(z_0)$.

Well-known second-order linear equations containing regular singular points include the following:

Bessel's Equation

$$z^2 \frac{d^2w}{dz^2} + z\frac{dw}{dz} + (z^2 - p^2)w = 0. \qquad (3.7.35a)$$

Legendre's Equation

$$(1 - z^2)\frac{d^2w}{dz^2} - 2z\frac{dw}{dz} + p(p+1)w = 0. \qquad (3.7.35b)$$

Hypergeometric Equation

$$z(1-z)\frac{d^2w}{dz^2} + [c - (a+b+1)z]\frac{dw}{dz} - abw = 0, \qquad (3.7.35c)$$

where p, a, b, c are constants.

We now return to questions involving nonlinear ODEs. In the late 19th and early 20th centuries, there were extensive studies undertaken by mathematicians in order to ennumerate those nonlinear ODEs that had poles as their only movable singularities. We say that ODEs possessing this property are of **Painlevé type** (named after one of the mathematicians of that time). Mathematically speaking, these equations are among the simplest possible because the solutions apart from their fixed singularities (which are known *a priori*) only have poles; in fact, they can frequently (perhaps always?) be linearized or solved exactly. It turns out that equations with this property arise frequently in physical applications, for example, fluid dynamics, quantum spin systems, relativity, etc. (See, for example, Ablowitz and Segur (1981), especially the sections on Painlevé equations.) The historical background and development is reviewed in the monograph of Ince (1956).

The simplest situation occurs with first-order nonlinear differential equations of the form

$$\frac{dw}{dz} = F(w, z) = \frac{P(w, z)}{Q(w, z)}, \tag{3.7.36}$$

where P and Q are polynomials in w and locally analytic functions of z. Then the *only equation* which is of Painlevé-type is

$$\frac{dw(z)}{dz} = A_0(z) + A_1(z)w + A_2(z)w^2. \tag{3.7.37}$$

Equation (3.7.37) is called a **Riccati equation**. Moreover, it can be linearized by the substitution

$$w(z) = \alpha(z)\left(\frac{\frac{d\psi}{dz}}{\psi}\right), \tag{3.7.38a}$$

where

$$\alpha(z) = -1/A_2(z) \tag{3.7.38b}$$

and $\psi(z)$ satisfies the linear equation

$$\frac{d^2\psi}{dz^2} = \left(A_1(z) + A_2'(z)/A_2(z)\right)\frac{d\psi}{dz} - A_0(z)A_2(z)\psi. \tag{3.7.38c}$$

Because Eq. (3.7.38c) linear, it has no movable singularities. But it does have movable zeroes; hence $w(z)$ from Eq. (3.7.38a) has movable poles.

Riccati equations are indeed special equations, and a large literature has been reserved for them. The above conclusions were first realized by Fuchs, but an extensive treatment was provided by the work of Painlevé. For equations (3.7.36),

Painlevé proved that the only movable singular points possible were algebraic; that is, no logarithmic or more exotic singular points arise in this case.

Painlevé also considered the question of enumerating those second nonlinear differential equations admitting poles as their only movable singularities. He studied equations of the form

$$\frac{d^2w}{dz^2} = F\left(w, \frac{dw}{dz}, z\right), \tag{3.7.39}$$

where F is rational in w and dw/dz, and whose coefficients are locally analytic in z. Painlevé and colleagues found (depending on how one counts) some fifty different types of equations, all of which were either reducible to (a) linear equations, (b) Riccati equations, (c) equations containing so-called elliptic functions, and (d) six "new" equations.

Elliptic functions are single-valued **doubly periodic** functions whose movable singularities are poles. We say $f(z)$ is a doubly-periodic function if there are two complex numbers ω_1 and ω_2 such that

$$\left.\begin{array}{rcl} f(z + \omega_1) & = & f(z), \\ f(z + \omega_2) & = & f(z) \end{array}\right\} \tag{3.7.40}$$

with a necessarily nonreal ratio: $\omega_2/\omega_1 = \gamma$, $\mathrm{Im}\,\gamma \neq 0$. There are no doubly-periodic functions with two real incommensurate periods and there are no triply-periodic functions. The numbers $m\omega_1 + n\omega_2$, with m, n integers, are periods of $f(z)$ and a lattice formed by the numbers $0, \omega_1, \omega_2$ and $\omega_1 + \omega_2$ as vertices is called the **period parallelogram** of $f(z)$.

An example of an elliptic function is the function defined by the convergent series

$$\mathcal{P}(z) = z^{-2} + \sum_{m,n=0}^{\infty}{}' \left[(z - \omega_{m,n})^{-2} - \omega_{m,n}^{-2}\right], \tag{3.7.41}$$

for $z \neq 0$, $z \neq \omega_{m,n}$ (if we wish $z = 0$ can be translated to $z = z_0$), where the prime means $(m, n) \neq (0, 0)$, and $\omega_{m,n} = m\omega_1 + n\omega_2$, and where ω_1 and ω_2 are the two periods of the elliptic function. The function $\mathcal{P}(z)$ satisfies a simple first-order equation. Writing $w = \mathcal{P}(z)$, we have

$$(w')^2 = 4w^3 - g_2w - g_3, \tag{3.7.42}$$

where the constants g_2, g_3 are given by

$$\left.\begin{array}{l} g_2 = 60 \displaystyle\sum_{m,n=0}^{\infty}{}' \omega_{m,n}^{-4}, \\[2mm] g_3 = 140 \displaystyle\sum_{m,n=0}^{\infty}{}' \omega_{m,n}^{-6}. \end{array}\right\} \tag{3.7.43}$$

Alternatively, by taking the derivative of Eq. (3.7.42), we see w satisfies

$$w'' = 6w^2 - \frac{g_2}{2}.$$

The function $w = \mathcal{P}(z)$, sometimes written as $w = \mathcal{P}(z, g_2, g_3)$, is called the **Weierstrass elliptic function**.

Another representation of elliptic functions is via the so-called **Jacobi elliptic functions**

$$w_1(z) = \text{sn}\,(z, k), \qquad w_2(z) = \text{cn}\,(z, k), \qquad w_3(z) = \text{dn}\,(z, k),$$

the first two of which are often referred to as the Jacobian sine and cosine. These functions satisfy

$$\frac{dw_1}{dz} = w_2 w_3, \qquad w_1(0) = 0; \tag{3.7.44a}$$

$$\frac{dw_2}{dz} = -w_1 w_3, \qquad w_2(0) = 1; \tag{3.7.44b}$$

$$\frac{dw_3}{dz} = -k^2 w_1 w_2, \qquad w_3(0) = 1. \tag{3.7.44c}$$

Multiplying Eq. (3.7.44a) by w_1 and Eq. (3.7.44b) by w_2, and adding, yields (in analogy with the trigonometric sine and cosine)

$$w_1^2(z) + w_2^2(z) = 1.$$

Similarly, from Eqs. (3.7.44a) and (3.7.44c),

$$k^2 w_1^2(z) + w_3^2(z) = 1,$$

whereupon we see from these equations that $w_1(z)$ satisfies a scalar first-order nonlinear ordinary differential equation:

$$\left(\frac{dw_1}{dz}\right)^2 = \left(1 - w_1^2\right)\left(1 - k^2 w_1^2\right). \tag{3.7.45}$$

Also note that by taking the derivative of Eq. (3.7.45), w_1 satisfies the second-order ODE

$$w_1'' = 2k^2 w_1^3 - (1 + k^2)w_1.$$

Using the substitution $u = w_1^2$ and changing variables, we can put Eq. (3.7.45) into the form Eq. (3.7.42). Indeed, the general form for an equation to have elliptic function solutions is

$$(w')^2 = (w - a)(w - b)(w - c)(w - d). \tag{3.7.46}$$

Equation (3.7.46) can also be transformed to either of the standard forms (3.7.42) or (3.7.45). (The "bilinear" transformation $w = (\alpha + \beta w_1)/(\gamma + \delta w_1)$, $\alpha\delta - \beta\gamma \neq 0$, can be used to transform Eq. (3.7.46) to Eqs. (3.7.42) or (3.7.45).) We also note that the autonomous second-order differential equation (meaning that the coefficients are independent of z)

$$\frac{d^2w}{dz^2} = w^3 + ew^2 + fw, \qquad \text{where } e\ f \text{ are constants,} \qquad (3.7.47)$$

can be solved by multiplying Eq. (3.7.47) by dw/dz and integrating. Then, by factorization we may put the result in the form (3.7.46).

The six new equations that Painlevé discovered are not reducible to "known" differential equations, but are referred to as the six Painlevé transcendents listed as P_I through P_{VI}. They are listed below (it is understood that $w' \equiv dw/dz$):

$$P_I : w'' = 6w^2 + z$$

$$P_{II} : w'' = 2w^3 + zw + a$$

$$P_{III} : w'' = \frac{(w')^2}{w} - \frac{w'}{z} + \frac{(aw^2 + b)}{z} + cw^3 + \frac{d}{w}$$

$$P_{IV} : w'' = \frac{(w')^2}{2w} + \frac{3w^3}{2} + 4zw^2 + 2(z^2 - a)w + \frac{b}{w}$$

$$P_V : w'' = \left(\frac{1}{2w} + \frac{1}{w-1}\right)(w')^2 - \frac{w'}{z} + \frac{(w-1)^2}{z^2}\left(aw + \frac{b}{w}\right)$$
$$+ \frac{cw}{z} + \frac{dw(w+1)}{w-1}$$

$$P_{VI} : w'' = \frac{1}{2}\left(\frac{1}{w} + \frac{1}{w-1} + \frac{1}{w-z}\right)(w')^2$$
$$- \left(\frac{1}{z} + \frac{1}{z-1} + \frac{1}{w-z}\right)w'$$
$$+ \frac{w(w-1)(w-2)}{z^2(z-1)^2}\left[a + \frac{bz}{w^2} + \frac{c(z-1)}{(w-1)^2} + \frac{dz(z-1)}{(w-z)^2}\right],$$

where a, b, c, d are arbitrary constants.

It turns out that the sixth equation contains the first five by a limiting procedure, carried out by first transforming w and z appropriately in terms of a suitable (small) parameter, and then taking limits of the parameter to zero. Recent research (see Fokas et al., 2006) has shown that these six equations can be linearized by transforming the equations via a somewhat complicated sequence of transformations into linear equations. The methods to understand these transformations and related solutions involve methods of complex analysis, as discussed in Chapter 7 of Ablowitz and Fokas (2003), on Riemann–Hilbert boundary value problems.

Second- and higher-order nonlinear equations need not have only poles or algebraic singularities. For example, the equation

$$\frac{d^2w}{dz^2} = \left(\frac{dw}{dz}\right)^2 \left(\frac{2w-1}{w^2+1}\right) \tag{3.7.48}$$

has the solution

$$w(z) = \tan(\log(az+b)), \tag{3.7.49}$$

where a and b are arbitrary constants. Hence the point $z = -b/a$ is a branch point and the function $w(z)$ has no limit as z approaches this point; it is also a cluster point of poles. Similarly, the equation

$$\frac{d^2w}{dz^2} = \frac{\alpha-1}{\alpha w}\left(\frac{dw}{dz}\right)^2 \tag{3.7.50}$$

has the solution

$$w(z) = c(z-d)^\alpha, \tag{3.7.51}$$

where c and d are arbitrary constants. Equation (3.7.51) has an algebraic branch point only if $\alpha = m/n$ where m and n are integers; otherwise the point $z = d$ is a transcendental branch point.

Third-order equations may possess even more exotic movable singular points. Indeed, motivated by Painlevé's work, Chazy (1911) showed that the equation

$$\frac{d^3w}{dz^3} = 2w\frac{d^2w}{dz^2} - 3\left(\frac{dw}{dz}\right)^2 \tag{3.7.52}$$

was solvable via a rather nontrivial transformation of coordinates. His solution shows that the general solution of Eq. (3.7.52) possesses a movable natural barrier. Indeed the barrier is a circle, whose center and radius depend on initial values. Interestingly enough, the solution w in Eq. (3.7.52) is related to the following system of equations, first considered in the case $\epsilon = -1$ by Darboux (1878) and then solved by Halphen (1881), which we refer to as the Darboux–Halphen system (when $\epsilon = 1$):

$$\left.\begin{array}{l} \dfrac{dw_1}{dz} = w_2w_3 + \epsilon w_1(w_2+w_3); \\[2mm] \dfrac{dw_2}{dz} = w_3w_1 + \epsilon w_2(w_3+w_1); \\[2mm] \dfrac{dw_3}{dz} = w_1w_2 + \epsilon w_3(w_1+w_2). \end{array}\right\} \tag{3.7.53}$$

In particular, when $\epsilon = -1$, Chazy's equation is related to the solutions of Eq. (3.7.53) by

$$w = -2(w_1 + w_2 + w_3).$$

When $\epsilon = 0$, Eqs. (3.7.53) are related (by scaling) to Eqs. (3.7.44a,b,c), and the solution may be written in terms of elliptic functions. Equations (3.7.53) with $\epsilon = -1$ arise in the study of relativity and integrable systems (Ablowitz and Clarkson, 1991).

In fact, Chazy's and the Darboux–Halphen system can be solved in terms of certain special functions which are generalizations of trigonometric and elliptic function; they are called automorphic functions which we will study further in Section 5.8. By direct calculation (whose details are outlined in the exercises) we can verify that the following (owing to Chazy) yields a solution to Eq. (3.7.52). Transform to a new independent variable

$$z(s) = \frac{\chi_2(s)}{\chi_1(s)} \tag{3.7.54a}$$

where χ_1 and χ_2 are two linearly independent solutions of the following hypergeometric equation (see Eq. (3.7.35c), where $a = b = \frac{1}{12}$ and $c = \frac{1}{2}$):

$$\frac{d^2\chi}{ds^2} = \alpha(s)\frac{d\chi}{ds} + \beta(s)\chi \tag{3.7.54b}$$

where

$$\alpha(s) = \left(\frac{\frac{7s}{6} - \frac{1}{2}}{s(1-s)}\right) \quad \text{and} \quad \beta(s) = \frac{1}{144s(1-s)};$$

that is,

$$s(1-s)\frac{d^2\chi}{ds^2} + \left(\frac{1}{2} - \frac{7s}{6}\right)\frac{d\chi}{ds} - \frac{\chi}{144} = 0. \tag{3.7.54c}$$

Then the solution w of Chazy's equation can be expressed as follows:

$$w(s(z)) = 6\frac{d}{dz}\log\chi_1 = \frac{6}{\chi_1}\frac{d\chi_1}{ds}\frac{ds}{dz} = \frac{6}{\chi_1}\frac{\frac{d\chi_1}{ds}}{(dz/ds)}$$

$$= \frac{6\chi_1}{\mathcal{W}(\chi_1, \chi_2)}\frac{d\chi_1}{ds}, \tag{3.7.55}$$

where $\mathcal{W}(\chi_1, \chi_2)$ is the Wronskian of χ_1 and χ_2 which satisfies $\mathcal{W}' = \alpha\mathcal{W}$, or

$$\mathcal{W}(\chi_1, \chi_2) = \chi_1\frac{d\chi_2}{ds} - \chi_2\frac{d\chi_1}{ds} = s^{-1/2}(1-s)^{-2/3}\mathcal{W}_0, \tag{3.7.56}$$

where $\mathcal{W}_0$ is an arbitrary constant. Although this yields, in principle, only a special solution to Eq. (3.7.52), the general solution can be obtained by making the transformation $\chi_1 \mapsto a\chi_1 + b\chi_2$, $\chi_2 \mapsto c\chi_1 + d\chi_2$, with $a, b, c,$ and d as arbitrary constants normalized to $ad - bc = 1$.

However, to understand the properties of the solution $w(z)$, we really need to understand the conformal map $z = z(s)$ and its inverse $s = s(z)$. (From $s(z)$ and $\chi_1(s(z))$ we find the solution $w(s(z))$.) Usually, this map is denoted by

$s = s(z; \alpha, \beta, \gamma)$, where α, β, γ are three parameters related to the hypergeometric equation (3.7.54c), which are in this case $\alpha = 0, \beta = \pi/2, \gamma = \pi/3$. This function is called a **Schwarzian triangle function**, and the map transforms the region defined inside a "circular triangle" (a triangle whose sides are either straight lines or circular arcs – at least one side being an arc) in the z-plane to the upper half s-plane. It turns out that by reflecting the triangle successively about any of its sides, and repeating this process infinitely, we can analytically continue the function $s(z, 0, \pi/2, \pi/3)$ everywhere inside a circle. For the solution normalized as in Eqs. (3.7.54a–c), this is a circle centered at the origin. The function $s = s(z; 0, \pi/2, \pi/3)$ is single valued and analytic inside the circle, but the circumference of the circle is a natural boundary – which in this case can be shown to consist of a dense set of essential singularities. The reader can find a further discussion of mappings of circular triangles and Schwarzian triangle functions in Section 5.8. Such functions are special cases of what are often called **automorphic functions**. Automorphic functions have the property that $s(\gamma(z)) = s(z)$, where $\gamma(z) = \frac{az+b}{cz+d}$, $ad - bc = 1$, and as such are generalizations of periodic functions, for example, elliptic functions.

It is worth remarking that the Darboux–Halphen system (3.7.53) can also be solved in terms of a Schwarzian triangle function. In fact, the solutions $\omega_1, \omega_2, \omega_3$, are given by the formulae

$$\omega_1 = -\frac{1}{2}\frac{d}{dz}\log\frac{s'(z)}{s}, \qquad \omega_2 = -\frac{1}{2}\frac{d}{dz}\log\frac{s'(z)}{1-s},$$
$$\omega_3 = -\frac{1}{2}\frac{d}{dz}\log\frac{s'(z)}{s(1-s)} \tag{3.7.57}$$

where $s(z)$ satisfies the equation

$$\{s, z\} = -\left(\frac{1}{s^2} + \frac{1}{(1-s)^2} + \frac{1}{s(1-s)}\right)\frac{(s'(z))^2}{2} \tag{3.7.58a}$$

and the term $\{s, z\}$ is the **Schwarzian derivative** defined by

$$\{s, z\} = \frac{s'''}{s'} - \frac{3}{2}\left(\frac{s''}{s'}\right)^2. \tag{3.7.58b}$$

Equation (3.7.58a) is obtained when we substitute ω_1, ω_2 and ω_3 given by Eq. (3.7.57) into Eq. (3.7.53), with $\epsilon = -1$. The function $s(z)$ is the "zero angle" Schwarzian triangle function: $s(z) = s(z, 0, 0, 0)$, which is discussed in Section 5.8. In the exercises, the following transformation involving Schwarzian derivatives is established:

$$\{z, s\} = \{s, z\}\frac{(-1)}{(s'(z))^2}. \tag{3.7.59}$$

Using Eqs. (3.7.58a–3.7.59), we obtain the equation

$$\{z, s\} = \frac{1}{2} \left(\frac{1}{s^2} + \frac{1}{(1-s)^2} + \frac{1}{s(1-s)} \right). \tag{3.7.60}$$

In Section 5.8 we show how to solve Eq. (3.7.60), and thereby find the inverse transformation $z = z(s)$ in terms of hypergeometric functions. We will not go further into these because it will take us too far outside the scope of this book.

3.7.1 Problems for Section 3.7

1. Discuss the nature of the singular points (location, fixed or movable) of the following differential equations and solve the differential equations.

   (a) $z\dfrac{dw}{dz} = 2w + z$;   (b) $z\dfrac{dw}{dz} = w^2$;

   (c) $\dfrac{dw}{dz} = a(z)w^3$,   where $a(z)$ is an entire function of z;

   (d) $z^2\dfrac{d^2w}{dz^2} + z\dfrac{dw}{dz} + w = 0$.

2. Solve the differential equation

$$\frac{dw}{dz} = w - w^2.$$

   Show that it has a pole as its only singularity.

3. Consider the equation

$$\frac{dw}{dz} = p(z)w^2 + q(z)w + r(z),$$

   where $p(z)$, $q(z)$, $r(z)$ are (for convenience) entire functions of z.

   (a) Let $w = \alpha(z)\phi'(z)/\phi(z)$. Show that taking $\alpha(z) = -1/p(z)$ eliminates the term $(\phi'/\phi)^2$, and find that $\phi(z)$ satisfies

$$\phi'' - \left(q(z) + \frac{p'(z)}{p(z)} \right) \phi' + p(z)r(z)\phi = 0.$$

   (b) Explain why the function $w(z)$ has, as its only movable singular points, poles. Where are they located? Can there be any fixed singular points? Explain.

4. Determine the indicial equation and the basic form of expansion representing the solution in the neighborhood of the regular singular points to the following equations:

(a) Bessel's Equation

$$z^2 \frac{d^2w}{dz^2} + z \frac{dw}{dz} + (z^2 - p^2)w = 0,$$

where p is not an integer;

(b) Legendre's Equation

$$(1 - z^2) \frac{d^2w}{dz^2} - 2z \frac{dw}{dz} + p(p + 1)w = 0,$$

where p is not an integer;

(c) Hypergeometric Equation

$$z(1 - z) \frac{d^2w}{dz^2} + [c - (a + b + 1)z] \frac{dw}{dz} - abw = 0,$$

one solution is satisfactory.

5. Suppose we are given the equation $\dfrac{d^2w}{dz^2} = 2w^3$.

(a) Let us look for a solution of the form

$$w = \sum_{n=0}^{\infty} a_n (z - z_0)^{n-r} = a_0 (z - z_0)^{-r} + a_1 (z - z_0)^{1-r} + \cdots ,$$

for z near z_0, $r > 0$. Substitute this into the equation to determine that $r = 1$ and $a_0 = \pm 1$.

(b) "Linearize" about the basic solution by letting $w = \pm 1/(z - z_0) + v$, and dropping quadratic terms in v to find

$$\frac{d^2v}{dz^2} = \frac{6v}{(z - z_0)^2}.$$

Solve this equation (said to be of Cauchy–Euler type) to find

$$v = A(z - z_0)^{-2} + B(z - z_0)^3.$$

(c) Explain why this indicates that all coefficients of subsequent powers in the following expansion (save possibly a_4),

$$w = \frac{\pm 1}{(z - z_0)} + a_1 + a_2(z - z_0) + a_3(z - z_0)^2 + a_4(z - z_0)^3 + \cdots ,$$

can be solved uniquely. Substitute the expansion into the equation for w, and find a_1, a_2, a_3, and establish the fact that a_4 is *arbitrary*. We obtain two arbitrary constants in this expansion: z_0 and a_4. The solution to $w'' = 2w^3$ can be expressed in terms of elliptic functions; its general solution is known to have only simple poles as its movable singular points.

(d) Show that a similar expansion works when we consider the equation

$$\frac{d^2w}{dz^2} = 2w^3 + zw.$$

(this is the second Painlevé equation, see Ince, 1956) and hence that the formal analysis indicates that the only movable algebraic singular points are poles. (Painlevé proved that there are no other singular points for this equation.)

(e) Show that this expansion *fails* when we consider

$$\frac{d^2w}{dz^2} = 2w^3 + z^2w,$$

because a_4 cannot be found. This indicates that a more general expansion is required. (In fact, another term of the form $b_4(z - z_0)^3 \log(z - z_0)$ must be added at this order, and further logarithmic terms must be added at all subsequent orders in order to obtain a consistent formal expansion.)

6. In this exercise we describe the verification that formulae (3.7.54a–3.7.56) indeed satisfy Chazy's equation.

(a) Use Eqs. (3.7.54a–3.7.55) to verify, by differentiation and resubstitution, the following formulae for the first three derivatives of w. Hint: use the linear equation (3.7.54b)

$$\frac{d^2\chi}{ds^2} = \alpha(s)\frac{d\chi}{ds} + \beta(s)\chi,$$

to re-substitute the second derivative $\chi''(s)$ in terms of the first derivative $\chi'(s)$ and the function $\chi(s)$ successively, thereby eliminating higher derivatives of $\chi(s)$.

$$\frac{dw}{dz} = 6\left(\chi_1^4\beta + \chi_1^2(\chi_1')^2\right)/\mathcal{W}^2, \tag{i}$$

$$\frac{d^2w}{dz^2} = 6\left[\chi_1^6(\beta' - 2\alpha\beta) + \chi_1^5\chi_1'6\beta + 2\chi_1^3(\chi_1')^3\right]/\mathcal{W}^3, \tag{ii}$$

$$\frac{d^3w}{dz^3} = 6\left[\chi_1^8\left(\beta'' - 2\alpha'\beta - 5\alpha\beta' + 6\alpha^2\beta + 6\beta^2\right)\right.$$
$$\left. + \chi_1^7\chi_1'(12\beta' - 24\alpha\beta) + 6\chi_1^4(\chi_1')^4 + 36\chi_1^6(\chi_1')^2\beta\right]/\mathcal{W}^4, \tag{iii}$$

where $\mathcal{W}$ is given by equation Eq. (3.7.56).

(b) By inserting (i)–(iii) into Chazy's equation (3.7.52) show that all terms cancel except for the following equation in α and β:

$$\beta'' - 2\alpha'\beta - 5\alpha\beta' + 6\alpha^2\beta + 24\beta^2 = 0. \tag{iv}$$

Show that the specific choices as in Eq. (3.7.54b),

$$\alpha(s) = \left(\frac{\frac{7s}{6} - \frac{1}{2}}{s(1-s)}\right) \quad \text{and} \quad \beta(s) = \frac{1}{144s(1-s)},$$

satisfy (iv) and hence verify Chazy's solution.

7. Consider an invertible function $s = s(z)$.

   (a) Show that the derivative d/dz transforms according to the relationship

   $$\frac{d}{dz} = \frac{1}{z'(s)}\frac{d}{ds}.$$

   (b) As in Eq. (3.7.58b), the Schwarzian derivative is defined as

   $$\{s, z\} = \left(\frac{s''}{s'}\right)' - \frac{1}{2}\left(\frac{s''}{s'}\right)^2.$$

   Show that

   $$\{s, z\} = \frac{1}{z'(s)}\frac{d^2}{ds^2}\left(\frac{1}{z'(s)}\right) - \frac{1}{2}\left(\frac{d}{ds}\left(\frac{1}{z'(s)}\right)\right)^2 = -\frac{1}{(z'(s))^2}\{z, s\}.$$

   (c) Consequently establish that

   $$\{z, s\} = \{s, z\}\left(-\frac{1}{(s'(z))^2}\right).$$

8. In this exercise we derive a different representation for the solution of Chazy's equation.

   (a) Show that

   $$s''(z) = \frac{d}{dz}(s'(z)) = s'(z)\frac{d}{ds}\left(\frac{1}{z'(s)}\right).$$

   (b) In the above formulae use $z(s) = \chi_2(s)/\chi_1(s)$, where χ_1 and χ_2 satisfy the hypergeometric equation (3.7.54a), and the Wronskian relation

   $$W(\chi_1, \chi_2) = (\chi_1\chi_2' - \chi_1'\chi_2) = W_0 s^{-1/2}(1-s)^{-2/3}$$

   to show that

   $$s''(z) = s'(z)\frac{d}{ds}\left(s^{\frac{1}{2}}(1-s)^{\frac{2}{3}}\chi_1^2(s)/W_0\right)$$

   $$= s'(z)\left(\frac{1}{2s}s'(z) - \frac{2}{3(1-s)}s'(z) + \frac{2\chi_1'}{\chi_1}s'(z)\right).$$

(c) Use Chazy's solution (3.7.55):

$$w = \frac{6\chi_1'}{\chi_1} s'(z)$$

to show that

$$w = 3\frac{s''}{s} - \frac{3}{2}\frac{s'}{s} + \frac{2s'}{1-s} = \frac{1}{2}\frac{d}{dz}\log\frac{(s')^6}{s^3(1-s)^4}.$$

(d) Note that here $s(z)$ is the Schwarzian triangle function with angles $0, \pi/2$, $\pi/3$; i.e.,

$$s(z) = s(z, 0, \pi/2, \pi/3).$$

The fact that Chazy's equations and the Darboux–Halphen system are related by the equation $w = -2(w_1 + w_2 + w_3)$ allows us to find a relation between the above Schwarzian, $s(z, 0, \pi/2, \pi/3)$ (for Chazy's equation), and the one used in the text for the solution of the Darboux–Halphen system with zero angles: $s(z, 0, 0, 0)$. Denote the latter Schwarzian by $\hat{s}(z)$; i.e., $\hat{s}(z) = s(z, 0, 0, 0)$. Show that Eq. (3.7.57) and $w = -2(w_1 + w_2 + w_3)$ yields the relationship

$$\frac{1}{2}\frac{d}{dz}\log\frac{(s')^6}{s^3(1-s)^4} = \frac{d}{dz}\log\frac{(\hat{s}')^3}{\hat{s}^2(1-\hat{s})^2},$$

or

$$\frac{(s')^6}{s^3(1-s)^4} = A\frac{(\hat{s}')^6}{\hat{s}^4(1-\hat{s})^4},$$

where A is a constant.

*3.8 Computational Methods

In this section we discuss some of the concrete aspects involving computation in the study of complex analysis. Our purpose here is not to be extensive in our discussion but rather to illustrate some basic ideas that can be readily implemented. We will discuss two topics: the evaluation of (a) Laurent series and (b) the solution of differential equations, both of which relate to our discussions in this chapter. We note that an extensive discussion of computational methods and theory can be found in Henrici (1977).

*3.8.1 Laurent series

In Section 3.3 we derived the Laurent series representation of a function analytic in an annulus, $R_1 \leq |z - z_0| \leq R_2$. It is given by the formulae (3.3.1) and (3.3.2), which we repeat here for the convenience of the reader:

$$f(z) = \sum_{n=-\infty}^{\infty} c_n (z - z_0)^n, \tag{3.8.1}$$

where

$$c_n = \frac{1}{2\pi i} \oint_C \frac{f(z)dz}{(z - z_0)^{n+1}}, \tag{3.8.2}$$

and C is any simple closed contour in the annulus which encloses the inner boundary $|z - z_0| = R_1$. We shall take C to be a circle of radius r. Accordingly the change of variables

$$z = z_0 + re^{i\theta}, \tag{3.8.3}$$

where r is the radius of a circle with $R_1 \leq r \leq R_2$, allows us to rewrite Eq. (3.8.2) as

$$\hat{c}_n = \frac{1}{2\pi} \int_{-\pi}^{\pi} f(\theta) e^{-in\theta} \, d\theta, \tag{3.8.4}$$

where $c_n = \hat{c}_n / r^n$. In fact, Eq. (3.8.4) gives the Fourier coefficients of the function

$$f(\theta) = \sum_{n=-\infty}^{\infty} \hat{c}_n e^{in\theta}, \tag{3.8.5}$$

with period 2π defined on the circle (3.8.3). Equation (3.8.4) can be used as a computational tool after discretization. We consider $2N$ points equally spaced along the circle, with $\theta_j = hj$, $j = -N, -N + 1, \ldots, N - 1$, and $\int_{-\pi}^{\pi} \to \sum_{j=-N}^{N-1}$ with $d\theta \to \Delta\theta = h = 2\pi/(2N) = \pi/N$; note that when $j = N$ then $\theta_N = \pi$.

The following discretization corresponds to what is usually called the discrete Fourier transform:

$$f(\theta_j) = \sum_{n=-N}^{N-1} \hat{c}_n e^{in\theta_j}, \tag{3.8.6}$$

where

$$\hat{c}_n = \frac{1}{2N} \sum_{j=-N}^{N-1} f(\theta_j) e^{-in\theta_j}. \tag{3.8.7}$$

We note that the formulae (3.8.6) and (3.8.7) can be calculated directly, at a "cost" of $O(N^2)$ multiplications. (The notation $O(N^2)$ means proportional to N^2; a formal definition can be found in Ablowitz and Fokas, 2003, Chapter 6.) Moreover, it is well known that in fact, the computational "cost" can be reduced significantly to $O(N \log N)$ multiplications by means of the Fast Fourier Transform (FFT): see, e.g., Henrici (1977).

Given a function at $2N$ equally spaced points on a circle, one can readily compute the discrete Fourier coefficients, $\hat{c}_n$. The approximate Laurent coefficients are then given by $c_n = \hat{c}_n/r^n$. (For all the numerical examples below we use $r = 1$.) As N increases, the approximation improves rapidly if the continuous function is expressible as a Laurent series. However, if the function $f(z)$ were analytic, we would find that the coefficients with negative indices would be zero (to a very good approximation).

Example 3.8.1 Consider the functions (a) $f(z) = 1/z$; and (b) $f(z) = e^{1/z}$. Note that with $z_0 = 0$ the exact answers are (a) $c_{-1} = 1$; and $c_n = 0$ for $n \neq 0$, (b) $c_n = 1/(-n)!$ for $n \leq 0$, and $c_n = 0$ for $n \geq 1$. The magnitude of the numerically computed coefficients, using $N = 16$, are shown in Figure 3.12 ($*$ represents the coefficient). Note that for part (a) we obtain only one significantly nonzero coefficient: $c_{-1} \approx 1$ and $c_n \approx 0$ to high accuracy. In part (b) we find that $c_n \approx 0$ for $n \geq 1$; all the coefficients agree with the exact values to a high degree of accuracy. Note that the coefficients decay rapidly for large negative n.

*3.8.2 Differential Equations

The solution of differential equations in the complex plane can be approximated by many of the computational methods often studied in numerical analysis. We shall discuss "time stepping" methods and series methods.

We consider the scalar differential equation

$$\frac{dy}{dz} = f(z, y), \tag{3.8.8}$$

with the initial condition $y(z_0) = y_0$, where f is analytic in both arguments in some domain D containing $z = z_0$. The key ideas are best illustrated by the explicit

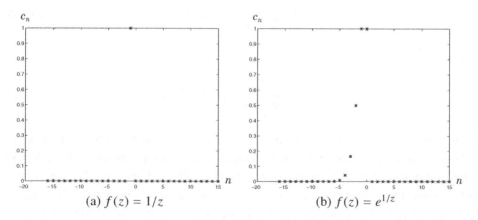

(a) $f(z) = 1/z$ (b) $f(z) = e^{1/z}$

Figure 3.12 Laurent coefficients c_n for two functions

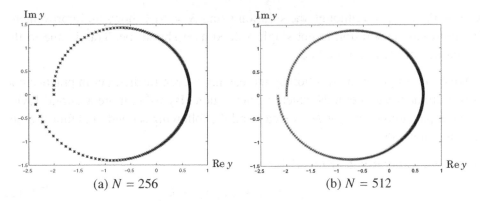

(a) $N = 256$ (b) $N = 512$

Figure 3.13 Explicit Euler's method, Example 3.8.2

Euler method. Here dy/dz is approximated by the difference $(y(z+h_n) - y(z))/h_n$. Write $z_{n+1} = z_n + h_n$, $y(z_n) = y_n$ and note that h_n is *complex*; that is, h_n can take any direction in the complex plane. Also note that we allow the step size, h_n, to vary from one time step to the next, which is necessary if, for instance, we want to integrate around the unit circle. In this application we keep $|h_n|$ constant. Hence, at $z = z_n$, we have the approximation

$$y_{n+1} = y_n + h_n f(z_n, y_n), \qquad n = 0, 1, \ldots \qquad (3.8.9)$$

with the initial condition $y(z_0) = y_0$. Using standard methods in numerical analysis, it can be shown that under suitable assumptions, Eq. (3.8.9) is an $O(h_n^2)$ approximation over every step and an $O(h_n)$ approximation if we integrate over a finite distance with $h_n \to 0$. Equation (3.8.9) is straightforward to apply as we now show.

Example 3.8.2 Approximate the solution of the equation

$$\frac{dy}{dz} = y^2, \qquad y(1) = -2$$

as z traverses along the contour C, where C is the unit circle in the complex plane. We discretize along the circle and take $z_n = e^{i\theta_n}$ where $\theta_n = 2\pi n/N$ and $h_n = z_{n+1} - z_n$, $n = 0, 1, \ldots N - 1$. The exact solution of $dy/dz = y^2$ is $y = 1/(A - z)$ where A is an arbitrary complex constant and we see that it has a pole at the location $z = A$. For the initial value $y(1) = -2$, $A = \frac{1}{2}$, and the pole is located at $z = \frac{1}{2}$.

Because we are taking a circuit around the unit circle, we never get close to the singular point. And because the solution is single valued we expect to return to the initial value after one circuit. We use the approximation (3.8.9) with $f(z_n, y_n) = y_n^2$, $y_0 = y(z_0) = -2$. The solutions using $N = 256$ and $N = 512$ are shown in Figure 3.13, where we plot the real part of y versus the imaginary part of y:

$$y(z_n) = y_R(z_n) + i y_I(z_n)$$

for $n = 0, 1, \ldots, N$. Although the solution using $N = 512$ shows an improvement, the approximate solution is not single valued as it should be. This is due to the inaccuracy of the Euler method.

We could improve the solution by increasing N even further, but in practice one uses more accurate methods which we now quote. By using more accurate Taylor series expansions of $y(z_n + h_n)$ we can find the following second- and fourth-order accurate methods:

(a) second-order Runge–Kutta (RK2)

$$y_{n+1} = y_n + \tfrac{1}{2} h_n (k_{n_1} + k_{n_2}) \tag{3.8.10}$$

where $k_{n_1} = f(z_n, y_n)$ and $k_{n_2} = f(z_n + h_n, y_n + h_n k_{n_1})$;
(b) fourth-order Runge–Kutta (RK4)

$$y_{n+1} = y_n + \frac{1}{6} h_n (k_{n_1} + 2k_{n_2} + 2k_{n_3} + k_{n_4}), \tag{3.8.11}$$

where

$$k_{n_1} = f(z_n, y_n), \qquad\qquad k_{n_2} = f\left(z_n + \frac{1}{2} h_n, y_n + \frac{1}{2} h_n k_{n_1}\right),$$

$$k_{n_3} = f\left(z_n + \frac{1}{2} h_n, y_n + \frac{1}{2} h_n k_{n_2}\right), \quad k_{n_4} = f(z_n + h_n, y_n + h_n k_{n_3}).$$

Example 3.8.3   We illustrate how the above methods, RK2 and RK4, work on the same problem as Example 3.8.2, choosing h_n in the same way as before. Using $N = 128$ we see in Figure 3.14 that the solution is indeed single valued (to numerical accuracy) as expected. It is clear that the solutions obtained by these methods are nearly single valued; they are a significant improvement over Euler's method. Moreover, RK4 is an improvement over RK2, although RK4 requires more function evaluations and more computer time.

Example 3.8.4   Consider the differential equation

$$\frac{dy}{dz} = \tfrac{1}{2} y^3,$$

with initial values

(a) $y(1) = 1$: the exact solution is $y(z) = 1/\sqrt{2 - z}$;
(b) $y(1) = 2i$: the exact solution is $y(z) = 2i/\sqrt{4z - 3}$.

The general solution is $y(z) = (z_0 - z)^{-1/2}$, where the proper branch of the square root is chosen to agree with the initial value. We integrate around the unit circle (choosing h_n as in the previous examples) using RK2 and RK4 for $N = 128$; the results are shown in Figure 3.15. For the initial value in part (a) the singularity lies

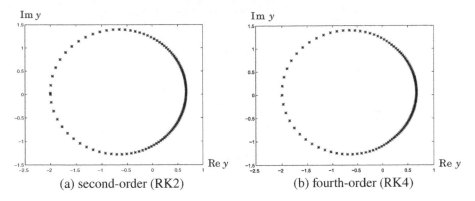

Figure 3.14 Runge–Kutta methods (Example 3.8.3)

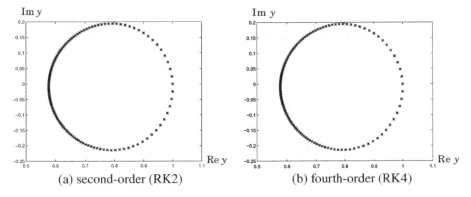

Figure 3.15 Part (a) of Example 3.8.4, using $y(1) = 1$ and $N = 128$

outside the unit circle and the numerical solutions are single valued. Figure 3.16 shows the logarithm (base 10) of the absolute value of the errors in the calculations graphed in Figure 3.15. Note that the error in RK4 is several orders of magnitude smaller than the error in RK2. For the initial value of part (b) the branch point is at $z = 3/4$ and thus lies inside the unit circle and the solutions are clearly not single valued. Numerically (see Figure 3.17) we find that the jump in the function $y(z)$ is approximately $4i$ as we traverse the circle from $\theta = 0$ to 2π, as expected from the exact solution.

As long as there are no singular points on or close to the integration contour there will be no difficulty in implementing the above time stepping algorithms. However, in practice one frequently has nearby singular points and the contour may need to be modified in order to analytically continue the solution. In this case it is sometimes useful to use series methods to approximate the solution of the differential equation and estimate the radius of convergence as the calculation proceeds. This is discussed next.

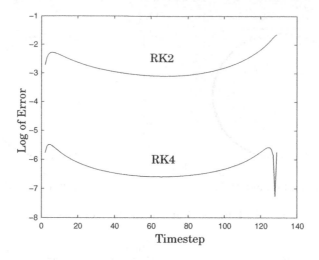

Figure 3.16 The error in the numerical solutions shown in Figure 3.15

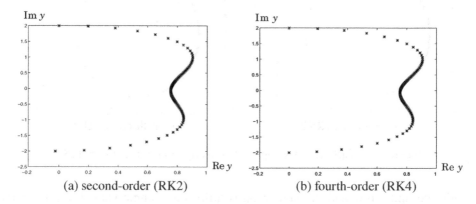

(a) second-order (RK2)   (b) fourth-order (RK4)

Figure 3.17 Part (b) of Example 3.8.4, using $y(1) = 2i$ and $N = 128$

Given Eq. (3.8.8) and noting Cauchy's Theorem 3.7.1 for differentiable equations, we can look for a series solution of the form

$$y = \sum_{n=0}^{\infty} A_n(z - z_0)^n$$

in the neighborhood of an analytic point of the equation. By inserting this series into the equation, we seek to develop a recursion relation between the coefficients; this can be difficult or unwieldy in complicated cases, but computationally speaking, it can almost always be accomplished. Having found such a recursion relation, we can evaluate the coefficients A_n and find an approximation to the radius of convergence: from the ratio test $R = |z - z_0| = \lim_{n\to\infty} |A_n/A_{n+1}|$ when this limit exists, or more generally vie the root test $R = [\lim_{n\to\infty} \sup_{m>n} |a_m|^{1/m}]^{-1}$ (see Section 3.2). As we proceed in the calculation we estimate the radius of convergence (for large n). We may need to modify our contour if the radius of convergence begins to shrink and move in a direction where the radius of convergence enlarges or remains acceptably large.

Example 3.8.5 Evaluate the series solution to the equation

$$\frac{dy}{dz} = y^2 + 1, \tag{3.8.12}$$

with $y(0) = 1$. The exact solution is obtained by integrating

$$\frac{dy}{1 + y^2} = dz$$

to yield $y = \tan(z + \pi/4)$. In order to obtain a recursion relation associated with the coefficients of the series solution $y = \sum_{n=0}^{\infty} A_n (z - z_0)^n$ it is useful to use the series product formula:

$$\sum_{n=0}^{\infty} A_n (z - z_0)^n \sum_{m=0}^{\infty} B_m (z - z_0)^m = \sum_{n=0}^{\infty} C_n (z - z_0)^n,$$

where $C_n = \sum_{p=0}^{n} A_p B_{n-p}$. The insertion of the series for y into Eq. (3.8.12) yields

$$\sum_{n=0}^{\infty} n A_n (z - z_0)^{n-1} = \left(\sum_{n=0}^{\infty} A_n (z - z_0)^n \right)^2 + 1.$$

Using the product formula and the transformation

$$\sum_{n=0}^{\infty} n A_n (z - z_0)^{n-1} = \sum_{n=0}^{\infty} (n + 1) A_{n+1} (z - z_0)^n,$$

we obtain the equation

$$\sum_{n=0}^{\infty} (n + 1) A_{n+1} (z - z_0)^n = \sum_{n=0}^{\infty} \left(\sum_{p=0}^{n} A_p A_{n-p} \right) (z - z_0)^n + 1$$

and hence the recursion relation

$$(n + 1) A_{n+1} = \sum_{p=0}^{n} A_p A_{n-p} + \delta_{n,0}, \tag{3.8.13}$$

where $\delta_{n,0}$ is the Kronecker delta function: $\delta_{n,0} = 1$ if $n = 0$, and 0 otherwise. Because we have posed the differential equation at $z = 0$, we begin with $z_0 = 0$ and $A_0 = y_0 = 1$. It is straightforward to compute the coefficients from this formula. Computing the ratios up to $n = 12$ (for example) we find that the final terms yield $\lim_{n \to \infty} A_n / A_{n+1} \approx A_{11} / A_{12} = 0.78539816$. It is clear that the series converges inside a radius of convergence R of approximately $\pi/4$ as it should. Suppose we use this series up to $z = 0.1$ in steps of 0.01. This means we use the recursion relation (3.8.13), but we use it repeatedly after each time step; that is, for each of the values $A_0 = y(z_j)$, with $z_j = 0, 0.01, 0.02, \ldots, 0.10$, we calculate the corresponding, successive coefficients, A_n, from Eq. (3.8.13) before we proceed to the next z_j. This means that in the series solution for y we are re-expanding about a new point $z_0 = z_j$. We obtain (still using $n = 12$ coefficients) $y(0.1) = 1.22305$ and an approximate radius of convergence $R = 0.6854$. Note that these values are very good approximations of the analytical values. We can also evaluate the series by moving into the complex plane. For example, if we expand around $z = 0.1$ and move in steps of $0.01i$ to $z = 0.1 + 0.1i$. We obtain $y(0.1 + 0.1i) = 1.1930 + 0.2457i$ and an approximate radius of convergence $R = 0.6967$. The series expansion is now seen to be valid in a larger region. This is true because we are now moving away from the singularity. The procedure can be repeated and we can analytically extend the solution by reexpanding the series about new points and employing the recursion relation to move into any region where the solution is analytic. In this way we can "internally" decide on how big a region of analyticity we wish to cover and always be sure to move into regions where the series solution is valid.

A detailed dicussion of series methods for solving ODEs appears in the work of Corliss and Chang (1982).

3.8.3 Problems for Section 3.8

1. Find the magnitude of the numerically computed Laurent coefficients with $z_0 = 0$ (using $N = 32$) for: (i) $f(z) = e^z$; (ii) $f(z) = \sqrt{z}$; (iii) $f(z) = 1/\sqrt{z}$; (iv) $f(z) = \tan 1/z$, and show that they agree with those in Figure 3.18.

   (a) Do the Laurent coefficients in Figure 3.18 correspond to what you would expect from analytical considerations? What is the true behavior of each function; i.e., what kind of singularities do these functions have?

   (b) Note that the coefficients decay at very different rates for the examples (i)–(iv). Explain why this is the case. (Hint: Relate it to the single-valuedness of the function.)

2. Consider the differential equation

$$\frac{dy}{dz} = y^2, \qquad y(z_0) = y_0.$$

   (a) Show that the analytical solution is given by

$$y(z) = \frac{y_0}{1 - y_0(z - z_0)}.$$

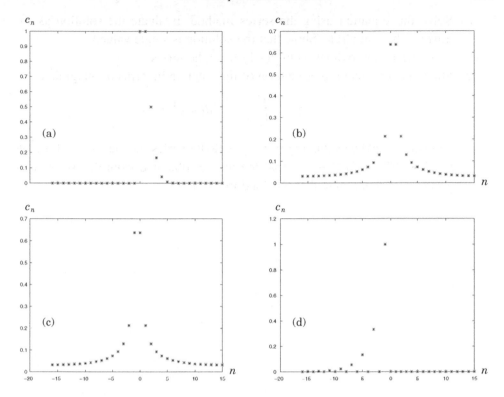

Figure 3.18 Laurent coefficients c_n for Problem 1: (a) $f(z) = e^z$; (b) $f(z) = z^{1/2}$; (c) $f(z) = z^{1/2}$; (d) $f(z) = \tan(1/z)$

(b) Write down the position of the singularity of the solution (a) above. What is the nature of the singularity?

(c) From (b) above note that the position of the singularity depends on the initial values, i.e., z_0 and y_0. Choose $z_0 = 1$ and find the values y_0 for which the singularity lies inside the unit circle.

(d) Use the "time stepping" numerical techniques discussed in this section (Euler, RK2 and RK4) to compute the solution on the unit circle $z = e^{i\theta}$ as θ varies from $\theta = 0$ to 4π.

3. Repeat Problem 2 above for the differential equation

$$\frac{dy}{dz} = \tfrac{1}{2}y^3, \qquad y(z_0) = y_0.$$

4. Consider the equation

$$\frac{dy}{dz} + 2zy = 1, \qquad y(1) = 1.$$

(a) Solve this equation using the series method. Evaluate the solution as we traverse the unit circle. Show that the solution is single valued.

(b) Evaluate an approximation to $y(-1)$ from the series.

(c) Show that an exact representation of the solution in terms of integrals is

$$y(z) = \int_1^z e^{t^2 - z^2} \, dt + e^{1 - z^2},$$

and verify that by evaluating $y(z)$ by a Taylor series (i.e., use $e^{t^2} = 1 + t^2 + t^4/2! + t^6/3! + t^8/4! + \cdots$) that the answer obtained from this series is a good approximation to that obtained in part (b).

Residue Calculus and Applications of Contour Integration

In this chapter we extend Cauchy's Theorem to cases where the integrand is not analytic, for example, when the integrand possesses isolated singular points. Each isolated singular point contributes a term proportional to what is called the residue of the singularity. This extension, called the residue theorem, is very useful in applications such as the evaluation of definite integrals of various types. The residue theorem provides a straightforward and sometimes the only method to compute these integrals. We also show how to use contour integration to compute the solutions of certain partial differential equations by the techniques of Fourier and Laplace transforms.

4.1 Cauchy Residue Theorem

Let $f(z)$ be analytic in the region D, defined by $0 < |z - z_0| < \rho$, and let $z = z_0$ be an isolated singular point of $f(z)$. The Laurent expansion of $f(z)$ (discussed in Section 3.3) in D is given by

$$f(z) = \sum_{n=-\infty}^{\infty} C_n(z - z_0)^n, \qquad (4.1.1)$$

with

$$C_n = \frac{1}{2\pi i} \oint_C \frac{f(z)dz}{(z - z_0)^{n+1}}, \qquad (4.1.2)$$

where C is a simple closed contour lying in D and enclosing z_0. The negative part of the series $\sum_{n=-\infty}^{-1} C_n(z - z_0)^n$ is referred to as the **principal part** of the series. The coefficient C_{-1} is called the **residue** of $f(z)$ at z_0, sometimes written as $C_{-1} = \text{Res } (f(z); z_0)$. We note that when $n = -1$, Eq. (4.1.2) yields

$$\oint_C f(z) \, dz = 2\pi i C_{-1}. \qquad (4.1.3)$$

Thus, Cauchy's Theorem is now seen to suitably generalize to functions $f(z)$ with one isolated singular point. Namely, we had previously proven that for $f(z)$ analytic in D the integral $\oint f(z)dz = 0$, where C was a closed contour in D. Equation (4.1.3) shows that the correct modification of Cauchy's Theorem, when $f(z)$ contains *one* isolated singular point at $z_0 \in D$, is that the integral be proportional to the residue (C_{-1}) of $f(z)$ at z_0. In fact, this concept is easily extended to functions with a finite number of isolated singular points. The result is often referred to as the **Cauchy Residue Theorem**, which we now state.

Theorem 4.1.1 (Cauchy Residue Theorem)   *Let $f(z)$ be analytic inside and on a simple closed contour C, except for a finite number of isolated singular points $z_1, \ldots, z_N$ located inside C. Then*

$$\oint_C f(z)\, dz = 2\pi i \sum_{j=1}^{N} a_j, \tag{4.1.4}$$

where a_j is the residue of $f(z)$ at $z = z_j$, denoted by $a_j = \text{Res}\ (f(z); z_j)$.

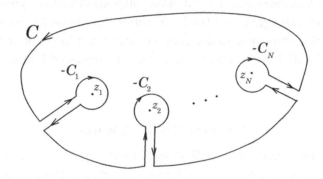

Figure 4.1 Proving Theorem 4.1.1

Proof   Consider Figure 4.1. We enclose each of the points z_j by small non-intersecting closed curves each of which lies within C : $C_1, C_2, \ldots C_N$ and is connected to the main closed contour by cross cuts. Because the integrals along the cross cuts vanish, we have that on the contour $\Gamma = C - C_1 - C_2 - \cdots - C_N$ (with each contour taken in the positive sense)

$$\int_\Gamma f(z)\, dz = 0,$$

which follows from Cauchy's Theorem. Thus

$$\oint_C f(z)\, dz = \sum_{j=1}^{N} \oint_{C_j} f(z)\, dz. \tag{4.1.5}$$

We now use the result (4.1.3) about each singular point. Because $f(z)$ has a Laurent expansion in the neighborhood of each singular point, $z = z_j$, Eq. (4.1.4) follows.   □

Some prototypical examples are described below.

Example 4.1.2   Evaluate

$$I_k = \frac{1}{2\pi i} \oint_{C_0} z^k \, dz, \qquad k \in \mathbf{Z},$$

where C_0 is the unit circle $|z| = 1$. Because z^k is analytic for $k = 0, 1, 2, \ldots$, we have $I_k = 0$ for $k = 0, 1, 2, \ldots$. Similarly, for $k = -2, -3, \ldots$ we find that the residue of z^k is zero, hence $I_k = 0$. For $k = -1$, the residue of z^{-1} is unity and thus $I_{-1} = 1$.

We write $I_k = \delta_{k,-1}$, where

$$\delta_{k,\ell} = \begin{cases} 1 & \text{when } k = \ell, \\ 0 & \text{otherwise} \end{cases}$$

is referred to as the Kronecker delta function.

Example 4.1.3

$$I = \frac{1}{2\pi i} \oint_{C_0} z \, e^{1/z} \, dz,$$

where C_0 is the unit circle $|z| = 1$. The function $f(z) = z e^{1/z}$ is analytic for all $z \neq 0$ inside C_0, and has the following Laurent expansion about $z = 0$;

$$z e^{1/z} = z \left(1 + \frac{1}{z} + \frac{1}{2! z^2} + \frac{1}{3! z^3} + \cdots \right).$$

Hence the residue $\mathrm{Res}\left(z e^{1/z}; 0 \right) = 1/2!$, and we have

$$I = \frac{1}{2}.$$

Example 4.1.4   Evaluate

$$I = \oint_{C_2} \frac{z+2}{z(z+1)} \, dz,$$

where C_2 is the circle $|z| = 2$.

We write the integrand as a partial fraction

$$\frac{z+2}{z(z+1)} = \frac{A}{z} + \frac{B}{z+1},$$

hence $z + 2 = A(z + 1) + Bz$, and we deduce (taking $z = 0$, $z = -1$) that $A = 2$ and $B = -1$. (In fact, the coefficients $A = 2$ and $B = -1$ are the residues of the function $\frac{z+2}{z(z+1)}$ at $z = 0$ and $z = -1$, respectively.) Thus

$$I = \oint_C \left(\frac{2}{z} - \frac{1}{z+1} \right) dz = 2\pi i(2 - 1) = 2\pi i,$$

where we note that the residue about $z = 0$ of $2/z$ is 2, and the residue of $1/(z + 1)$ about $z = -1$ is 1.

So far we have evaluated the residue by expanding $f(z)$ in a Laurent expansion about the point $z = z_j$. Indeed, if $f(z)$ has an essential singular point at $z = z_0$, then expansion in terms of a Laurent expansion is the only general method to evaluate the residue. If, however, $f(z)$ has a pole in the neighborhood of z_0, then there is a simple formula, which we now give.

Let $f(z)$ be defined by

$$f(z) = \frac{\phi(z)}{(z - z_0)^m}, \tag{4.1.6}$$

where $\phi(z)$ is analytic in the neighborhood of $z = z_0$, m is a positive integer, and if $\phi(z_0) \neq 0$ f has a pole of order m. Then the residue of $f(z)$ at z_0 is given by

$$C_{-1} = \frac{1}{(m-1)!} \left(\frac{d^{m-1}}{dz^{m-1}} \phi \right) (z = z_0)$$

$$= \frac{1}{(m-1)!} \left(\frac{d^{m-1}}{dz^{m-1}} (z - z_0)^m f(z) \right) (z = z_0). \tag{4.1.7}$$

(This means that one first computes the $(m - 1)^{st}$ derivative of $\phi(z)$ and then evaluates it at $z = z_0$.)

The derivation of this formula follows from the fact that if $f(z)$ has a pole of order m at $z = z_0$ then it can be written in the form (4.1.6). Because $\phi(z)$ is analytic in the neighborhood of z_0,

$$\phi(z) = \phi(z_0) + \phi'(z_0)(z - z_0) + \cdots + \frac{\phi^{(m-1)}(z_0)}{(m-1)!} (z - z_0)^{m-1} + \cdots .$$

Dividing this expression by $(z - z_0)^m$ it then follows that the coefficient of the $(z - z_0)^{-1}$ term, denoted by C_{-1}, is given by (4.1.7). (From the derivation, it also follows that Eq. (4.1.7) holds even if the order of the pole is overestimated; e.g. Eq. (4.1.7) holds even if $\phi(z_0) = 0$, $\phi'(z_0) \neq 0$, which implies the order of the pole is $m - 1$.)

For a simple pole we have $m = 1$, hence the formula

$$C_{-1} = \phi(z_0) = \lim_{z \to z_0} ((z - z_0) f(z))$$

$$\text{(simple pole)}. \tag{4.1.8}$$

Suppose our function is given by a ratio of two functions $N(z)$ and $D(z)$, where both are analytic in the neighborhood of $z = z_0$,

$$f(z) = \frac{N(z)}{D(z)}.$$ (4.1.9)

Then if $D(z)$ has a zero of order m at z_0, we may write $D(z) = (z - z_0)^m \tilde{D}(z)$, where $\tilde{D}(z_0) \neq 0$ and $\tilde{D}(z)$ is analytic near $z = z_0$. Hence $f(z)$ takes the form (4.1.6), where $\phi(z) = N(z)/\tilde{D}(z)$ and Eq. (4.1.7) applies. In the special case of a simple pole, $m = 1$, so from the Taylor series of $N(z)$ and $D(z)$, we have $N(z) = N(z_0) + (z - z_0)N'(z_0) + \cdots$, and $\tilde{D}(z) = D'(z_0) + (z - z_0)\frac{D''(z_0)}{2!} + \cdots$, whereupon $\phi(z_0) = \frac{N(z_0)}{D'(z_0)}$ and Eq. (4.1.8) yields

$$C_{-1} = \frac{N(z_0)}{D'(z_0)} \qquad \text{with } D'(z) \neq 0.$$ (4.1.10)

Special cases such as $N(z_0) = D'(z_0) = 0$ can be derived in similarly.

In the following examples we illustrate the use of formulae (4.1.7) and (4.1.10).

Example 4.1.5 Evaluate

$$I = \frac{1}{2\pi i} \oint_{C_2} \left(\frac{3z + 1}{z(z - 1)^3} \right) dz,$$

where C_2 is the circle $|z| = 2$.

The function

$$f(z) = \frac{3z + 1}{z(z - 1)^3}$$

has the form (4.1.6) near $z = 0$, $z = 1$. We have

$$\text{Res } (f(z); 0) = \left(\frac{3z + 1}{(z - 1)^3} \right)_{z=0} = -1,$$

$$\text{Res } (f(z); 1) = \frac{1}{2!} \left(\frac{d^2}{dz^2} \left(\frac{3z + 1}{z} \right) \right)_{z=1} = \frac{1}{2!} \left(\frac{d^2}{dz^2} \left(3 + \frac{1}{z} \right) \right)_{z=1}$$

$$= +1,$$

hence $I = 0$.

Example 4.1.6 Evaluate

$$I = \frac{1}{2\pi i} \oint_{C_0} \cot z \, dz,$$

where C_0 is the unit circle $|z| = 1$.

The function $\cot z = \cos z / \sin z$ is a ratio of two analytic functions, whose singularities occur at the zeroes of $\sin z : z = n\pi, n = 0, \pm 1, \pm 2, \ldots ..$ Because the contour C_0 encloses only the singularity $z = 0$, we can use formula (4.1.10) to find

$$I = \lim_{z \to 0} \frac{\cos z}{(\sin z)'} = 1.$$

Sometimes it is useful to work with the residue at infinity. The residue at infinity, Res $(f(z); \infty)$, in analogy with the case of finite isolated singular points (see Eq. (4.1.5)) is given by the formula

$$\text{Res } (f(z); \infty) = \frac{1}{2\pi i} \oint_{C_\infty} f(z) \, dz, \qquad (4.1.11a)$$

where C_∞ denotes the limit $R \to \infty$ of a circle C_R with radius $|z| = R$. For example, if $f(z)$ is analytic at infinity with $f(\infty) = 0$, it has the expansion

$$f(z) = \frac{a_{-1}}{z} + \frac{a_{-2}}{z^2} + \cdots .$$

Then we have

$$\text{Res } (f(z); \infty) = \frac{1}{2\pi i} \oint_{C_\infty} f(z) \, dz$$

$$= \lim_{R \to \infty} \frac{1}{2\pi i} \int_0^{2\pi} \left(\frac{a_{-1}}{Re^{i\theta}} + \frac{a_{-2}}{(Re^{i\theta})^2} + \cdots \right) i Re^{i\theta} \, d\theta$$

$$= a_{-1}. \qquad (4.1.11b)$$

In fact (4.1.11a,b) hold even when $f(\infty) \neq 0$, as long as $f(z)$ has a Laurent series in the neighborhood of $z = \infty$.

As mentioned earlier, it is sometimes convenient, when analyzing the behavior of a function near infinity, to make the change of variables $z = 1/t$. Using $dz = -\frac{1}{t^2} dt$ and noting that the counterclockwise (positive direction) of $C_R: z = Re^{i\theta}$ transforms to a clockwise rotation (negative direction) in $t: t = \frac{1}{z} = \frac{1}{R} e^{-i\theta} = \epsilon e^{-i\theta}$, with $\epsilon = \frac{1}{R}$, we have

$$\text{Res } (f(z); \infty) = \frac{1}{2\pi i} \oint_{C_\infty} f(z) \, dz = \frac{1}{2\pi i} \oint_{C_\epsilon} \left(\frac{1}{t^2} \right) f \left(\frac{1}{t} \right) dt, \qquad (4.1.12)$$

where C_ϵ is the limit as $\epsilon \to 0$ of a small circle ($\epsilon = 1/R$) around the origin in the t-plane. Hence the residue at ∞ is given by

$$\text{Res } (f(z); \infty) = \text{Res } \left\{ \frac{1}{t^2} f \left(\frac{1}{t} \right) ; 0 \right\};$$

that is, the right-hand side is the coefficient of t^{-1} in the expansion of $f(1/t)/t^2$ near $t = 0$; the left-hand side is the coefficient of z^{-1} in the expansion of $f(z)$ at $z = \infty$. Sometimes we write

$$\text{Res } (f(z); \infty) = \lim_{z \to \infty} (z f(z)) \qquad \text{when } f(\infty) = 0. \qquad (4.1.13)$$

The concept of residue at infinity is quite useful when we integrate rational functions. Rational functions have only isolated singular points in the extended plane and are analytic elsewhere. Let $z_1, z_2, \ldots, z_N$ denote the finite singularities. Then, *for every rational function,*

$$\sum_{j=1}^{N} \text{Res} \, (f(z); z_j) = \text{Res} \, (f(z); \infty). \tag{4.1.14}$$

This follows from an application of the residue theorem. We know that

$$\frac{1}{2\pi i} \lim_{R \to \infty} \oint_{C_R} f(z) \, dz = \sum_{j=1}^{N} \text{Res} \left(f(z); z_j \right)$$

because $f(z)$ has poles at $\{z_j\}_{j=1}^{N}$. On the other hand, because $f(z)$ is a rational function it has a Laurent series near infinity, hence we have Res $(f(z); \infty) = \frac{1}{2\pi i} \lim_{R \to \infty} \oint_{C_R} f(z) \, dz$.

We illustrate the use of the residue at infinity in the following examples.

Example 4.1.7   We reconsider Example 4.1.5, but we now use Res $(f(z); \infty)$.

We note that all the singularities of $f(z)$ lie inside C_2, and the integrand is a rational function with $f(\infty) = 0$. Thus $I = \text{Res} \, (f(z); \infty)$. Because $f(z) = 3/z^3 + \cdots$ as $z \to \infty$, we use Eq. (4.1.13) to find

$$\text{Res} \, (f(z); \infty) = \lim_{z \to \infty} \frac{(3z + 1)}{z(z - 1)^3} = 0.$$

Hence $I = 0$, as we had found by a somewhat longer calculation!

We illustrate this idea with another example.

Example 4.1.8   Evaluate

$$I = \frac{1}{2\pi i} \oint_C \frac{a^2 - z^2}{a^2 + z^2} \frac{dz}{z}$$

where C is any simple closed contour enclosing the points $z = 0$ and $z = \pm ia$.

The function

$$f(z) = \frac{a^2 - z^2}{a^2 + z^2} \frac{1}{z}$$

is a rational function with $f(\infty) = 0$, hence it has only isolated singular points. Note also that $f(z) = -1/z + \cdots$ as $z \to \infty$. Hence

$$I = \text{Res} \, (f(z); \infty).$$

We again use Eq. (4.1.13) to obtain

$$I = \lim_{z \to \infty} (z f(z)) = -1.$$

The value $w(z_j)$, defined by

$$w(z_j) = \frac{1}{2\pi i} \oint_C \frac{dz}{z - z_j} = \frac{1}{2\pi i} \left[\log(z - z_j)\right]_C = \frac{\Delta\theta_j}{2\pi}, \qquad (4.1.15)$$

is called the **winding number** of the curve C around the point z_j. Here, $\Delta\theta_j$ is the total change in the argument of $z - z_j$ when z traverses the curve C around the point z_j. The value $w(z_j)$ represents the number of times (positive means counterclockwise) that C encircles or winds around z_j.

By the process of deformation of contours, including the introduction of cross cuts and the like, one can generalize the Cauchy Residue Theorem (4.1.1) to

$$\oint_C f(z)\,dz = 2\pi i \sum_{j=1}^{N} w(z_j)a_j, \qquad a_j = \mathrm{Res}\,(f(z); z_j), \qquad (4.1.16)$$

where the hypothesis of Theorem 4.1.1 remains intact except for allowing the contour C to be nonsimple – hence the need for introducing the winding numbers $w(z_j)$ at every point $z = z_j$ with residue $a_j = \mathrm{Res}\,(f(z); z_j)$.

In applications it is usually clear how to break up a nonsimple contour into a series of simple ones; we shall not go through the formal proof in the general case. Instead, we illustrate the procedure with an example.

Example 4.1.9   Use Eq. (4.1.16) to evaluate

$$I = \oint_C \frac{dz}{z^2 + a^2},$$

for $a > 0$, where C is the nonsimple contour of Figure 4.2.

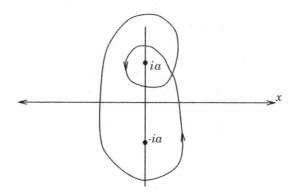

Figure 4.2 Nonsimple curve, for Example 4.1.9

The residue of $1/(z^2 + a^2)$ is

$$\mathrm{Res}\left(\frac{1}{z^2 + a^2}; \pm ia\right) = \left(\frac{1}{2z}\right)_{\pm ia} = \pm\frac{1}{2ai}.$$

We see from Figure 4.2 that the winding numbers are $w(ia) = +2$ and $w(-ia) = +1$.
Thus

$$I = 2\pi i \left[2\left(\frac{1}{2ai}\right) - \frac{1}{2ai} \right] = \frac{\pi}{a}.$$

More generally, corresponding to any two closed curves C_1 and C_2, we have

$$\oint_{C_1} \frac{dz}{z^2 + a^2} = \oint_{C_2} \frac{dz}{z^2 + a^2} + \frac{N\pi}{a} \tag{4.1.17}$$

where N is an appropriate integer related to the winding numbers of C_1 and C_2.
Note that $N\pi/a$ is intimately related to the function

$$\Phi(z) = \int_{z_0}^{z} \frac{du}{u^2 + a^2} = \frac{1}{a} \tan^{-1} \frac{z}{a} + \Phi_0,$$

where

$$\Phi_0 = \frac{-1}{a} \tan^{-1} \left(\frac{z_0}{a}\right)$$

or $z = a \tan a(\Phi - \Phi_0)$. Because z is periodic, with period $N\pi/a$, changing Φ by
$N\pi/a$ yields the same value for z, because the period of $\tan x$ is π.

Incorporating the winding numbers in Cauchy's Residue Theorem shows that, in
the general case, the difference between two contours, C_1, C_2, of a function $f(z)$
analytic inside these contours, save for a finite number of isolated singular points,
is given by

$$\left(\oint_{C_1} - \oint_{C_2} \right) f(z) dz = 2\pi i \sum_{j=1}^{N} w_j a_j.$$

The points $a_j = \text{Res}\,(f(z); z_j)$ are the periods of the inverse function $z = z(\Phi)$,
defined by

$$\Phi(z) = \int_{z_0}^{z} f(z)\, dz; \qquad z = z(\Phi).$$

4.1.1 Problems for Section 4.1

1. Evaluate the integrals $\frac{1}{2\pi i} \oint_C f(z)\, dz$, where C is the unit circle centered at the
   origin, and $f(z)$ is given by:

   (a) $\dfrac{z+1}{2z^3 - 3z^2 - 2z}$; (b) $\dfrac{\cosh(1/z)}{z}$; (c) $\dfrac{e^{-\cosh z}}{4z^2 + \pi^2}$;

   (d) $\dfrac{\log(z+2)}{2z+1}$, principal branch; (e) $\dfrac{(z + 1/z)}{z(2z - 1/2z)}$.

2. Evaluate the integrals $\frac{1}{2\pi i} \oint_C f(z) \, dz$ where C is the unit circle centered at the origin with $f(z)$ given below. Do these problems by both: (i) enclosing the singular points inside C; and (ii) enclosing the singular points outside C (by including the point at infinity). Show that you obtain the same result in both cases.

$$\text{(a)} \ \frac{z^2 + 1}{z^2 - a^2}, \quad a^2 < 1; \qquad \text{(b)} \ \frac{z^2 + 1}{z^3}; \qquad \text{(c)} \ z^2 e^{-1/z}.$$

3. Determine the type of singular point each of the following functions have at $z = \infty$:

   (a) z^m, m a positive integer;   (b) $z^{1/3}$;   (c) $(z^2 + a^2)^{1/2}, a^2 > 0$;

   (d) $\log z$;   (e) $\log(z^2 + a^2), a^2 > 0$;   (f) e^z;

   (g) $z^2 \sin \dfrac{1}{z}$;   (h) $\dfrac{z^2}{z^3 + 1}$;   (i) $\sin^{-1} z$;   (j) $\log\left(1 - e^{1/z}\right)$.

4. Let $f(z)$ be analytic outside a circle C_R enclosing the origin.

   (a) Show that

   $$\frac{1}{2\pi i} \oint_{C_R} f(z) \, dz = \frac{1}{2\pi i} \oint_{C_\rho} f\left(\frac{1}{t}\right) \frac{dt}{t^2},$$

   where C_ρ is a circle of radius $1/R$ enclosing the origin. For $R \to \infty$ conclude that the integral can be computed to be Res $(f(1/t)/t^2; 0)$.

   (b) Suppose $f(z)$ has the convergent Laurent expansion

   $$f(z) = \sum_{j=-\infty}^{-1} A_j z^j.$$

   Show that the integral above equals A_{-1}; see also Eq. (4.1.11a,b).

5. Show

   (a)

   $$\frac{1}{2\pi i} \oint_C \frac{e^{2z}}{(z - 1)^2} \, dz = 2e^2,$$

   where C is a circle radius $R > 1$ centered at $z = 0$;

   (b)

   $$\frac{1}{2\pi i} \oint_{C_0} \frac{P(z)}{z(z^2 + 4)^2} \, dz = \frac{P(0)}{16},$$

   where $P(z)$ is analytic inside C_0 which is the unit circle (unit radius, centered at $z = 0$);

(c)

$$\frac{1}{2\pi i} \oint_C \frac{\sin z}{(z - \pi/2)^3} \, dz = -1/2,$$

where C is a circle radius r, enclosing $\pi/2$.

6. (a) The following identity for Bessel functions is valid:

$$\exp\left(\frac{w}{2}(z - 1/z)\right) = \sum_{n=-\infty}^{\infty} J_n(w)z^n.$$

Show that

$$J_n(w) = \frac{1}{2\pi i} \oint_C \exp\left(\frac{w}{2}(z - 1/z)\right) \frac{dz}{z^{n+1}},$$

where C is the unit circle centered at the origin.

(b) Use

$$\exp\left(\frac{w}{2}(z - 1/z)\right) = \exp\left(\frac{w}{2}z\right)\exp\left(-\frac{w}{2z}\right),$$

multiply the two series for exponentials to compute the following series representation for the Bessel function of "zeroth" order:

$$J_0(w) = \sum_{k=0}^{\infty} (-1)^k \frac{w^{2k}}{2^{2k}(k!)^2}.$$

7. Suppose we know that, everywhere outside the circle C_R, radius R centered at the origin, the functions $f(z)$, $g(z)$ are analytic, with $\lim_{z \to \infty} f(z) = C_1$ and $\lim_{z \to \infty}(zg(z)) = C_2$, where C_1 and C_2 are constant. Show that

$$\frac{1}{2\pi i} \oint_{C_R} g(z)e^{f(z)} \, dz = C_2 e^{C_1}.$$

8. Consider the following integral:

$$I_R = \oint_{C_R} \frac{dz}{z^2 \cosh z},$$

where C_R is a square centered at the origin whose sides lie along the lines $x = \pm(R + 1)\pi$ and $y = \pm(R + 1)\pi$, where R is a positive integer. Evaluate this integral both by residues and by direct evaluation of the line integral and show that in either case $\lim_{R \to \infty} I_R = 0$, where the limit is taken over the integers. (In the direct evaluation, use estimates of the integrand. Hint: See Examples 4.2.7, 4.2.8.)

9. Suppose $f(z)$ is a meromorphic function (i.e., $f(z)$ is analytic everywhere in the finite z-plane except at isolated points where it has poles) with N simple zeroes (i.e., $f(z_0) = 0$, $f'(z_0) \neq 0$) and M simple poles inside a circle C. Show that

$$\frac{1}{2\pi i} \oint_C \frac{f'(z)}{f(z)} \, dz = N - M.$$

4.2 Evaluation of Certain Definite Integrals

We begin this section by developing methods to evaluate real integrals of the form

$$I = \int_{-\infty}^{\infty} f(x) \, dx, \tag{4.2.1}$$

where $f(x)$ is a real-valued function that will be specified later. Integrals with infinite endpoints converge depending on the existence of a limit; namely, we say that I **converges** if the two limits in

$$I = \lim_{L \to \infty} \int_{-L}^{\alpha} f(x) \, dx + \lim_{R \to \infty} \int_{\alpha}^{R} f(x) \, dx, \qquad \text{for } \alpha \text{ finite}, \tag{4.2.2}$$

exist. When evaluating integrals in complex analysis it is useful (as we will see) to consider a more restrictive limit by taking $L = R$, and this is sometimes referred to as the **Cauchy Principal Value at Infinity**, I_p:

$$I_p = \lim_{R \to \infty} \int_{-R}^{R} f(x) \, dx. \tag{4.2.3}$$

If Eq. (4.2.2) is convergent, then $I = I_p$ by simply taking as a special case $L = R$.

It is possible for I_p to exist but not the more general limit (4.2.2). For example, if $f(x)$ is odd and nonzero at infinity (e.g. $f(x) = x$), then $I_p = 0$ but I will not exist. In applications one frequently checks the convergence of I by using the usual tests of calculus and *then* one evaluates the integral via Eq. (4.2.3).

In what follows, unless otherwise explicitly stated, we shall only consider integrals with infinite limits whose convergence can be established in the sense of Eq. (4.2.2).

We first show how to evaluate integrals of the form

$$I = \int_{-\infty}^{\infty} f(x) \, dx,$$

where $f(x) = N(x)/D(x)$, where $N(x)$ and $D(x)$ are real polynomials (that is, $f(x)$ is a rational function), $D(x) \neq 0$ for $x \in \mathbb{R}$, and $D(x)$ is at least two degrees greater than the degree of $N(x)$; the latter hypothesis implies convergence of the integral. The method is to consider the integral

$$\oint_C f(z) \, dz = \int_{-R}^{R} f(x) \, dx + \int_{C_R} f(z) \, dz \tag{4.2.4}$$

(see Figure 4.3) in which C_R is a large semicircle and the contour C encloses all the singularities of $f(z)$, namely, those locations where $D(z) = 0$; that is, $z_1, z_2, \ldots, z_N$. We use Cauchy's Residue Theorem and suitable analysis showing that $\lim_{R \to \infty} \int_{C_R} f(z)\, dz = 0$ (this is true owing to the assumptions on $f(x)$ and is proven in Theorem 4.2.2), in which case from (4.2.4) we have, as $R \to \infty$,

$$\int_{-\infty}^{\infty} f(x)\, dx = 2\pi i \sum_{j=1}^{N} \operatorname{Res}\, (f(z); z_j). \tag{4.2.5}$$

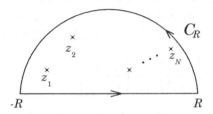

Figure 4.3 Evaluating (4.2.5), contour in the upper half plane

The integral can also be evaluated by using the closed contour in the lower half plane, shown in Figure 4.4. Note that because $D(x)$ is a real polynomial, its complex zeroes come in complex conjugate pairs.

Figure 4.4 Evaluating (4.2.5), contour in the lower half plane

We illustrate the method first by an example.

Example 4.2.1   Evaluate

$$I = \int_{-\infty}^{\infty} \frac{x^2}{x^4 + 1}\, dx.$$

We begin by establishing that the contour integral along the semicircular arc described in Eq. (4.2.4) vanishes as $R \to \infty$. Using $f(z) = z^2/(z^4 + 1)$, $z = Re^{i\theta}$, $dz = iRe^{i\theta}\, d\theta$, and $|dz| = R\, d\theta$, we have

$$\left| \int_{C_R} f(z)\, dz \right| \le \int_{\theta=0}^{\pi} \frac{|z|^2}{|z^4 + 1|}\, |dz| \le \int_{\theta=0}^{\pi} \frac{|z|^2}{|z|^4 - 1}\, |dz|$$

$$= \frac{\pi R^3}{R^4 - 1} \xrightarrow[R \to \infty]{} 0.$$

These inequalities follow from $|z^4 + 1| \ge |z|^4 - 1$, which implies $1/|z^4 + 1| \le 1/(R^4 - 1)$; we have used the integral inequalities of Chapter 2 (see, for example, Theorem 2.4.6). Thus we have shown how Eq. (4.2.5) is arrived at in this example.

The residues of the function $f(z)$ are easily calculated from Eq. (4.1.10) by noting that all poles are simple; they may be found by solving $z^4 = -1 = e^{i\pi}$, and hence there is one pole located in each of the four quadrants. We shall use the contour in Figure 4.3 so we need only the zeroes in the first and second quadrants: $z_1 = e^{i\pi/4}$ and $z_2 = e^{i(\pi/4+\pi/2)} = e^{3i\pi/4}$. Thus Eq. (4.2.5) yields

$$I = 2\pi i \left[\left(\frac{z^2}{4z^3} \right)_{z_1} + \left(\frac{z^2}{4z^3} \right)_{z_2} \right]$$

$$= \frac{2\pi i}{4} \left(e^{-i\pi/4} + e^{-3i\pi/4} \right) = \frac{\pi}{2} \left(e^{i\pi/4} + e^{-i\pi/4} \right)$$

$$= \pi \cos(\pi/4) = \pi/\sqrt{2},$$

where we have used $i = e^{i\pi/2}$. We also note that if we used the contour depicted in Figure 4.4 and evaluated the residues in the third and fourth quadrants, we would arrive at the same result – as we must.

More generally, we have the following theorem.

Theorem 4.2.2   *Let $f(z) = N(z)/D(z)$ be a rational function such that the degree of $D(z)$ exceeds the degree of $N(z)$ by at least 2. Then*

$$\lim_{R \to \infty} \int_{C_R} f(z)\, dz = 0.$$

Proof   We write

$$f(z) = \frac{a_n z^n + a_{n-1} z^{n-1} + \cdots + a_1 z + a_0}{b_m z^m + b_{m-1} z^{m-1} + \cdots + b_1 z + b_0}.$$

Then, using the same ideas as in Example 4.2.1,

$$\left| \int_C R f(z)\, dz \right| \le \int_0^{\pi} (R\, d\theta) \frac{|a_n|\,|z|^n + |a_{n-1}|\,|z|^{n-1} + \cdots + |a_1|\,|z| + |a_0|}{|b_m|\,|z|^m - |b_{m-1}|\,|z|^{m-1} - \cdots - |b_1|\,|z| - |b_0|}$$

$$= \frac{\pi R\, (\,|a_n| R^n + \cdots + |a_0|\,)}{|b_m| R^m - |b_{m-1}| R^{m-1} - \cdots - |b_0|} \xrightarrow[R \to \infty]{} 0,$$

because $m \ge n + 2$.   □

Integrals that are closely related to the one described above are of the form

$$I_1 = \int_{-\infty}^{\infty} f(x) \cos kx \, dx, \qquad I_2 = \int_{-\infty}^{\infty} f(x) \sin kx \, dx,$$

$$I_{3\pm} = \int_{-\infty}^{\infty} f(x) e^{\pm ikx} \, dx, \qquad k > 0,$$

where $f(x)$ is a rational function satisfying the conditions in Theorem 4.2.2. These integrals may be evaluated by a method similar to the ones described earlier. When evaluating integrals such as I_1 or I_2, we first replace them by integrals of the form I_3. We evaluate, say I_{3+}, by using the contour in Eq. (4.2.1). Again, we need to evaluate the integral along the upper semicircle. Because $e^{ikz} = e^{ikx} e^{-ky}$, where $z = x + iy$, we have $|e^{ikz}| \le 1$, for $y > 0$, and

$$\left| \int_{C_R} f(z) e^{ikz} \, dz \right| \le \int_0^{\pi} |f(z)| \, |dz| \xrightarrow[R \to \infty]{} 0,$$

from the results of Theorem 4.2.2. Thus, using,

$$I_{3+} = \int_{-\infty}^{\infty} f(x) e^{ikx} \, dx$$

$$= \int_{-\infty}^{\infty} f(x) \cos kx \, dx + i \int_{-\infty}^{\infty} f(x) \sin kx \, dx,$$

we have from (4.2.5), suitably modified,

$$I_{3+} = I_1 + iI_2 = 2\pi i \sum_{j=1}^{N} \text{Res} \left(f(z) e^{ikz}; z_j \right), \tag{4.2.6}$$

and hence by taking real and imaginary parts of Eq. (4.2.6), we can compute I_1 and I_2.

It should be noted that to evaluate I_{3-}, we use a semicircular contour in the lower half of the plane; that is, Figure 4.4. The calculations are similar to those before, save for the fact that we need to compute the residues in the lower half plane and we find that $I_{3-} = I_1 - iI_2 = -2\pi i \sum_{j=1}^{N} \text{Res}\ (f(z); z_j)$.

We note that in other applications one might need to consider integrals $\int_{C_R} e^{-kz} f(z) \, dz$, where C_R is a semicircle in the *right half plane*, and/or $\int_{C_L} e^{kz} f(z) \, dz$, where C_L is a semicircle in the *left half plane*. The methods needed to show that such integrals are zero as $R \to \infty$ are similar to those presented above, hence there is no need to elaborate further.

Example 4.2.3 Evaluate

$$I = \int_{-\infty}^{\infty} \frac{\cos kx}{(x+b)^2 + a^2} \, dx, \qquad k > 0,\ a > 0,\ b \text{ real}.$$

We consider

$$I_+ = \int_{-\infty}^{\infty} \frac{e^{ikx}}{(x+b)^2 + a^2} \, dx$$

and use the contour in Figure 4.3 and Theorem 4.2.2 to find

$$I_+ = 2\pi i \, \text{Res} \left(\frac{e^{ikz}}{(z+b)^2 + a^2} ; z_0 = ia - b \right), \qquad a > 0,$$

$$= 2\pi i \left(\frac{e^{ikz}}{2(z+b)} \right)_{z_0 = ia - b} = \frac{\pi}{a} e^{-ka} e^{-ibk} \quad a, k > 0.$$

Using

$$I_+ = \int_{-\infty}^{\infty} \frac{\cos kx}{(x+b)^2 + a^2} \, dx + i \int_{-\infty}^{\infty} \frac{\sin kx}{(x+b)^2 + a^2} \, dx,$$

we have

$$I = \frac{\pi}{a} e^{-ka} \cos bk$$

and

$$J = \int_{-\infty}^{\infty} \frac{\sin kx}{(x+b)^2 + a^2} \, dx = \frac{-\pi}{a} e^{-ka} \sin bk.$$

If $b = 0$, the latter formula reduces to $J = 0$, which also follows directly from the fact that the integrand is odd. The reader can also verify that

$$\left| \int_{C_R} \frac{e^{ikz}}{(z+b)^2 + a^2} dz \right| \le \int_{C_R} \frac{|dz|}{|z|^2 - 2|b| \, |z| - a^2 - b^2}$$

$$= \frac{\pi R}{R^2 - 2bR - (a^2 + b^2)} \xrightarrow[R \to \infty]{} 0.$$

In applications we frequently wish to evaluate integrals like $I_{3\pm}$ involving $f(x)$ for which all that is known is $f(x) \to 0$ as $|x| \to \infty$. From calculus we know that in these cases the integral still converges, conditionally, but our estimates in leading to Eq. (4.2.6) must be made more carefully. We say that $f(z) \to 0$ **uniformly** as $R \to \infty$ in C_R if $|f(z)| \le K_R$, where K_R depends only on R (not on $\arg z$) and $K_R \to 0$ as $R \to \infty$. We have the following result, called Jordan's Lemma.

Lemma 4.2.4 (Jordan's Lemma)   *Suppose that on the circular arc C_R in Figure 4.3 we have $f(z) \to 0$ uniformly as $R \to \infty$. Then*

$$\lim_{R \to \infty} \int_{C_R} e^{ikz} f(z) \, dz = 0, \qquad k > 0.$$

Proof With $|f(z)| \leq K_R$, where K_R is independent of θ and $K_R \to 0$ as $R \to \infty$, we have

$$I = \left| \int_{C_R} e^{ikz} f(z)\, dz \right| \leq \int_0^\pi e^{-ky} K_R\, R\, d\theta.$$

Using $y = R \sin \theta$, and $\sin(\pi - \theta) = \sin \theta$, we obtain

$$\int_0^\pi e^{-ky}\, d\theta = \int_0^\pi e^{-kR \sin \theta}\, d\theta = 2 \int_0^{\pi/2} e^{-kR \sin \theta}\, d\theta.$$

But in the region $0 \leq \theta \leq \pi/2$ we also have the estimate $\sin \theta \geq 2\theta/\pi$ (see Figure 4.5).

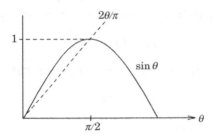

Figure 4.5 Jordan's Lemma

Thus

$$I \leq 2K_R R \int_0^{\pi/2} e^{-2kR\theta/\pi}\, d\theta = \frac{2K_R R\pi}{2kR} \left(1 - e^{-kR} \right)$$

and $I \to 0$ as $R \to \infty$, because $K_R \to 0$. $\qquad\qquad\square$

We note that if $k < 0$, a similar result holds for the contour in Figure 4.4:

Moreover, by suitably rotating the contour, Jordan's Lemma applies to the cases $k = i\ell$ for $\ell \neq 0$; see, for example, the contours given in Figure 4.23 in Section 4.5. Consequently, Eq. (4.2.6) follows whenever Jordan's Lemma applies.

Jordan's Lemma is used in the following example.

Example 4.2.5 Evaluate

$$I = 2 \int_{-\infty}^{\infty} \frac{x \sin \alpha x \cos \beta x}{x^2 + \gamma^2}\, dx, \qquad \gamma > 0, \alpha, \beta \text{ real.}$$

The trigonometric formula

$$\sin \alpha x \cos \beta x = \frac{1}{2} \left[\sin(\alpha - \beta)x + \sin(\alpha + \beta)x \right]$$

motivates the introduction of the integrals

$$J = \int_{-\infty}^{\infty} \frac{xe^{i(\alpha-\beta)x}}{x^2 + \gamma^2}\, dx + \int_{-\infty}^{\infty} \frac{xe^{i(\alpha+\beta)x}}{x^2 + \gamma^2}\, dx$$

$$= J_1 + J_2.$$

Jordan's Lemma applies because the function $f(z) = z/(z^2 + \gamma^2) \to 0$ uniformly as $z \to \infty$ and we note that,

$$|f| \le \frac{R}{(R^2 - \gamma^2)} \equiv K_R.$$

In this case it is necessary to appeal to Jordan's Lemma because the denominator is only one degree higher than the numerator. If $\alpha - \beta > 0$, then we close our contour in the upper half plane and the only residue is $z = i\gamma$, with $\gamma > 0$, hence

$$J_1 = i\pi e^{-(\alpha-\beta)\gamma}.$$

On the other hand, if $\alpha - \beta < 0$, we close in the lower half plane and find

$$J_1 = -i\pi e^{(\alpha-\beta)\gamma}.$$

Combining these results gives

$$J_1 = i\pi\, \text{sgn}\,(\alpha - \beta)e^{-|\alpha-\beta|\gamma}.$$

Similarly, for I_2 we find

$$J_2 = i\pi\, \text{sgn}\,(\alpha + \beta)e^{-|\alpha+\beta|\gamma}.$$

Thus

$$J = i\pi\left[\text{sgn}\,(\alpha - \beta)e^{-|\alpha-\beta|\gamma} + \text{sgn}\,(\alpha + \beta)e^{-|\alpha+\beta|\gamma}\right],$$

and, by taking the imaginary part,

$$I = \pi\left[\text{sgn}\,(\alpha - \beta)e^{-|\alpha-\beta|\gamma} + \text{sgn}\,(\alpha + \beta)e^{-|\alpha+\beta|\gamma}\right].$$

If we take $\text{sgn}\,(0) = 0$ then the case $\alpha = \beta$ is incorporated in this result. This could either be established directly using $\sin \alpha x \cos \alpha x = \frac{1}{2}\sin 2\alpha x$, or by noting that $J_1 = 0$ due to the oddness of the integrand. This is a consequence of employing the Cauchy Principal Value integral. (Note that the integral I is convergent.)

We now consider a class of real integrals of the following type:

$$I = \int_0^{2\pi} f(\sin\theta, \cos\theta)\, d\theta,$$

where $f(x, y)$ is a rational function of x, y. We make the substitution

$$z = e^{i\theta}, \qquad dz = ie^{i\theta}\, d\theta.$$

Then, using $\cos\theta = (e^{i\theta} + e^{-i\theta})/2$ and $\sin\theta = (e^{i\theta} - e^{-i\theta})/2i$, we have

$$\cos\theta = (z + 1/z)/2, \qquad \sin\theta = (z - 1/z)/2i. \qquad (4.2.7)$$

Thus

$$\int_0^{2\pi} d\theta\, f(\sin\theta, \cos\theta) = \oint_{C_0} \frac{dz}{iz}\, f\left(\frac{z - 1/z}{2i}, \frac{z + 1/z}{2}\right),$$

where C_0 is the unit circle $|z| = 1$. Using the residue theorem gives

$$I = \oint_{C_0} f\left(\frac{z - 1/z}{2i}, \frac{z + 1/z}{2}\right)\left(\frac{dz}{iz}\right)$$

$$= 2\pi i \sum_{j=1}^{N} \mathrm{Res}\left(\frac{f\left(\frac{z-1/z}{2i}, \frac{z+1/z}{2}\right)}{iz}; z_j\right).$$

The fact that $f(x, y)$ is a rational function of x, y implies that the residue calculation amounts to finding the zeroes of a polynomial.

Example 4.2.6 Evaluate

$$I = \int_0^{2\pi} \frac{d\theta}{A + B\sin\theta}, \qquad A^2 > B^2, \quad A > 0.$$

Employing the substitution (4.2.7), with C_0 the unit circle $|z| = 1$, and assuming, for now, that $B \ne 0$, yields

$$I = \oint_{C_0} \frac{dz}{iz}\, \frac{1}{\left(A + B\left(\frac{z-1/z}{2i}\right)\right)} = \oint_{C_0} \frac{2\,dz}{2iAz + B(z^2 - 1)}$$

$$= \frac{2}{B} \oint_{C_0} \frac{dz}{z^2 + 2i\frac{A}{B}z - 1}.$$

The roots of the denominator z_1 and z_2 which satisfy $(z - z_1)(z - z_2) = z^2 + 2iAz/B - 1 = 0$ are given by

$$z_1 = -i\frac{A}{B} + i\sqrt{\left(\frac{A}{B}\right)^2 - 1} = \frac{-iA + i\sqrt{A^2 - B^2}}{B},$$

$$z_2 = -i\frac{A}{B} - i\sqrt{\left(\frac{A}{B}\right)^2 - 1} = \frac{-iA - i\sqrt{A^2 - B^2}}{B}.$$

Because $z_1 z_2 = -1$, we find that $|z_1|\,|z_2| = 1$; hence if one root is inside C_0, the other is outside. Because $A^2 - B^2 > 0$, and $A > 0$, it follows that $|z_1| < |z_2|$;

hence z_1 lies inside. Thus, computing the residue of the integral, we have, from Eq. (4.1.8),

$$I = 2\pi i \left(\frac{2}{B}\right)\left(\frac{1}{z_1 - z_2}\right)$$

$$= \frac{4\pi i}{B}\frac{B}{2i\sqrt{A^2 - B^2}} = \frac{2\pi}{\sqrt{A^2 - B^2}};$$

the value of I when $B = 0$ is $2\pi/A$. We note that we also have computed

$$I = \int_0^{2\pi} \frac{d\theta}{A + B\cos\theta},$$

simply by making the substitution $\theta = \pi/2 + \phi$ inside the original integral.

As another illustration of the residue theorem and calculation of integrals, we describe how to obtain a "pole" expansion of a function via a contour integral.

Example 4.2.7   Evaluate

$$I = \frac{1}{2\pi i}\oint_C \frac{\pi\cot\pi\zeta}{z^2 - \zeta^2}\,d\zeta, \qquad z^2 \neq 0, 1^2, 2^2, 3^2, \ldots,$$

where C is the contour given by the rectangle $(-N - \frac{1}{2}) \leq x \leq (N + \frac{1}{2}), -N \leq y \leq N$ (see Figure 4.6). We will show that it implies

$$\pi\cot\pi z = z\sum_{n=-\infty}^{\infty}\frac{1}{z^2 - n^2}$$

$$= z\left(\frac{1}{z^2} + \frac{2}{z^2 - 1^2} + \frac{2}{z^2 - 2^2} + \cdots\right) \qquad z \neq 0, \pm 1, \pm 2, \ldots. \qquad (4.2.8)$$

We take N sufficiently large so that z lies inside C. The poles are located at $\zeta = n = 0, \pm 1, \pm 2, \ldots, \pm N$, and at $\zeta = \pm z$; hence

$$I = \sum_{n=-N}^{N}\pi\left(\frac{\cos\pi\zeta}{\pi\cos\pi\zeta}\frac{1}{z^2 - \zeta^2}\right)_{\zeta=n}$$

$$+ \pi\left(\frac{\cot\pi\zeta}{-2\zeta}\right)_{\zeta=z} + \pi\left(\frac{\cot\pi\zeta}{-2\zeta}\right)_{\zeta=-z}$$

$$= \sum_{n=-N}^{N}\frac{1}{z^2 - n^2} - \pi\frac{\cot\pi z}{z}.$$

Next we estimate the contour integral on the vertical sides, $\zeta = \pm(N + \frac{1}{2}) + i\eta$. Here the integrand satisfies

$$\left|\frac{\pi\cot\pi\zeta}{z^2 - \zeta^2}\right| \leq \frac{\pi|\tanh\pi\eta|}{|\zeta|^2 - |z|^2} \leq \frac{\pi}{N^2 - |z|^2}$$

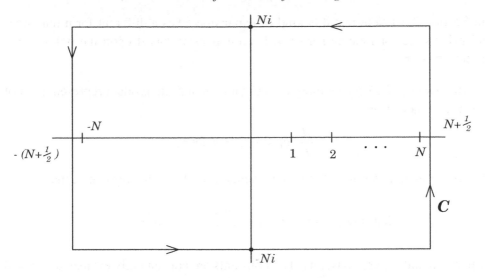

Figure 4.6 Rectangular contour C for Example 4.2.7

because $|\zeta| > N$, $|\tanh \eta| \le 1$ and

$$|\cot \pi \zeta| = \left| \frac{\sin\left[\pi \left(N + \frac{1}{2}\right)\right] (\sinh \pi \eta)(-i)}{\sin\left[\pi \left(N + \frac{1}{2}\right)\right] (\cosh \pi \eta)} \right|.$$

On the horizontal sides $\zeta = \xi \pm iN$, and the integral satisfies

$$\left| \frac{\pi \cot \pi \zeta}{z^2 - \zeta^2} \right| \le \frac{\pi \coth \pi N}{|\zeta|^2 - |z|^2} \le \frac{\pi \coth \pi N}{N^2 - |z|^2}$$

because $|\zeta| > N$ and

$$|\cot \pi \zeta| = \left| \frac{e^{\mp \pi N} e^{i\pi \xi} + e^{\pm \pi N} e^{-i\pi \xi}}{e^{\mp \pi N} e^{i\pi \xi} - e^{\pm \pi N} e^{-i\pi \xi}} \right|$$

$$\le \frac{e^{\pi N} + e^{-\pi N}}{e^{\pi N} - e^{-\pi N}} = \coth \pi N.$$

Thus

$$|I| \le \frac{1}{2\pi i} \oint_C \left| \frac{\pi \cot \pi \zeta}{z^2 - \zeta^2} \right| |d\zeta|$$

$$\le \frac{1}{2\pi} \frac{2(2N)\pi}{N^2 - |z|^2} + \frac{1}{2\pi} \frac{2(2N+1)\pi \coth \pi N}{N^2 - |z|^2} \xrightarrow[N \to \infty]{} 0,$$

since $\coth \pi N \to 1$ as $N \to \infty$. Hence we recover Eq. (4.2.8) in the limit $N \to \infty$. Formula (4.2.8) is referred to as a Mittag-Leffler expansion of the function $\pi \cot \pi z$. (The interested reader will find a discussion of Mittag-Leffler expansions in Section 3.6.) Note that this kind of expansion takes a different form than does a Taylor series or Laurent series. It is an expansion based upon the poles of the function $\cot \pi z$.

The result (4.2.8) can be integrated to yield an infinite product representation of $\sin \pi z$. Namely, from

$$\frac{d}{dz} \log \sin \pi z = \pi \cot \pi z$$

it follows by integration (taking the principal branch for the logarithm) that

$$\log \sin \pi z = \log z + A_0 + \sum_{n=1}^{\infty} \left(\log(z^2 - n^2) - A_n \right),$$

where A_0 and A_n are constants. The constants are conveniently evaluated at $z = 0$ by noting that $\lim_{z \to 0} \log \frac{\sin \pi z}{z} = \log \pi$. Thus $A_0 = \log \pi$, and $A_n = \log(-n^2)$; hence taking the exponential yields

$$\frac{\sin \pi z}{\pi} = z \prod_{n=1}^{\infty} \left(1 - \frac{z^2}{n^2} \right). \tag{4.2.9}$$

This is an example of the so-called Weierstrass Factor Theorem, discussed in Section 3.6.

It turns out that Eq. (4.2.8) can also be obtained by evaluation of a different integral, a fact that is not immediately apparent. We illustrate this in the following example.

Example 4.2.8   Evaluate

$$I = \frac{1}{2\pi i} \oint_C \pi \cot \pi \zeta \left(\frac{1}{\zeta} - \frac{1}{\zeta - z} \right) d\zeta \qquad z \neq 0, \pm 1, \pm 2, \dots,$$

where C is the same contour as in Example 4.2.7 and is depicted in Figure 4.6.

Residue calculation yields

$$I = ((-)\pi \cot \pi \zeta)_{\zeta=z} + \sum_{n=-N}^{N} {}' \left\{ \frac{\pi \cos \pi \zeta}{\pi \cos \pi \zeta} \left(\frac{1}{\zeta} - \frac{1}{\zeta - z} \right) \right\}_{\zeta=n, \, n\neq0}$$

$$+ \left(\frac{\pi \cos \pi \zeta}{\pi \cos \pi \zeta} \frac{(-)}{\zeta - z} \right)_{\zeta=0}$$

$$= -\pi \cot \pi z + \sum_{n=-N}^{N} {}' \left(\frac{1}{z-n} + \frac{1}{n} \right) + \frac{1}{z},$$

where $\sum_{n=-N}^{N}{}'$ means we omit the $n = 0$ contribution. We also note that the contribution from the double pole at $\zeta = 0$ vanishes, because $\cot \pi\zeta/\zeta \sim 1/(\pi\zeta^2) - \pi/3 + \cdots$ as $\zeta \to 0$.

Finally, we estimate the integral I on the boundary in the same manner as in Example (4.2.7) to find

$$|I| \le \frac{1}{2\pi} \oint_C |\pi \cot \pi\zeta| \frac{|z|}{|\zeta|(|\zeta| - |z|)} |d\zeta|$$

$$\le \frac{|z|}{2\pi} \left(\frac{4N\pi}{N(N - |z|)} + \frac{2(2N + 1)\pi \coth \pi N}{N(N - |z|)} \right)$$

$$\to 0 \quad \text{as} \quad N \to \infty.$$

Hence

$$\pi \cot \pi z = \sum_{n=-\infty}^{\infty}{}' \left(\frac{1}{z - n} + \frac{1}{n} \right) + \frac{1}{z}. \qquad (4.2.10)$$

Note that the expansion (4.2.10) has a suitable "convergence factor" $(1/n)$ inside the sum, otherwise it would diverge. When we combine the terms appropriately for $n = \pm 1, \pm 2, \ldots$, we find

$$\pi \cot \pi z = \frac{1}{z} + \frac{1}{z - 1} + \frac{1}{1} + \frac{1}{z + 1} - \frac{1}{1} + \frac{1}{z - 2} + \frac{1}{2} + \frac{1}{z + 2} - \frac{1}{2} + \cdots$$

$$= \frac{1}{z} + \frac{2z}{z^2 - 1} + \frac{2z}{z^2 - 2^2} + \cdots = \sum_{n=-\infty}^{\infty} \frac{z}{z^2 - n^2},$$

which is Eq. (4.2.8).

When employing contour integration sometimes it is necessary to employ special properties of the integrand, as is illustrated next.

Example 4.2.9 Evaluate

$$I = \int_0^\infty \frac{dx}{x^3 + a^3} \qquad a > 0.$$

Because we have an integral on $(0, \infty)$, we cannot immediately use a contour like that of Figure 4.3. If the integral was $\int_0^\infty f(x)\, dx$ where $f(x)$ was an even function, i.e., $f(x) = f(-x)$, then we would have

$$\int_0^\infty f(x)\, dx = \frac{1}{2} \int_{-\infty}^\infty f(x)\, dx.$$

However, in this case the integrand is not even, and, for $x < 0$, has a singularity. Nevertheless there is a symmetry that can be employed, namely, $(xe^{2\pi i/3})^3 = x^3$. This suggests using the contour of Figure 4.7, where C_R is the sector $R\, e^{i\theta} : 0 \le \theta \le 2\pi/3$.

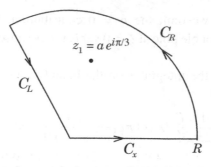

Figure 4.7 Contour for Example 4.2.9

We therefore have

$$\oint_C \frac{dz}{z^3 + a^3} = \left(\int_{C_L} + \int_{C_x} + \int_{C_R} \right) \frac{dz}{z^3 + a^3}$$

$$= 2\pi i \sum_j \operatorname{Res} \left(\frac{1}{z^3 + a^3} ; z_j \right).$$

The only pole inside C satisfies $z^3 = -a^3 = a^3 e^{i\pi}$ and is given by $z_1 = a e^{i\pi/3}$. The residue is obtained from

$$\operatorname{Res} \left(\frac{1}{z^3 + a^3} ; z_1 \right) = \left(\frac{1}{3z^2} \right)_{z_1} = \frac{1}{3a^2 e^{2\pi i/3}} = \frac{1}{3a^2} e^{-2\pi i/3}.$$

The integral on C_R tends to zero because of Theorem 4.2.2. Alternatively, by direct calculation,

$$\left| \int_{C_R} \frac{dz}{z^3 + a^3} \right| \leq \frac{2\pi R}{3(R^3 - a^3)} \to 0, \quad R \to \infty.$$

The integral on C_L is evaluated by making the substitution $z = e^{2\pi i/3} r$ (where the orientation is taken into account):

$$\int_{C_L} \frac{dz}{z^3 + a^3} = \int_{r=R}^0 \frac{e^{2\pi i/3}}{r^3 + a^3} dr = -e^{2\pi i/3} I.$$

Thus taking into account the contributions from C_x ($0 \leq z = x \leq R$) and from C_L, we have

$$I \left(1 - e^{2\pi i/3} \right) = \lim_{R \to \infty} \int_0^R \frac{dr}{r^3 + a^3} \left(1 - e^{2\pi i/3} \right) = \frac{2\pi i}{3a^2} e^{-2\pi i/3}.$$

Thus,

$$I = \frac{2\pi i}{3a^2} \frac{e^{-2\pi i/3}}{1 - e^{2\pi i/3}} = \frac{\pi}{3a^2} \left(\frac{2i}{e^{-i\pi/3} - e^{i\pi/3}} \right) e^{-i\pi} = \frac{\pi}{3a^2 \sin \pi/3}$$

$$= \frac{2\pi}{3\sqrt{3}a^2}.$$

The following example, similar in spirit to Eq. (4.2.7), allows us to calculate the following conditionally convergent integrals:

$$C = \int_0^\infty \cos(tx^2)\, dx, \qquad (4.2.11)$$

$$S = \int_0^\infty \sin(tx^2)\, dx. \qquad (4.2.12)$$

Example 4.2.10   Evaluate

$$I = \int_0^\infty e^{itx^2}\, dx.$$

For convenience we take $t > 0$. Consider the contour depicted in Figure 4.8 where the contour C_R is the sector $R e^{i\theta} : 0 \le \theta \le \pi/4$. Because e^{itz^2} is analytic inside $C = C_x + C_R + C_L$, we have

$$\oint_C e^{itz^2}\, dz = \left(\int_{C_L} + \int_{C_x} + \int_{C_R} \right) e^{itz^2}\, dz = 0.$$

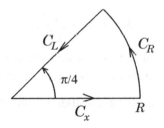

Figure 4.8   Contour for Example 4.2.10

The integral on C_R is estimated using the same idea as in Jordan's Lemma, namely $\sin \theta \ge 2\theta/\pi$ for $0 \le \theta \le \pi/2$:

$$\left| \int_{C_R} e^{itz^2}\, dz \right| = \left| \int_0^{\pi/4} e^{it R^2 (\cos 2\theta + i \sin 2\theta)} R e^{i\theta} i\, d\theta \right|$$

$$\le \int_0^{\pi/4} R e^{-t R^2 \sin 2\theta}\, d\theta \le \int_0^{\pi/4} R e^{-t R^2 \frac{4\theta}{\pi}}\, d\theta$$

$$= \frac{\pi}{4tR} \left(1 - e^{-tR^2} \right),$$

where we used $\sin x \ge \frac{2x}{\pi}$ for $0 < x < \frac{\pi}{2}$. Thus $\left| \int_{C_R} e^{itz^2}\, dz \right| \to 0$ as $R \to \infty$.

Hence on C_x, $z = x$, and on C_L, $z = r\,e^{i\pi/4}$;

$$\int_{C_x} e^{itz^2}\, dz = \int_0^R e^{itx^2}\, dx,$$

$$\int_{C_L} e^{itz^2}\, dz = \int_R^0 e^{-tr^2}\, dr\, e^{i\pi/4}.$$

Thus

$$I = \int_0^\infty e^{itx^2}\, dx = e^{i\pi/4} \int_0^\infty e^{-tr^2}\, dr,$$

and this transforms I to a well-known real definite integral which can be evaluated directly. We use polar coordinates

$$J^2 = \left(\int_0^\infty e^{-tx^2}\, dx \right)^2 = \int_0^\infty \int_0^\infty e^{-t(x^2+y^2)}\, dx\, dy = \int_{\theta=0}^{\pi/2} \int_{\rho=0}^\infty e^{-t\rho^2} \rho\, d\rho\, d\theta = \frac{\pi}{4t}.$$

Thus, taking $R \to \infty$,

$$J = \int_0^\infty e^{-tx^2}\, dx = \frac{1}{2}\sqrt{\frac{\pi}{t}} \qquad (4.2.13)$$

and

$$I = e^{i\pi/4} \frac{1}{2}\sqrt{\frac{\pi}{t}} = \left(\cos\frac{\pi}{4} + i\sin\frac{\pi}{4} \right) \frac{1}{2}\sqrt{\frac{\pi}{t}}.$$

Hence Eqs. (4.2.11) and (4.2.12) are found to be

$$S = C = \frac{1}{2}\sqrt{\frac{\pi}{2t}}. \qquad (4.2.14)$$

Incidentally, it should be noted that we cannot evaluate the integral I in the same way (via polar coordinates) we do on J, because I is not absolutely convergent.

The following example exhibits yet another variant of contour integration.

Example 4.2.11   Evaluate

$$I = \int_{-\infty}^\infty \frac{e^{px}}{1 + e^x}\, dx$$

for $0 < \operatorname{Re} p < 1$. The condition on p is required for convergence of the integral. Consider the contour depicted in Figure 4.9.

We have

$$\oint_C \frac{e^{pz}}{1 + e^z}\, dz = \left(\int_{C_x} + \int_{C_{SR}} + \int_{C_{SL}} + \int_{C_T} \right) \frac{e^{pz}}{1 + e^z}\, dz$$

$$= 2\pi i \sum_j \operatorname{Res} \left(\frac{e^{pz}}{1 + e^z}; z_j \right).$$

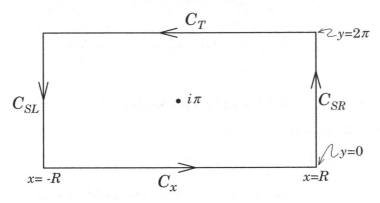

Figure 4.9 Contour for Example 4.2.11

The only poles of the function $e^{pz}/(1 + e^z)$ occur when $e^z = -1$ or, by taking the logarithm, when $z = i(\pi + 2n\pi)$, with $n = 0, \pm 1, \pm 2, \dots$. The contour is chosen such that $z = x + iy$, for $0 \le y \le 2\pi$; hence the only pole inside the contour is $z = i\pi$, with the residue

$$\text{Res} \left(\frac{e^{pz}}{1 + e^z}; i\pi \right) = \left(\frac{e^{pz}}{e^z} \right)_{z=i\pi} = e^{(p-1)i\pi}.$$

The integrals along the sides are readily estimated and shown to vanish as $R \to \infty$. Indeed on C_{SR} we have $z = R + iy$, $0 \le y \le 2\pi$

$$\left| \int_{C_{SR}} \frac{e^{pz}}{1 + e^z} \, dz \right| = \left| \int_0^{2\pi} \frac{e^{p(R+iy)}}{1 + e^{R+iy}} \, i \, dy \right| \le \frac{e^{pR}}{e^R - 1} 2\pi \to 0, \ R \to \infty, \ \text{Re } p < 1.$$

On C_{SL} we have $z = -R + iy$, with $0 \le y \le 2\pi$, hence

$$\left| \int_{C_{SL}} \frac{e^{pz}}{1 + e^z} \, dz \right| = \left| \int_{2\pi}^0 \frac{e^{p(-R+iy)}}{1 + e^{-R+iy}} \, i \, dy \right| \le \frac{e^{-pR}}{1 - e^{-R}} 2\pi \to 0, \ R \to \infty, \ \text{Re } p > 0.$$

The integral at the top has $z = x + 2\pi i$, $e^z = e^x$, so

$$\int_{C_T} \frac{e^{pz}}{1 + e^z} \, dz = e^{2\pi i p} \int_{+R}^{-R} \frac{e^{px}}{1 + e^x} \, dx.$$

Hence, combining the above computations we have, as $R \to \infty$,

$$\int_{-\infty}^{\infty} \frac{e^{px}}{1 + e^x} \, dx \left(1 - e^{2\pi i p} \right) = 2\pi i \, e^{(p-1)i\pi}$$

or

$$\int_{-\infty}^{\infty} \frac{e^{px}}{1 + e^x} \, dx = 2\pi i \frac{e^{-i\pi}}{e^{-ip\pi} - e^{ip\pi}} = \frac{\pi}{\sin p\pi}.$$

4.2.1 Problems for Section 4.2

1. Evaluate the following real integrals.

(a) $\displaystyle\int_0^{\infty} \frac{dx}{x^2 + a^2}, \quad a^2 > 0.$

Verify your answer by using the usual antiderivatives.

(b) $\displaystyle\int_0^{\infty} \frac{dx}{(x^2 + a^2)^2}, \quad a^2 > 0.$

(c) $\displaystyle\int_0^{\infty} \frac{dx}{(x^2 + a^2)(x^2 + b^2)}, \quad a^2, b^2 > 0.$

(d) $\displaystyle\int_0^{\infty} \frac{dx}{x^6 + 1}.$

2. Evaluate the following real integrals by residue integration.

(a) $\displaystyle\int_{-\infty}^{\infty} \frac{x \sin x}{(x^2 + a^2)} \, dx, \quad a^2 > 0.$

(b) $\displaystyle\int_{-\infty}^{\infty} \frac{\cos kx}{(x^2 + a^2)(x^2 + b^2)} \, dx, \quad a^2, b^2, k > 0.$

(c) $\displaystyle\int_{-\infty}^{\infty} \frac{x \cos kx}{x^2 + 4x + 5} \, dx, \quad k > 0.$

(d) $\displaystyle\int_0^{\infty} \frac{\cos kx}{x^4 + 1} \, dx, \quad k \text{ real.}$

(e) $\displaystyle\int_0^{\infty} \frac{x^3 \sin kx}{x^4 + a^4} \, dx, \quad k \text{ real, } a^4 > 0.$

(f) $\displaystyle\int_0^{2\pi} \frac{d\theta}{1 + \cos^2 \theta}.$

(g) $\displaystyle\int_0^{\pi/2} \sin^4 \theta \, d\theta.$

(h) $\displaystyle\int_0^{2\pi} \frac{d\theta}{(5 - 3 \sin \theta)^2}.$

(i) $\displaystyle\int_{-\infty}^{\infty} \frac{\cos kx \cos mx}{(x^2 + a^2)} \, dx, \quad a^2 > 0, k, m \text{ real.}$

3. Show

$$\int_0^{2\pi} \cos^{2n} \theta \, d\theta = \int_0^{2\pi} \sin^{2n} \theta \, d\theta = \frac{2\pi}{2^{2n}} B_n, \qquad n = 1, 2, 3, \ldots,$$

where

$$B_n = 2^{2n} \frac{1 \cdot 3 \cdot 5 \cdots (2n - 1)}{2 \cdot 4 \cdot 6 \cdots (2n)}.$$

Hint: use the fact that in the binomial expansion of $(1 + w)^{2n}$ the coefficient of the term w^n is B_n.

4. Show that

$$\int_0^\infty \frac{\cosh ax}{\cosh \pi x} \, dx = \frac{1}{2} \sec \left(\frac{a}{2} \right), \qquad |a| < \pi.$$

Hint: use a rectangular contour with corners at $\pm R$ and $\pm R + i$.

5. Consider a rectangular contour with corners at $b \pm iR$ and $b + 1 \pm iR$. Use this contour to show that

$$\lim_{R \to \infty} \frac{1}{2\pi i} \int_{b-iR}^{b+iR} \frac{e^{az}}{\sin \pi z} \, dz = \frac{1}{\pi(1 + e^{-a})},$$

where $0 < b < 1$, $|\operatorname{Im} a| < \pi$.

6. Consider a rectangular contour C_R with corners at $(\pm R, 0)$ and $(\pm R, a)$. Show that

$$\oint_{C_R} e^{-z^2} \, dz = \int_{-R}^R e^{-x^2} \, dx - \int_R^R e^{-(x+ia)^2} \, dx + J_R = 0,$$

where

$$J_R = \int_0^a e^{-(R+iy)^2} i \, dy - \int_0^a e^{-(-R+iy)^2} i \, dy.$$

Show that $\lim_{R \to \infty} J_R = 0$, whereupon we have

$$\int_{-\infty}^\infty e^{-(x+ia)^2} \, dx = \int_{-\infty}^\infty e^{-x^2} \, dx = \sqrt{\pi},$$

and consequently deduce that

$$\int_{-\infty}^\infty e^{-x^2} \cos 2ax \, dx = \sqrt{\pi} e^{-a^2}.$$

7. Use a sector contour with radius R as in Figure 4.8, centered at the origin with angle $0 \le \theta \le \frac{2\pi}{5}$, to find, for $a > 0$,

$$\int_0^\infty \frac{dx}{x^5 + a^5} = \frac{\pi}{5a^4 \sin \frac{\pi}{5}}.$$

8. Consider the contour integral

$$I(N) = \frac{1}{2\pi i} \oint_{C(N)} \frac{\pi \csc \pi \zeta}{z^2 - \zeta^2} \, d\zeta,$$

where the contour $C(N)$ is the rectangular contour depicted in Figure 4.6 (see also Example 4.2.7).

(a) Show that calculation of the residues implies that

$$I(N) = \sum_{n=-N}^{N} \frac{(-1)^n}{z^2 - n^2} - \frac{\pi \csc \pi z}{z}.$$

(b) Estimate the line integral along the boundary and show that $\lim_{N \to \infty} I(N) = 0$, and consequently that

$$\pi \csc \pi z = z \sum_{n=-\infty}^{\infty} \frac{(-1)^n}{z^2 - n^2}.$$

(c) Use the result of part (b) to obtain the following representation of π:

$$\pi = 2 \sum_{n=-\infty}^{\infty} \frac{(-1)^n}{1 - 4n^2}.$$

9. Consider a rectangular contour with corners $(N + \frac{1}{2})(\pm 1 \pm i)$, to evaluate

$$\frac{1}{2\pi i} \oint_{C(N)} \frac{\pi \cot \pi z \coth \pi z}{z^3} \, dz,$$

and in the limit as $N \to \infty$, show that

$$\sum_{n=1}^{\infty} \frac{\coth n\pi}{n^3} = \frac{7}{180} \pi^3.$$

Hint: note

$$\text{Res} \left(\frac{\pi \cot \pi z \coth \pi z}{z^3} ; 0 \right) = -\frac{7\pi^3}{45}.$$

4.3 Indented Contours, Principal Value Integrals, and Integrals with Branch Points

4.3.1 *Principal Value integrals*

In Section 4.2 we introduced the notion of the Cauchy Principal Value Integral at infinity (see Eq. (4.2.3)). Frequently in applications we are also interested in integrals with integrands which have singularities at a finite location. Consider the integral $\int_a^b f(x) \, dx$, where $f(x)$ has a singularity at x_0, $a < x_0 < b$. Convergence of such an integral depends on the existence of the following limit, where $f(x)$ has a singularity at $x = x_0$:

$$I = \lim_{\epsilon \to 0^+} \int_a^{x_0 - \epsilon} f(x) \, dx + \lim_{\delta \to 0^+} \int_{x_0 + \delta}^b f(x) \, dx. \tag{4.3.1}$$

We say the integral $\int_a^b f(x)\,dx$ is **convergent** if and only if Eq. (4.3.1) exists and is finite; otherwise we say it is **divergent**. The integral might exist even if the $\lim_{x \to x_0} f(x)$ is infinite or is divergent. For example, the integral $\int_0^2 dx/(x-1)^{1/3}$ is convergent, whereas the integral $\int_0^2 dx/(x-1)^2$ is divergent. Sometimes by restricting the definition (4.3.1) we can make sense of a divergent integral. In this respect the so-called Cauchy Principal Value integral, denoted $\fint_a^b$ (where $\delta = \epsilon$ in Eq. (4.3.1)),

$$\fint_a^b f(x)\,dx = \lim_{\epsilon \to 0^+} \left(\int_a^{x_0-\epsilon} + \int_{x_0+\epsilon}^b \right) f(x)\,dx \qquad (4.3.2)$$

is quite useful. (Here the Cauchy Principal Value integral is required because of the singularity at $x = x_0$. We usually do not explicitly refer to where the singularity occurs, unless there is a special reason to do so, such as when the singularity is at infinity.) We say the Cauchy Principal Value integral exists if and only if the limit (4.3.2) exists. For example, the integral

$$\int_{-1}^2 \frac{1}{x}\,dx = \lim_{\epsilon \to 0^+} \int_{-1}^{-\epsilon} \frac{1}{x}\,dx + \lim_{\delta \to 0^+} \int_\delta^2 \frac{1}{x}\,dx$$
$$= \lim_{\epsilon \to 0^+} \ln|\epsilon| - \lim_{\delta \to 0^+} \ln|\delta| + \ln 2$$

does not exist, whereas

$$\fint_{-1}^2 \frac{1}{x}\,dx = \lim_{\epsilon \to 0}(\ln|\epsilon| - \ln|\epsilon|) + \ln 2 = \ln 2$$

does exist.

More generally, in applications we are sometimes interested in functions on an infinite interval with many points $\{x_i\}_{i=1}^N$ for which $\lim_{x \to x_i} f(x)$ is either infinite or does not exist. We say the following Cauchy Principal Value integral exists if and only if for $a < x_1 < x_2 < \cdots < x_N < b$

$$\fint_{-\infty}^\infty f(x)\,dx = \lim_{R \to \infty} \left(\int_{-R}^a + \int_b^R \right) f(x)\,dx + \lim_{\epsilon_1, \epsilon_2, \ldots, \epsilon_N \to 0^+} \left(\int_a^{x_1-\epsilon_1} \right.$$
$$\left. + \int_{x_1+\epsilon_1}^{x_2-\epsilon_2} + \int_{x_2+\epsilon_2}^{x_3-\epsilon_3} + \cdots + \int_{x_{N-1}+\epsilon_{N-1}}^{x_N-\epsilon_N} + \int_{x_N+\epsilon_N}^b \right) f(x)\,dx$$
$$(4.3.3)$$

exists. In practice we usually combine the integrals and consider the double limit $R \to \infty$ and $\epsilon_i \to 0^+$, for example, $\int_{-R}^{x_1-\epsilon_1}$ in Eq. (4.3.3), and do not bother to partition the integrals into intermediate values with a, b inserted. Examples will serve to clarify this point. Hereafter we consider $\epsilon_i > 0$, and $\lim_{\epsilon_i \to 0}$ means $\lim_{\epsilon_i \to 0^+}$.

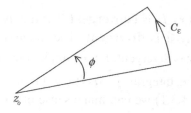

Figure 4.10 Small circular arc C_ϵ

The following theorems will be useful later. We consider integrals on a small circular arc with radius ϵ, center $z = z_0$, with the arc subtending an angle ϕ (see Figure 4.10). There are two important cases:

(a) $(z - z_0)f(z) \to 0$ uniformly (independent of the angle θ along C_ϵ) as $\epsilon \to 0$; and

(b) $f(z)$ possessing a simple pole at $z = z_0$.

Theorem 4.3.1 (a) *Suppose that on the contour C_ϵ, depicted in Figure 4.10, we have $(z - z_0)f(z) \to 0$ uniformly as $\epsilon \to 0$. Then*

$$\lim_{\epsilon \to 0} \int_{C_\epsilon} f(z)\, dz = 0.$$

(b) *Suppose $f(z)$ has a simple pole at $z = z_0$ with residue* Res $(f(z); z_0) = C_{-1}$. *Then for the contour C_ϵ,*

$$\lim_{\epsilon \to 0} \int_{C_\epsilon} f(z)\, dz = i\phi C_{-1}, \tag{4.3.4}$$

where the integration is carried out in the positive (counterclockwise) sense.

Proof (a) The hypothesis, $(z - z_0)f(z) \to 0$ uniformly as $\epsilon \to 0$, means that on C_ϵ, $|(z - z_0)f(z)| \le K_\epsilon$, where K_ϵ depends on ϵ, not on arg $(z - z_0)$ (recall, if $z - z_0 = re^{i\theta}$ we define θ to be arg$(z - z_0)$), and $K_\epsilon \to 0$ as $\epsilon \to 0$. Estimating the integral $(z = z_0 + \epsilon e^{i\theta}, \theta = \arg(z - z_0))$ using $|f(z)| \le K_\epsilon/\epsilon$, $\phi = \max |\arg(z - z_0)|$, gives

$$\left| \int_{C_\epsilon} f(z)\, dz \right| \le \int_{C_\epsilon} |f(z)|\, |dz| \le \frac{K_\epsilon}{\epsilon} \int_0^\phi \epsilon\, d\theta$$

$$= K_\epsilon \phi \to 0, \qquad \epsilon \to 0.$$

(b) If $f(z)$ has a simple pole with Res $(f(z); z_0) = C_{-1}$, then, from the Laurent expansion of $f(z)$ in the neighborhood of $z = z_0$,

$$f(z) = \frac{C_{-1}}{z - z_0} + g(z),$$

where $g(z)$ is analytic in the neighborhood of $z = z_0$. Thus

$$\lim_{\epsilon \to 0} \int_{C_\epsilon} f(z)\, dz = \lim_{\epsilon \to 0} C_{-1} \int_{C_\epsilon} \frac{dz}{z - z_0} + \lim_{\epsilon \to 0} \int_{C_\epsilon} g(z)\, dz.$$

The first integral on the right-hand side is evaluated, using $z = z_0 + \epsilon\, e^{i\theta}$ and taking θ from 0 to ϕ, to find

$$\int_{C_\epsilon} \frac{dz}{z - z_0} = \int_0^\phi \frac{i\epsilon\, e^{i\theta}\, d\theta}{\epsilon\, e^{i\theta}} = i\phi.$$

In the second integral, $|g(z)| \le M$, which is a constant in the neighborhood of $z = z_0$ because $g(z)$ is analytic there; hence we can apply part (a) of this theorem to find that the second integral vanishes in the limit $\epsilon \to 0$, and we recover Eq. (4.3.4). □

As a first example we show

$$\int_{-\infty}^{\infty} \frac{\sin ax}{x}\, dx = \operatorname{sgn}(a)\, \pi, \tag{4.3.5}$$

where

$$\operatorname{sgn}(a) = \begin{cases} -1 & a < 0, \\ 0 & a = 0, \\ 1 & a > 0. \end{cases} \tag{4.3.6}$$

Example 4.3.2   Evaluate

$$I = \int_{-\infty}^{\infty} \frac{e^{iax}}{x}\, dx, \qquad a \text{ real}. \tag{4.3.7}$$

Let us first consider $a > 0$ and the contour depicted in Figure 4.11. Because there are no poles enclosed by the contour, we have

$$\oint_C \frac{e^{iaz}}{z}\, dz = \left(\int_{-R}^{-\epsilon} + \int_{\epsilon}^{R} \right) \frac{e^{iax}}{x}\, dx + \int_{C_\epsilon} \frac{e^{iaz}}{z}\, dz + \int_{C_R} \frac{e^{iaz}}{z}\, dz = 0.$$

Because $a > 0$, the integral $\int_{C_R} e^{iaz}\, dz/z$ satisfies Jordan's Lemma (i.e., Lemma 4.2.4); hence it vanishes as $R \to \infty$. Similarly, $\int_{C_\epsilon} e^{iaz}\, dz/z$ is calculated using Theorem 4.3.1(b) to find

$$\lim_{\epsilon \to 0} \int_{C_\epsilon} \frac{e^{iaz}}{z}\, dz = -i\pi,$$

where we note that on C_ϵ the angle subtended is π and the residue of the function e^{iaz}/z is unity. The minus sign is a result of the direction being *clockwise*. Taking the limit $R \to \infty$ we have

$$I = \int_{-\infty}^{\infty} \frac{\cos ax}{x}\, dx + i \int_{-\infty}^{\infty} \frac{\sin ax}{x}\, dx = i\pi.$$

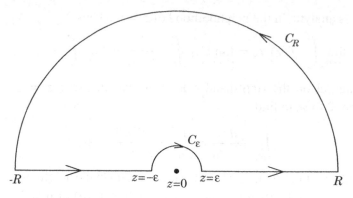

Figure 4.11 Contour of integration, Example 4.3.2

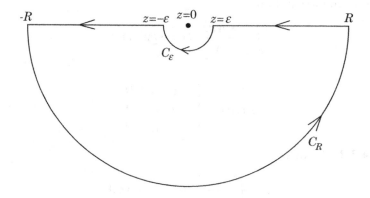

Figure 4.12 Alternative contour, Example 4.3.2

Thus, by setting real and imaginary parts equal we obtain $\oint (\cos ax)/x \, dx = 0$ (which is consistent with the fact that $(\cos ax)/x$ is odd) and Eq. (4.3.5) with $a > 0$. The case $a < 0$ follows because $\frac{\sin ax}{x}$ is odd in a. The same result could be obtained by using the contour in Figure 4.12. We also note that there is no need for the principal value in the integral (4.3.5), because it is a (weakly) convergent integral.

The following example is similar except that there are two locations where the integral has principal value contributions.

Example 4.3.3   Evaluate

$$I = \int_{-\infty}^{\infty} \frac{\cos x - \cos a}{x^2 - a^2} \, dx, \qquad a \in \mathbb{R}.$$

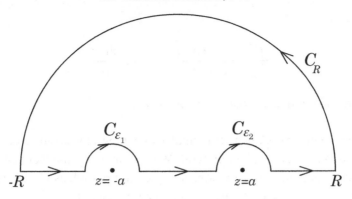

Figure 4.13 Contour C, Example 4.3.3

We note that the integral is convergent and well-defined at $x = \pm a$, because l'Hôpital's rule shows

$$\lim_{x \to \pm a} \frac{\cos x - \cos a}{x^2 - a^2} = \lim_{x \to \pm a} \frac{-\sin x}{2x} = \frac{-\sin a}{2a}.$$

We evaluate I by considering

$$J = \oint_C \frac{e^{iz} - \cos a}{z^2 - a^2}\, dz,$$

where the contour C is depicted in Figure 4.13. Because there are no poles enclosed by C, we have

$$0 = \oint_C \frac{e^{iz} - \cos a}{z^2 - a^2}\, dz$$

$$= \left\{ \int_{-R}^{-a-\epsilon_1} + \int_{-a+\epsilon_1}^{a-\epsilon_2} + \int_{a+\epsilon_2}^{R} + \int_{C_{\epsilon_1}} + \int_{C_{\epsilon_2}} + \int_{C_R} \right\} \frac{e^{iz} - \cos a}{z^2 - a^2}\, dz.$$

Along C_R we find, by Theorem 4.2.2 and Jordan's Lemma 4.2.4, that

$$\lim_{R \to \infty} \left| \int_{C_R} \frac{e^{iz} - \cos a}{z^2 - a^2}\, dz \right| = 0.$$

Similarly, from Theorem 4.3.1 we find (note that the directions of C_{ϵ_1} and C_{ϵ_2} are clockwise; that is, in the negative direction)

$$\lim_{\epsilon_1 \to 0} \int_{C_{\epsilon_1}} \frac{e^{iz} - \cos a}{z^2 - a^2}\, dz = -i\pi \left(\frac{e^{iz} - \cos a}{2z} \right)_{z=-a} = \frac{\pi \sin a}{2a}$$

and

$$\lim_{\epsilon_2 \to 0} \int_{C_{\epsilon_2}} \frac{e^{iz} - \cos a}{z^2 - a^2}\, dz = -i\pi \left(\frac{e^{iz} - \cos a}{2z} \right)_{z=a} = \frac{\pi \sin a}{2a}.$$

Thus, as $R \to \infty$,

$$\fint_{-\infty}^{\infty} \frac{e^{ix} - \cos a}{x^2 - a^2} \, dx = -\pi \frac{\sin a}{a},$$

and hence, by taking the real part, $I = -\pi(\sin a)/a$.

Again we note that the Cauchy Principal Value integral was only a device used to obtain a result for a well-defined integral. We also mention the fact that in practice one frequently calculates contributions along contours such as C_{ϵ_i} by carrying out the calculation directly without resorting to Theorem 4.3.1.

Our final illustration of Cauchy Principal Values is the evaluation of an integral similar to that of Example 4.2.11.

Example 4.3.4   Evaluate

$$I = \int_{-\infty}^{\infty} \frac{e^{px} - e^{qx}}{1 - e^x} \, dx,$$

where $0 < p, q < 1$.

We observe that this integral is convergent and well defined. We evaluate two separate integrals:

$$I_1 = \fint_{-\infty}^{\infty} \frac{e^{px}}{1 - e^x} \, dx$$

and

$$I_2 = \fint_{-\infty}^{\infty} \frac{e^{qx}}{1 - e^x} \, dx,$$

noting that $I = I_1 - I_2$.

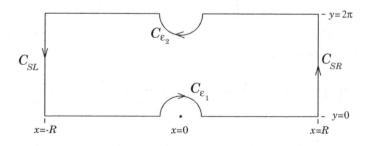

Figure 4.14 Contour of integration for Example 4.3.4

In order to evaluate I_1, we consider the contour depicted in Figure 4.14,

$$
\begin{aligned}
J &= \oint_C \frac{e^{pz}}{1 - e^z}\, dz \\
&= \left(\int_{-R}^{-\epsilon_1} + \int_{\epsilon_1}^{R} \right) \frac{e^{px}}{1 - e^x}\, dx + \left(\int_{C_{SR}} + \int_{C_{SL}} \right) \frac{e^{pz}}{1 - e^z}\, dz \\
&\quad + \left(\int_{R}^{-\epsilon_2} + \int_{\epsilon_2}^{-R} \right) \frac{e^{2\pi i p}\, e^{px}}{1 - e^x}\, dx + \left(\int_{C_{\epsilon_1}} + \int_{C_{\epsilon_2}} \right) \frac{e^{pz}}{1 - e^z}\, dz.
\end{aligned}
$$

Along the top path line we take $z = x + 2\pi i$. The integral $J = 0$ because no singularities are enclosed. The estimates of Example 4.2.11 show that the integrals along the sides C_{SL} and C_{SR} vanish. From Theorem 4.3.1 we have

$$
\lim_{\epsilon_1 \to 0} \int_{C_{\epsilon_1}} \frac{e^{pz}}{1 - e^z}\, dz = -i\pi \left(\frac{e^{pz}}{-e^z} \right)_{z=0} = i\pi
$$

and

$$
\lim_{\epsilon_2 \to 0} \int_{C_{\epsilon_2}} \frac{e^{pz}}{1 - e^z}\, dz = -i\pi \left(\frac{e^{pz}}{-e^z} \right)_{z=2\pi i} = i\pi e^{2\pi i p}.
$$

Hence, taking $R \to \infty$,

$$
I_1 = \int_{-\infty}^{\infty} \frac{e^{px}}{1 - e^x}\, dx = -i\pi \frac{1 + e^{2\pi i p}}{1 - e^{2\pi i p}} = \pi \cot \pi p.
$$

Clearly a similar analysis is valid for I_2 where p is replaced by q. Thus, combining these results, we find

$$
I = \pi(\cot \pi p - \cot \pi q).
$$

4.3.2 Integrals with Branch Points

In the remainder of this section we consider integrands that involve branch points. To evaluate the integrals, we introduce suitable branch cuts associated with the relevant multivalued functions. The procedure will be illustrated by a variety of examples.

Before working out examples we prove a theorem that will be useful in providing estimates for cases where Jordan's Lemma is not applicable.

Theorem 4.3.5 *If on a circular arc C_R of radius R and center $z = 0$, $z f(z) \to 0$ uniformly as $R \to \infty$, then*

$$
\lim_{R \to \infty} \int_{C_R} f(z)\, dz = 0.
$$

Proof   Let $\phi > 0$ be the angle enclosed by the arc C_R. Then

$$\left| \int_{C_R} f(z)\, dz \right| \leq \int_0^\phi |f(z)| R\, d\theta \leq K_R \phi.$$

Because $zf(z) \to 0$ *uniformly*, it follows that $|zf(z)| = R|f(z)| \leq K_R,\ K_R \to 0$, as $R \to \infty$.   □

Example 4.3.6   Use contour integration to evaluate

$$I = \int_0^\infty \frac{dx}{(x+a)(x+b)}, \qquad a, b > 0.$$

Because the integrand is not even, we cannot extend our integration region to the entire real line. Hence the methods of Section 4.2 will not work directly. Instead, we consider the contour integral

$$J = \oint_C \frac{\log z}{(z+a)(z+b)}\, dz,$$

where C is the "keyhole" contour depicted in Figure 4.15. We take $\log z$ to be on its principal branch $z = re^{i\theta} : 0 \leq \theta < 2\pi$, and choose a branch cut along the x-axis, $0 \leq x < \infty$.

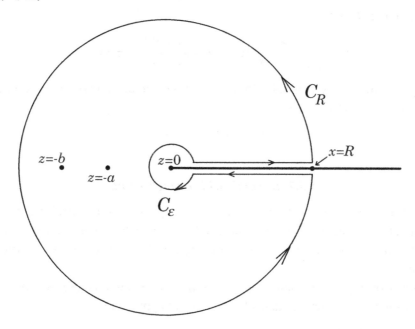

Figure 4.15 "Keyhole" contour, Example 4.3.6

An essential ingredient of the method is that owing to the location of the branch cut, the sum of the integrals on each side of the cut do not cancel.

$$J = \int_{\epsilon}^{R} \frac{\log x}{(x+a)(x+b)}\,dx + \int_{R}^{\epsilon} \frac{\log(xe^{2\pi i})}{(x+a)(x+b)}\,dx$$
$$+ \left(\int_{C_{\epsilon}} + \int_{C_R}\right) \frac{\log z}{(z+a)(z+b)}\,dz$$
$$= 2\pi i \left\{ \left(\frac{\log z}{z+a}\right)_{z=-b} + \left(\frac{\log z}{z+b}\right)_{z=-a} \right\}.$$

Theorems 4.3.1a and 4.3.5 show that

$$\lim_{\epsilon \to 0} \int_{C_{\epsilon}} \frac{\log z}{(z+a)(z+b)}\,dz = 0$$

and

$$\lim_{R \to \infty} \int_{C_R} \frac{\log z}{(z+a)(z+b)}\,dz = 0.$$

Using $\log(x\,e^{2\pi i}) = \log x + 2\pi i$, $\log(-a) = \log|a| + i\pi$, for $a > 0$, to simplify the expression for J, we have

$$-2\pi i \int_{0}^{\infty} \frac{dx}{(x+a)(x+b)} = 2\pi i \left(\frac{\log b - i\pi}{a-b} + \frac{\log a - i\pi}{b-a}\right)$$

hence

$$I = \left(\frac{\log b/a}{b-a}\right).$$

We worked this example for illustrative purposes and, of course, could have evaluated this integral by elementary methods because

$$I = \left(\frac{1}{b-a}\right) \int_{0}^{\infty} \left(\frac{1}{x+a} - \frac{1}{x+b}\right)\,dx = \left(\frac{1}{b-a}\right)\left[\ln\left(\frac{x+a}{x+b}\right)\right]_{0}^{\infty}$$
$$= \frac{\log b/a}{b-a}.$$

Another example is the following.

Example 4.3.7 Evaluate

$$I = \int_{0}^{\infty} \frac{\log^2 x}{x^2 + 1}\,dx.$$

Consider

$$J = \oint_{C} \frac{\log^2 z}{z^2 + 1}\,dz$$

where C is the contour depicted in Figure 4.16, and we take the principal branch of $\log z : z = r\,e^{i\theta}, 0 \le \theta < 2\pi$. Hence there is a branch cut along $x > 0$.

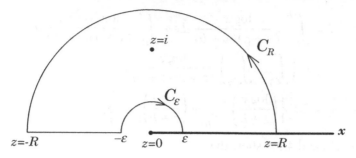

Figure 4.16 Contour of integration, Example 4.3.7

We have

$$
J = \int_R^\epsilon \frac{[\log(re^{i\pi})]^2}{(re^{i\pi})^2 + 1} e^{i\pi}\, dr + \int_\epsilon^R \frac{\log^2 x}{x^2 + 1}\, dx
$$

$$
+ \left(\int_{C_R} + \int_{C_\epsilon} \right) \frac{\log^2 z}{z^2 + 1}\, dz
$$

$$
= 2\pi i \left(\frac{\log^2 z}{2z} \right)_{z=i=e^{i\pi/2}}.
$$

Theorems 4.3.1a and 4.3.5 show that $\int_{C_\epsilon} \to 0$ as $\epsilon \to 0$, and $\int_{C_R} \to 0$ as $R \to \infty$. Thus the above equation simplifies, and

$$
2 \int_0^\infty \frac{\log^2 x}{x^2 + 1}\, dx + 2i\pi \int_0^\infty \frac{\log x}{x^2 + 1}\, dx - \pi^2 \int_0^\infty \frac{dx}{x^2 + 1}
$$

$$
= 2\pi i \frac{(i\pi/2)^2}{2i} = -\frac{\pi^3}{4}.
$$

However, the last integral can also be evaluated by contour integration, using the method of Section 4.2 (with the contour C_R depicted in Figure 4.3) to find

$$
\int_0^\infty \frac{dx}{x^2 + 1} = \frac{1}{2} \int_{-\infty}^\infty \frac{dx}{x^2 + 1} = \frac{1}{2} 2\pi i \left(\frac{1}{2z} \right)_{z=i} = \frac{\pi}{2}.
$$

Hence, we have

$$
2 \int_0^\infty \frac{\log^2 x\, dx}{x^2 + 1} + 2\pi i \int_0^\infty \frac{\log x\, dx}{x^2 + 1} = \frac{\pi^3}{4},
$$

whereupon, by taking the real and imaginary parts,

$$
I = \frac{\pi^3}{8} \quad \text{and} \quad \int_0^\infty \frac{\log x}{x^2 + 1}\, dx = 0.
$$

An example which uses some of the ideas of both Examples 4.3.6 and 4.3.7 is the following.

Example 4.3.8 Evaluate

$$I = \int_0^\infty \frac{x^{m-1}}{x^2 + 1} \, dx, \qquad 0 < m < 2.$$

The condition on m is required for the convergence of the integral.

Consider the contour integral

$$J = \oint_C \frac{z^{m-1}}{z^2 + 1} \, dz,$$

where, like in Example 4.3.6, C is the keyhole contour in Figure 4.15 with a branch cut along $x > 0$. We take the principal branch of $z^m : z = re^{i\theta}$, for $0 \le \theta < 2\pi$. The residue theorem yields

$$\begin{aligned}
J &= \int_\epsilon^R \frac{x^{m-1}}{x^2 + 1} \, dx + \int_R^\epsilon \frac{(xe^{2\pi i})^{m-1}}{x^2 + 1} \, dx \\
&\quad + \left(\int_{C_\epsilon} + \int_{C_R} \right) \frac{z^{m-1}}{z^2 + 1} \, dz \\
&= 2\pi i \left[\left(\frac{z^{m-1}}{2z} \right)_{z=i=e^{i\pi/2}} + \left(\frac{z^{m-1}}{2z} \right)_{z=-i=e^{3i\pi/2}} \right].
\end{aligned}$$

Theorems 4.3.1a and 4.3.5 imply that $\int_{C_\epsilon} \to 0$ as $\epsilon \to 0$, and $\int_{C_R} \to 0$ as $R \to \infty$. Therefore the above equation simplifies to

$$\begin{aligned}
\int_0^\infty \frac{x^{m-1}}{x^2 + 1} \, dx \left(1 - e^{2\pi i m} \right) &= 2\pi i \left(\frac{e^{i\pi(m-1)/2}}{2i} - \frac{e^{3i\pi(m-1)/2}}{2i} \right) \\
&= -\pi i \left(e^{im\pi/2} + e^{3im\pi/2} \right).
\end{aligned}$$

Hence,

$$I = -\pi i \, e^{im\pi/2} \left(\frac{1 + e^{im\pi}}{1 - e^{2im\pi}} \right) = \pi \frac{\cos \frac{m\pi}{2}}{\sin m\pi} = \frac{\pi}{2 \sin \frac{m\pi}{2}}.$$

The following example illustrates how we deal with more complicated multivalued functions and their branch cut structure. The reader is encouraged to review Section 2.3 before considering the next example.

Example 4.3.9 Evaluate

$$I = \int_{-1}^1 \frac{\sqrt{1 - x^2}}{1 + x^2} \, dx,$$

where the square root function takes on a positive value (i.e., $\sqrt{1} = +1$) in the range $-1 < x < 1$.

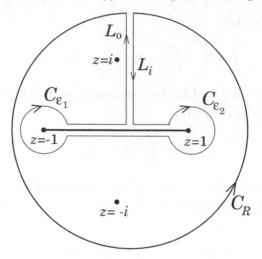

Figure 4.17 Contour C for Example 4.3.9

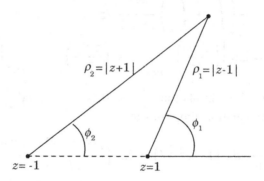

Figure 4.18 Polar coordinates for Example 4.3.9

We consider the contour integral

$$J = \oint_C \frac{(z^2 - 1)^{1/2}}{1 + z^2} \, dz$$

where the contour C and the relevant branch cut structure are depicted in Figure 4.17.

Before calculating the various contributions to J we first discuss the multivalued function $\sqrt{z^2 - 1}$ (see Figure 4.18). Using polar coordinates, we find

$$\left(z^2 - 1\right)^{\frac{1}{2}} = \sqrt{\rho_1 \rho_2} \, e^{i(\phi_1 + \phi_2)/2}, \qquad 0 \le \phi_1, \phi_2 < 2\pi,$$

where

$$(z - 1)^{\frac{1}{2}} = \sqrt{\rho_1} \, e^{i\phi_1/2}, \qquad \rho_1 = |z - 1|$$

and

$$(z + 1)^{\frac{1}{2}} = \sqrt{\rho_2} e^{i\phi_2/2}, \qquad \rho_2 = |z + 1|.$$

With this choice of branch, we find

$$(z^2 - 1)^{\frac{1}{2}} = \begin{cases} \sqrt{x^2 - 1}, & 1 < x < \infty, \\ -\sqrt{x^2 - 1}, & -\infty < x < -1, \\ i\sqrt{1 - x^2}, & -1 < x < 1, \quad y \to 0^+, \\ -i\sqrt{1 - x^2}, & -1 < x < 1, \quad y \to 0^-. \end{cases}$$

Thus we have a branch cut along $-1 \le x \le 1$. Using these expressions in the contour integral J, it follows that

$$J = \int_{-1+\epsilon_1}^{1-\epsilon_2} \frac{i\sqrt{1 - x^2}}{1 + x^2} \, dx + \int_{1-\epsilon_2}^{-1+\epsilon_1} \frac{-i\sqrt{1 - x^2}}{1 + x^2} \, dx$$

$$+ \left(-\int_{C_{\epsilon_1}} - \int_{C_{\epsilon_2}} + \int_{C_R} \right) \frac{(z^2 - 1)^{1/2}}{1 + z^2} \, dz$$

$$= 2\pi i \left[\left(\frac{(z^2 - 1)^{1/2}}{2z} \right)_{z=e^{i\pi/2}} + \left(\frac{(z^2 - 1)^{1/2}}{2z} \right)_{z=e^{3i\pi/2}} \right].$$

We note that the crosscut integrals vanish,

$$\left(\int_{L_0} + \int_{L_i} \right) \frac{(z^2 - 1)^{1/2}}{1 + z^2} \, dz = 0,$$

because L_0 and L_i are chosen in a region where $(z^2 - 1)^{1/2}$ is continuous and single-valued, and L_0 and L_i are arbitrarily close to each other. Theorem 4.3.1a shows that $\int_{C_{\epsilon_i}} \to 0$ as $\epsilon_i \to 0$; that is,

$$\left| \int_{C_{\epsilon_i}} \frac{(z^2 - 1)^{1/2}}{1 + z^2} \, dz \right| \le \int_0^{2\pi} \frac{\left(|(z - 1)(z + 1)|^{1/2} \right)}{\left| 2 + \epsilon_i^2 e^{2i\theta} + 2\epsilon_i e^{i\theta} \right|} \epsilon_i \, d\theta$$

$$\le \int_0^{2\pi} \frac{\sqrt{\epsilon_i^2 + 2\epsilon_i}}{2 - 2\epsilon_i - \epsilon_i^2} \epsilon_i \, d\theta \xrightarrow[\epsilon_i \to 0]{} 0.$$

The contribution on C_R is calculated as follows:

$$\int_{C_R} \frac{(z^2 - 1)^{1/2}}{1 + z^2} \, dz = \int_0^{2\pi} \frac{\left(R^2 e^{2i\theta} - 1 \right)^{1/2}}{1 + R^2 e^{2i\theta}} iR \, e^{i\theta} \, d\theta,$$

where we note that $(R^2 e^{2i\theta} - 1)^{1/2} \approx R e^{i\theta}$ as $R \to \infty$ because the chosen branch implies $\lim_{z \to \infty} (z^2 - 1)^{1/2} = z$. Hence,

$$\lim_{R \to \infty} \int_{C_R} \frac{(z^2 - 1)^{1/2}}{1 + z^2} \, dz = 2\pi i.$$

Calculation of the residues requires computing the correct branch of $(z^2 - 1)^{1/2}$.

$$\left(\frac{(z^2 - 1)^{1/2}}{2z}\right)_{z=e^{i\pi/2}} = \frac{\sqrt{2}e^{i(3\pi/4+\pi/4)/2}}{2i} = \frac{1}{\sqrt{2}},$$

$$\left(\frac{(z^2 - 1)^{1/2}}{2z}\right)_{z=e^{3i\pi/2}} = \frac{\sqrt{2}e^{i(5\pi/4+7\pi/4)/2}}{-2i} = \frac{1}{\sqrt{2}}.$$

Taking $\epsilon_i \to 0$, $R \to \infty$, and substituting the above results in the expression for J, we find

$$2i \int_{-1}^{1} \frac{\sqrt{1 - x^2}}{1 + x^2} \, dx = 2\pi i(\sqrt{2} - 1)$$

hence

$$I = \pi(\sqrt{2} - 1).$$

It should be noted that the contribution along C_R is proportional to the residue at infinity. Namely, calling

$$f(z) = \frac{(z^2 - 1)^{1/2}}{1 + z^2}$$

it follows that

$$f(z) = \frac{\left(z^2(1 - \frac{1}{z^2})\right)^{\frac{1}{2}}}{z^2\left(1 + \frac{1}{z^2}\right)} = \frac{1}{z}\left(1 - \frac{1}{2z^2} + \cdots\right)\left(1 - \frac{1}{z^2} + \cdots\right).$$

Thus, the coefficient of $\frac{1}{z}$ is unity, hence, from the definition (4.1.11a,b),

$$\text{Res } (f(z); \infty) = 1$$

and

$$\lim_{R \to \infty} \int_{C_R} \frac{(z^2 - 1)^{1/2}}{1 + z^2} = 2\pi i \, \text{Res} \left(\frac{(z^2 - 1)^{1/2}}{1 + z^2}; \infty\right) = 2\pi i.$$

The above analysis shows that calculating I follows from

$$2I = 2\pi i \sum_{j=1}^{2} \text{Res} \left(f(z); z_j\right) - 2\pi i \text{Res} \left(f(z); z_\infty\right),$$

where the three residues are calculated at

$$z_1 = e^{i\pi/2}, \qquad z_2 = e^{3i\pi/2}, \qquad z_\infty = \infty.$$

The minus sign is due to the orientation of the contour C_R with respect to z_∞.

The calculation involving the residue at infinity can be carried out for a large class of functions. In practice one usually computes the contribution at infinity by evaluating an integral along a large circular contour, C_R, in the same manner as we have done here.

4.3.3 Problems for Section 4.3

1. (a) Use principal value integrals to show that
$$\int_0^\infty \frac{\cos kx - \cos mx}{x^2}\, dx = \frac{-\pi}{2}(|k| - |m|), \qquad k, m, \text{ real.}$$

Hint: note that the function $f(z) = (e^{ikz} - e^{imz})/z^2$ has a simple pole at the origin.

(b) Setting $k = 2, m = 0$, deduce that
$$\int_0^\infty \frac{\sin^2 x}{x^2}\, dx = \frac{\pi}{2}.$$

2. Show that
$$\int_0^\infty \frac{\sin x}{x(x^2 + 1)}\, dx = \frac{\pi}{2}\left(1 - \frac{1}{e}\right).$$

3. Show that
$$\int_{-\infty}^\infty \frac{(\cos x - 1)}{x^2(x^2 + a^2)}\, dx = -\frac{\pi}{a^2} + \frac{\pi}{a^3}(1 - e^{-a}), \qquad a > 0.$$

4. Use a rectangular contour with corners at $\pm R$ and $\pm R + i\pi/k$, with an appropriate indentation, to show that
$$\int_0^\infty \frac{x}{\sinh kx}\, dx = \frac{\pi^2}{4k|k|} \qquad \text{for } k \neq 0, \text{ and } k \text{ real.}$$

5. Projection operators can be defined as follows. Consider a function $F(z)$:
$$F(z) = \frac{1}{2\pi i} \int_C \frac{f(\zeta)}{\zeta - z}\, d\zeta,$$

where C is a contour, typically infinite (e.g. the real axis) or closed (e.g. a circle), and z lies off the contour. Then, the "plus" and "minus" projections of $F(z)$ at $z = \zeta_0$ are defined by the following limit:
$$F^\pm(\zeta_0) = \lim_{z \to \zeta_0^\pm} \left[\frac{1}{2\pi i} \int_C \frac{f(\zeta)}{\zeta - z}\, d\zeta\right],$$

where $\zeta_0^\pm$ are points just inside (+) or outside (−) a closed contour (i.e., $\lim_{z \to \zeta_0^+}$ denotes the limit from points z inside the contour C), or to the left (+), right (−) of an infinite contour. Note: the "+" region lies to the *left* of the contour, where we take the standard orientation for a contour, i.e., the contour is taken with counterclockwise orientation. To simplify the analysis, we will assume that $f(x)$ can be analytically extended in the neighborhood of the curve C.

(a) Show that

$$F^\pm(\zeta_0) = \frac{1}{2\pi i} \fint_C \frac{f(\zeta)}{\zeta - \zeta_0} \, d\zeta \pm \frac{1}{2} f(\zeta_0),$$

where $\fint_C$ is the principal value integral which omits the point $\zeta = \zeta_0$.

(b) Suppose that $f(\zeta) = 1/(\zeta^2 + 1)$, and the contour C is the real axis (infinite), with orientation take from $-\infty$ to ∞; find $F^\pm(\zeta_0)$.

(c) Suppose that $f(\zeta) = 1/(\zeta^2 + a^2)$, for $a^2 > 1$, and the contour C is the unit circle centered at the origin with counterclockwise orientation; find $F^\pm(\zeta_0)$.

6. An important application of complex variables is to solve equations for functions analytic in a certain region given a relationship on a boundary (see Ablowitz and Fokas, 2003, Chapter 7). A simple example of this is the following. Solve for the function $\psi^+(z)$ analytic in the upper half plane, and $\psi^-(z)$ analytic in the lower half plane, given the following relationship on the real axis (where Re $z = x$):

$$\psi^+(x) - \psi^-(x) = f(x),$$

where, say, $f(x)$ is differentiable and absolutely integrable. A solution to this problem for $\psi^\pm(z) \to 0$ as $|z| \to \infty$ is given by

$$\psi^\pm(x) = \lim_{\epsilon \to 0^+} \frac{1}{2\pi i} \int_{-\infty}^{\infty} \frac{f(\zeta)}{\zeta - (x \pm i\epsilon)} \, d\zeta.$$

To simplify the analysis, we will assume that $f(x)$ can be analytically extended in the neighborhood of the real axis.

(a) Explain how this solution could be formally obtained by introducing the "projection" operators

$$P^\pm = \lim_{\epsilon \to 0^+} \frac{1}{2\pi i} \int_{-\infty}^{\infty} \frac{d\zeta}{\zeta - (x \pm i\epsilon)}$$

and in particular why it should be true that

$$P^+\psi^+ = \psi^+, \qquad P^-\psi^- = -\psi^-, \qquad P^+\psi^- = P^-\psi^+ = 0.$$

(b) Verify the results in part (a) for the example

$$\psi^+(x) - \psi^-(x) = \frac{1}{x^4 + 1}$$

and find $\psi^\pm(z)$ in this example.

(c) Show that

$$\psi^{\pm}(x) = \frac{1}{2\pi i} \int_{-\infty}^{\infty} \frac{f(\zeta)}{\zeta - x} \, d\zeta \pm \frac{1}{2} f(x).$$

In operator form the first term is usually denoted as $H(f(x))/2i$, where $H(f(x))$ is called the Hilbert transform. Then show that

$$\psi^{\pm}(x) = \frac{1}{2i} H(f(x)) \pm \frac{1}{2} f(x)$$

or, in operator form,

$$\psi^{\pm}(x) = \frac{1}{2} (\pm 1 - iH) f(x).$$

7. Use the "keyhole" contour of Figure 4.15 to show that on the principal branch of x^k:

(a) $I(a) = \displaystyle\int_0^{\infty} \frac{x^{k-1}}{(x+a)} \, dx = \frac{\pi}{\sin k\pi} a^{k-1}, \qquad 0 < k < 1, \quad a > 0;$

(b) $\displaystyle\int_0^{\infty} \frac{x^{k-1}}{(x+1)^2} \, dx = \frac{(1-k)\pi}{\sin k\pi}, \qquad 0 < k < 2.$

Verify this result by evaluating $I'(1)$ in part (a).

8. Use the technique described in Problem 7 above, using $\oint_C f(z)(\log z)^2 \, dz$, to establish that

(a) $I(a) = \displaystyle\int_0^{\infty} \frac{\log x}{x^2 + a^2} \, dx = \frac{\pi}{2a} \log a, a > 0;$

(b) $\displaystyle\int_0^{\infty} \frac{\log x}{(x^2 + 1)^2} \, dx = -\frac{\pi}{4}.$

Verify this by computing $I'(1)$ in part (a).

9. Use the keyhole contour in Figure 4.15 to establish that

$$\int_0^{\infty} \frac{x^{-k}}{x^2 + 2x \cos \phi + 1} \, dx = \frac{\pi}{\sin k\pi} \frac{\sin(k\phi)}{\sin \phi},$$

for $0 < k < 1, 0 < \phi < \pi$.

10. By using a large semicircular contour, enclosing the left half plane with a suitable keyhole contour, show that

$$\frac{1}{2\pi i} \int_{a-i\infty}^{a+i\infty} \frac{e^{zt}}{\sqrt{z}} \, dz = \frac{1}{\sqrt{\pi t}} \qquad \text{for } a, t > 0.$$

This is the inverse Laplace transform of the function $1/\sqrt{z}$. (The Laplace transform and its inverse will be discussed in Section 4.5.)

11. Consider the integral

$$I_R = \frac{1}{2\pi i} \oint_{C_R} \frac{dz}{(z^2 - a^2)^{1/2}}, \qquad a^2 > 0,$$

where C_R is a circle of radius R centered at the origin enclosing the points $z = \pm a$. Take the principal value of the square root.

(a) Evaluate the residue of the integrand at infinity and show that $I_R = 1$.

(b) Evaluate the integral by defining the contour around the branch points and along the branch cuts between $z = -a$ to $z = a$, to find (see Section 2.3) that

$$I_R = \frac{2}{\pi} \int_0^a \frac{dx}{\sqrt{a^2 - x^2}}.$$

Use the well-known indefinite integral

$$\int \frac{dx}{\sqrt{a^2 - x^2}} = \sin^{-1} x/a + \text{constant},$$

to obtain the same result as in part (a).

12. Use the transformation $t = (x - 1)/(x + 1)$ on the principal branch of the following functions, to show that

$$\text{(a)} \int_{-1}^1 \left(\frac{1+t}{1-t} \right)^{k-1} dt = \frac{2(1-k)\pi}{\sin k\pi}, \qquad 0 < k < 2;$$

$$\text{(b)} \int_{-1}^1 \log\left(\frac{1+t}{1-t} \right) \frac{dt}{1-at} = \frac{1}{2a} \log^2\left(\frac{1+a}{1-a} \right), \qquad 0 < a < 1.$$

13. Use the keyhole contour of Figure 4.15 to show that for the principal branch of $x^{1/2}$ and $\log x$,

$$\int_0^\infty \frac{x^{1/2} \log x}{(1 + x^2)} dx = \frac{\pi^2}{2\sqrt{2}}$$

and

$$\int_0^\infty \frac{x^{1/2}}{(1 + x^2)} dx = \frac{\pi}{\sqrt{2}}.$$

14. Consider the following integral with the principal branch of the square root:

$$\int_0^1 \sqrt{x(1 - x)} \, dx.$$

Use the contour integral

$$I_R = \frac{1}{2\pi i} \oint_{C_R} (z(z - 1))^{1/2} \, dz,$$

where C_R is the outside part of the "two-keyhole" or "dogbone" contour, similar to Figure 4.17, this time enclosing $z = 0$ and $z = 1$. Take the branch cut on the real axis between $z = 0$ and $z = 1$.

(a) Show that the behavior at $z = \infty$ implies that

$$\lim_{R \to \infty} I_R = \frac{-1}{8}.$$

(b) Use the principal branch of this function to show

$$I_R = \frac{1}{\pi} \int_0^1 \sqrt{x(1-x)} \, dx,$$

and conclude that

$$\int_0^1 \sqrt{x(1-x)} \, dx = \frac{\pi}{8}.$$

(c) Use the same method to show

$$\int_0^1 x^n \sqrt{x(1-x)} \, dx = -\pi b_{n+2},$$

where b_{n+2} is the coefficient of the term x^{n+2} in the binomial expansion of $(1-x)^{1/2}$; i.e.,

$$(1-x)^{\frac{1}{2}} = 1 - \frac{x}{2} - \frac{x^2}{8} - \frac{x^3}{16} - \cdots .$$

15. In Problem 12 of Section 2.6.4 we derived the formula

$$v(r, \varphi) = v(r = 0) + \frac{1}{2\pi} \int_0^{2\pi} u(\theta) \frac{2r \sin(\varphi - \theta)}{1 - 2r \cos(\varphi - \theta) + r^2} \, d\theta,$$

where $u(\theta)$ is given on the unit circle and the harmonic conjugate to $u(r, \varphi)$, $v(r, \varphi)$, is determined by the formula above. (In that problem $u(r, \varphi)$ was also derived.) Let $\zeta = re^{i\varphi}$. Show that as $r \to 1$, $\zeta \to e^{i\varphi}$ with $z = e^{i\theta}$, the above formula may be written as (use the trigonometric identity (I) below)

$$v(\varphi) = v(0) - \frac{1}{2\pi i} \int_0^{2\pi} u(\theta) \frac{e^{i\varphi} + e^{i\theta}}{e^{i\varphi} - e^{i\theta}} \, d\theta,$$

where the integral is taken as the Cauchy principal value. Show that

$$\frac{e^{i\varphi} + e^{i\theta}}{e^{i\varphi} - e^{i\theta}} = i \left(\frac{\sin(\theta - \varphi)}{1 - \cos(\theta - \varphi)} \right) = i \cot \left(\frac{\theta - \varphi}{2} \right), \tag{I}$$

using

$$\cos x = 2\cos^2 \frac{x}{2} - 1 = 1 - 2\sin^2 \frac{x}{2}, \qquad \sin x = 2 \sin \frac{x}{2} \cos \frac{x}{2},$$

and therefore deduce that

$$v(\varphi) = v(0) + \frac{1}{2\pi}\int_0^{2\pi} u(\theta)\cot\left(\frac{\varphi - \theta}{2}\right)\,d\theta.$$

This formula relates the boundary values, on the circle, between the real and imaginary parts of a function $f(z) = u + iv$ which is analytic inside the circle.

4.4 The Argument Principle, Rouché's Theorem

The Cauchy Residue Theorem can be used to obtain a useful result regarding the number of zeroes and poles of a meromorphic function. In what follows, we refer to second-order, third-order, ..., poles as "poles of order 2, 3,...."

Theorem 4.4.1 (Argument Principle)   *Let $f(z)$ be a meromorphic function defined inside and on a simple closed contour C, with no zeroes or poles on C. Then*

$$I = \frac{1}{2\pi i}\oint_C \frac{f'(z)}{f(z)}\,dz = N - P = \frac{1}{2\pi}[\arg f(z)]_C, \qquad (4.4.1)$$

where N and P are the number of zeroes and poles, respectively, of $f(z)$ inside C, where a multiple zero or pole is counted according to its multiplicity, and where $\arg f(z)$ is the argument of $f(z)$; that is, $f(z) = |f(z)|\exp(i\arg f(z))$ and $[\arg f(z)]_C$ denotes the change in the argument of $f(z)$ over C.

Proof   Suppose $z = z_i$ is a zero/pole of order n_i. Then

$$\frac{f'(z)}{f(z)} = \frac{\pm n_i}{z - z_i} + \phi(z), \qquad (4.4.2)$$

where ϕ is analytic in the neighborhood of z and the plus/minus sign stands for the zero/pole case, respectively. Formula (4.4.2) follows from the fact that if $f(z)$ has a zero of order n_{iz}, we can write $f(z)$ as $f(z) = (z - z_i)^{n_{iz}}g(z)$, where $g(z_i) \neq 0$ and $g(z)$ is analytic in the neighborhood of z_i; whereas if $f(z)$ has a pole of order n_{ip} then $f(z) = g(z)/(z - z_i)^{n_{ip}}$. Equation (4.4.2) then follows by differentiation with $\phi(z) = g'(z)/g(z)$. Applying the Cauchy Residue Theorem yields

$$I = \frac{1}{2\pi i}\oint_C \frac{f'(z)}{f(z)}\,dz = \sum_{i_z=1}^{M_z} n_{iz} - \sum_{i_p=1}^{M_p} n_{ip} = N - P,$$

where M_z (M_p) is the number of zero (pole) locations z_{iz} (respectively z_{ip}) with multiplicity n_{iz} (respectively n_{ip}).

In order to show that $I = [\arg f(z)]_C/2\pi$, we parametrize C as in Section 2.4. That is, we let $z = z(t)$ on $a \leq t \leq b$, where $z(a) = z(b)$. Thus we have, for I, the line integral

$$I = \frac{1}{2\pi i} \int_a^b \frac{f'(z(t))}{f(z(t))} z'(t)\, dt = \frac{1}{2\pi i} [\log f(z(t))]_{t=a}^b$$

$$= \frac{1}{2\pi i} [\log |f(z(t))| + i \arg f(z(t))]_{t=a}^b = \frac{1}{2\pi} [\arg f(z)]_C \qquad (4.4.3)$$

where we have taken the principal branch of the logarithm. $\qquad\qquad\qquad\square$

Geometrically, Eq. (4.4.3) corresponds to the following. Consider Figure 4.19.

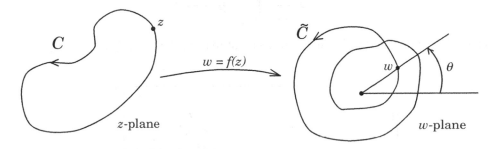

Figure 4.19 The mapping w

Let $w = f(z)$ be the image of the point z under the mapping $w = f(z)$, and let $\theta = \arg f(z)$ be the angle that the ray from the origin to w makes with respect to the horizontal; then $w = |f(z)| \exp(i\theta)$. Equation (4.4.3) corresponds to the number of times the point w winds around the origin on the image curve $\tilde{C}$ when z moves around C. Under the transformation $w = f(z)$, we find

$$\frac{1}{2\pi i} \oint_C \frac{f'(z)}{f(z)}\, dz = \frac{1}{2\pi i} \oint_{\tilde{C}} \frac{dw}{w} = \frac{1}{2\pi} [\arg w]_{\tilde{C}}. \qquad (4.4.4)$$

The quantity $\frac{1}{2\pi}[\arg w]_{\tilde{C}}$ is called the **winding number** of $\tilde{C}$ about the origin.

The following extension of Theorem 4.4.1 is useful. Suppose that $f(z)$ and C satisfy the hypothesis of Theorem 4.4.1, and let $h(z)$ be analytic inside and on C. Then

$$J = \frac{1}{2\pi i} \oint_C \frac{f'(z)}{f(z)} h(z)\, dz$$

$$= \sum_{i_z=1}^{M_z} n_{iz} h(z_{iz}) - \sum_{i_p=1}^{M_p} n_{ip} h(z_{ip}), \qquad (4.4.5)$$

where $h(z_{iz})$ and $h(z_{ip})$ are the values of $h(z)$ at the locations z_{iz} and z_{ip}. Formula (4.4.5) follows from Eq. (4.4.2) by evaluating the contour integral associated with

$$\frac{f'(z)}{f(z)} h(z) = \frac{\pm n_i}{z - z_i} h(z) + h(z)\phi(z).$$

This amounts to obtaining the residue

$$\text{Res}\left(\frac{f'(z)}{f(z)}h(z); z_i\right) = \pm n_i\, h(z_i).$$

If the zeroes/poles are simple, then in Eq. (4.4.5) we take $n_{iz} = n_{ip} = 1$.

Example 4.4.2   Consider the following integral:

$$I(\zeta) = \frac{1}{2\pi i}\oint_{C_N}\frac{\pi\cot\pi z}{\zeta^2 - z^2}\,dz,$$

where C_N is depicted in Figure 4.20. Deduce that $\pi\cot\pi\zeta = \zeta\sum_{n=-\infty}^{\infty}\frac{1}{\zeta^2-n^2}$.

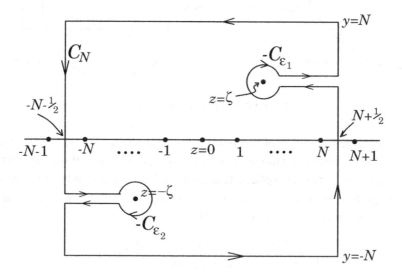

Figure 4.20 The contour C_N

We have an integral of the form (4.4.5), where $f(z) = \sin\pi z$, $h(z) = \frac{1}{\zeta^2-z^2}$, $z_i = n$, and $n = 0, \pm 1, \pm 2, \ldots, \pm N$. Deforming the contour and using Eq. (4.4.5) with $n_{iz} = 1$, $n_{ip} = 0$, we find

$$\frac{1}{2\pi i}\left(\oint_{C_{NR}} - \oint_{C_{\epsilon_1}} - \oint_{C_{\epsilon_2}}\right)\frac{\pi\cot\pi z}{\zeta^2 - z^2}\,dz = \sum_{n=-N}^{N}\frac{1}{\zeta^2 - n^2},$$

where C_{NR} denotes the rectangular contour alone, without crosscuts and circles, around $z = \pm\zeta$. Taking $N \to \infty$ and $\epsilon_i \to 0$, we have, after computing the residues about C_{ϵ_i}, we find

$$\lim_{N\to\infty}\frac{1}{2\pi i}\oint_{C_{NR}}\frac{\pi\cot\pi z}{\zeta^2 - z^2}\,dz - \left[\left(\frac{\pi\cot\pi z}{-2z}\right)_{z=\zeta} + \left(\frac{\pi\cot\pi z}{-2z}\right)_{z=-\zeta}\right] = \sum_{n=-\infty}^{\infty}\frac{1}{\zeta^2 - n^2}.$$

It was shown in Example 4.2.7 that $\oint_{C_{NR}} \to 0$ as $N \to \infty$; hence we have the series representation

$$\pi \cot \pi \zeta = \zeta \sum_{n=-\infty}^{\infty} \frac{1}{\zeta^2 - n^2}. \tag{4.4.6}$$

The Argument Principle is also frequently used to determine the number of zeroes located in a given region of the plane.

Example 4.4.3 Determine the number of zeroes located inside the first quadrant of the function

$$f(z) = z^3 + 1.$$

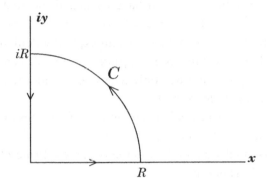

Figure 4.21 Contour in first quadrant for Example 4.4.3

Consider the contour in Figure 4.21. We begin at $z = 0$ where $f(z) = 1$. We take

$$\arg f(z) = \phi, \qquad \tan \phi = \frac{\mathrm{Im}\, f(x, y)}{\mathrm{Re}\, f(x, y)},$$

with the principal branch $\phi = 0$ corresponding to $f(z) = 1$. On the circle $|z| = R$,

$$z = R\, e^{i\theta}, \qquad f(z) = R^3 e^{3i\theta}\left(1 + \frac{1}{R^3 e^{3i\theta}}\right).$$

On the real axis $z = x$, $f(x, y) = x^3 + 1$ is positive and continuous; hence at $x = R$, we have $y = 0$, $\theta(R, 0) = 0$. Next we follow the $\arg f(x, y)$ around the circular contour. For large R, $1 + 1/(R^3 e^{3i\theta})$ is near 1; hence $\arg f(z) \approx 3\theta$, and at $z = iR$, $\theta = \pi/2$; hence $\arg f(z) \approx 3\pi/2$. Now along the imaginary axis $z = iy$

$$f(x, y) = -iy^3 + 1.$$

As y traverses from $y \to \infty$ to $y \to 0^+$, $\tan \phi$ goes from $-\infty$ to 0^- (note $\tan \phi \approx -y^3$ for $y \to \infty$); that is, ϕ goes from $\frac{3\pi}{2}$ to 2π. Note also that we are in the fourth quadrant because $\mathrm{Im}\, f < 0$ and $\mathrm{Re}\, f > 0$.

Thus the Argument Principle gives

$$N = \frac{1}{2\pi}[\arg f]_C = 1$$

because arg f changed by 2π over this circuit. Note the number of poles $P = 0$ because $f(z)$ is a polynomial, hence analytic in C. Thus, we have one zero located in the first quadrant. Because complex roots for a real polynomial must have a complex conjugate root, it follows that the other complex root occurs in the fourth quadrant. Cubic equations with real coefficients have at least one real root. Because it is not on the positive real axis (remember $x^3 + 1 > 0$ for $x > 0$), this root is necessarily on the negative axis.

In this particular example we could have evaluated the roots directly; that is, $z^3 = -1 = e^{i\pi} = e^{(2n+1)\pi i}$ for $n = 0, 1, 2$; or $z_1 = e^{i\pi/3}$, $z_2 = e^{i\pi} = -1$, and $z_3 = e^{5i\pi/3}$. In more complicated examples an explicit calculation of the roots is usually impossible. In such examples the Argument Principle can be a very effective tool. However, our purpose here was only to give the reader the basic ideas of the method as applied to a simple example.

As we have seen, one must calculate the number of changes of sign in Im $f(x, y)$ and Re $f(x, y)$ in order to compute arg $f(x, y)$.

A result that is essentially a corollary to the Argument Principle is the following, often termed Rouché's Theorem.

Theorem 4.4.4 (Rouché)   *Let $f(z)$ and $g(z)$ be analytic on and inside a simple closed contour C. If $|f(z)| > |g(z)|$ on C, then $f(z)$ and $[f(z) + g(z)]$ have the same number of zeroes inside the contour C.*

In Theorem 4.4.4, multiple zeroes are enumerated in the same manner as in the Argument Principle.

Proof   Because $|f(z)| > |g(z)| \geq 0$ on C, then $|f(z)| \neq 0$; hence $f(z) \neq 0$ on C. Thus, writing

$$w(z) = \frac{f(z) + g(z)}{f(z)},$$

it follows that the contour integral $\frac{1}{2\pi i} \oint_C \frac{w'(z)}{w(z)} \, dz$ is well defined (there are no poles on C). Moreover, $(w(z) - 1) = g/f$, whereupon

$$|w(z) - 1| < 1, \tag{4.4.7}$$

and hence all points $w(z)$ in the w-plane lie within the circle of unit radius centered at $(1, 0)$. Thus, we conclude that the origin $w = 0$ cannot be enclosed by $\tilde{C}$ (see Figure 4.22). Here $\tilde{C}$ is the image curve in the w-plane (if $w = 0$ were enclosed then $|w - 1| = 1$ somewhere on $\tilde{C}$). Hence $[\arg w(z)]_C = 0$ and $N = P$ for $w(z)$.

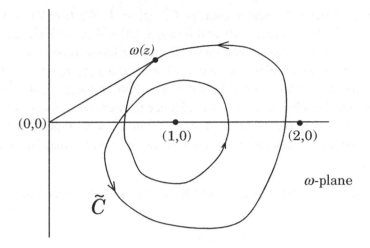

Figure 4.22 The contour $\tilde{C}$, proof of Rouché's Theorem

Therefore the number of zeroes of $f(z)$ (poles of $w(z)$) equals the number of zeroes of $f(z) + g(z)$. □

Rouché's theorem can be used to prove the fundamental theorem of algebra (also discussed in Section 2.6). Namely, every polynomial

$$P(z) = z^n + a_{n-1}z^{n-1} + a_{n-2}z^{n-2} + \cdots + a_0$$

has n and only n roots counting multiplicities: $P(z_i) = 0$, for $i = 1, 2, \ldots, n$. We write $f(z) = z^n$ and $g(z) = a_{n-1}z^{n-1} + a_{n-2}z^{n-2} + \cdots + a_0$. For $|z| > 1$ we find

$$|g(z)| \leq |a_{n-1}||z|^{n-1} + |a_{n-2}||z|^{n-2} + \cdots + |a_0|$$
$$\leq (|a_{n-1}| + |a_{n-2}| + \cdots + |a_0|)\,|z|^{n-1}.$$

If our contour C is taken to be a circle with radius R greater than unity, then $|f(z)| = R^n > |g(z)|$ whenever

$$R > \max\left(1, |a_{n-1}| + |a_{n-2}| + \cdots + |a_0|\right).$$

Hence $P(z) = f(z) + g(z)$ has the same number of roots as $f(z) = z^n = 0$ which is n. Moreover, all the roots of $P(z)$ are contained inside the circle $|z| < R$ because, by the above estimate for R,

$$|P(z)| = |z^n + g(z)| \geq R^n - |g(z)| > 0$$

and therefore $|P(z)|$ does not vanish for $|z| \geq R$.

Example 4.4.5 Show that all the roots of $P(z) = z^8 - 4z^3 + 10$ lie between $1 \leq |z| \leq 2$.

First we consider the circular contour C_1: $|z| = 1$. We take $f(z) = 10$, and $g(z) = z^8 - 4z^3$. Thus, $|f(z)| = 10$, and $|g(z)| \leq |z|^8 + 4|z|^3 = 5$. Hence $|f| > |g|$, which implies that $P(z)$ has no roots on C_1. Because f has no roots inside C_1, neither does $P(z) = (f + g)(z)$. Next we take $f(z) = z^8$ and $g(z) = -4z^3 + 10$. On the circular contour C_2 we have $|z| = 2$, $|f(z)| = 2^8 = 256$, and $|g(z)| \leq 4|z|^3 + 10 = 42$, so $|f| > |g|$ and $P(z)$ has no roots on C_2. Hence the number of roots of $(f + g)(z)$ equals the number of zeroes of $f(z) = z^8$. Thus $z^8 = 0$ implies that there are eight roots inside C_2. Because they cannot be inside or on C_1, they lie in the region $1 < |z| < 2$.

Example 4.4.6   Show that there is exactly one root inside the contour C_1: $|z| = 1$, for

$$h(z) = e^z - 4z - 1.$$

We take $f(z) = -4z$ and $g(z) = e^z - 1$ on C_1

$$|f(z)| = |4z| = 4, \qquad |g(z)| = |e^z - 1| \leq |e^z| + 1 < e + 1 < 4.$$

Thus $|f| > |g|$ on C_1, and hence $h(z) = (f + g)(z)$ has the same number of roots as $f(z) = 0$, which is 1.

4.4.1 Problems for Section 4.4

1. Verify the Argument Principle, Theorem 4.4.1, for the following functions. Take the contour C to be a unit circle centered at the origin.

   (a) z^n, for n an integer (positive or negative);
   (b) e^z;
   (c) $\coth 4\pi z$;
   (d) $P(z)/Q(z)$, where $P(z)$ and $Q(z)$ are polynomials of degree N and M, respectively, and have all their zeroes inside C.
   (e) What happens if we consider $f(z) = e^{1/z}$?

2. Show

$$\frac{1}{2\pi i} \oint_C \frac{f'(z)}{(f(z) - f_0)} \, dz = N,$$

   where N is the number of points z where $f(z) = f_0$ (a constant) inside C; $f'(z)$ and $f(z)$ are analytic inside and on C, and $f(z) \neq f_0$ on the boundary of C.

3. Use the Argument Principle to show that

   (a) $f(z) = z^5 + 1$ has one zero in the first quadrant;
   (b) $f(z) = z^7 + 1$ has two zeroes in the first quadrant.

4. Suppose $f(z)$ is a function which is analytic inside and on a simple closed contour C with no zeroes on C. If the image $\tilde{C}$ of C under the transformation $w = f(z)$ looks like the right-hand figure in Figure 4.19, find $[\arg f(c)]_C$ and the number of zeroes of $f(z)$ inside C.

5. Show that there are no zeroes of $f(z) = z^4 + z^3 + 5z^2 + 2z + 4$ in the first quadrant. Use the fact that on the imaginary axis, $z = iy$, the argument of the function for large y starts with a certain value which corresponds to a quadrant of the argument of $f(z)$. Each change in sign of Re $f(iy)$ and Im $f(iy)$ corresponds to a suitable change of quadrant of the argument.

6. (a) Show that $e^z - (4z^2 + 1) = 0$ has exactly two roots for $|z| < 1$. Hint: in Rouché's Theorem use $f(z) = -4z^2$, $g(z) = e^z - 1$, so that for C being the unit circle,

$$|f(z)| = 4 \quad \text{and} \quad |g(z)| = |e^z - 1| \le |e^z| + 1.$$

   (b) Show that the improved estimate $|g(z)| \le e - 1$ can be deduced from $e^z - 1 = \int_0^z e^w \, dw$, and this allows us to establish that $e^z - (2z + 1) = 0$, has exactly one root for $|z| < 1$.

7. (a) Consider the mapping $w = z^3$. When we encircle the origin in the z-plane one time, how many times do we encircle the origin in the w-plane? Explain why this agrees with the Argument Principle.

   (b) Suppose we consider $w = z^3 + a_2 z^2 + a_1 z + a_0$ for three constants a_0, a_1, a_2. If we encircle the origin in the z-plane once on a very large circle, how many times do we encircle the w-plane?

   (c) Suppose we have a mapping $w = f(z)$ where $f(z)$ is analytic inside and on a simple closed contour C in the z-plane. Let us define $\tilde{C}$ as the (non-simple) image in the w-plane of the contour C in the z-plane. If we deduce that it encloses the origin ($w = 0$) N times, and encloses the point $w = 1$ M times, what is the change in arg w over the contour $\tilde{C}$?

8. Suppose that $f(z)$ is analytic in a region containing a simple closed contour C. Let $|f(z)| \le M$ on C; show via Rouché's Theorem that $|f(z)| \le M$ inside C. (The maximum of an analytic function is attained on its boundary; this provides an alternate proof of the maximum modulus result in Section 2.6.) Hint: suppose there is a value of $f(z)$, say f_0, such that $|f_0| > M$. Consider the two functions $-f_0$ and $f(z) - f_0$. Then Rouché's Theorem implies that functions $-f_0, f(z) - f_0$ have the same number of zeroes inside C.

4.5 Fourier and Laplace Transforms

One of the most valuable tools in mathematics, physics, and engineering is making use of the properties a function takes on in a so-called *transform* (or *dual*) space.

In suitable function spaces, defined below, the Fourier transform pair is given by the following relations:

$$f(x) = \frac{1}{2\pi} \int_{-\infty}^{\infty} \hat{F}(k)e^{ikx} \, dk, \qquad -\infty < x < \infty, \tag{4.5.1}$$

$$\hat{F}(k) = \int_{-\infty}^{\infty} f(x)e^{-ikx} \, dx, \qquad -\infty < k < \infty. \tag{4.5.2}$$

The function $\hat{F}(k)$ is called the **Fourier transform** of $f(x)$. The integral in Eq. (4.5.1) is referred to as the **inverse Fourier transform**. In mathematics, the study of Fourier transforms is central in fields like harmonic analysis. In physics and engineering, applications of Fourier transforms are crucial, for example, in the study of quantum mechanics, wave propagation, and signal processing. In this section we introduce the basic notions and give a heuristic derivation of Eqs. (4.5.1)–(4.5.2). In the next section we apply these concepts to solve some of the classical partial differential equations.

In what follows we make some general remarks about the relevant function spaces for $f(x)$ and $\hat{F}(k)$. However, in our calculations we will apply complex variable techniques and will not use any deep knowledge of function spaces. Relations (4.5.1)–(4.5.2) always hold if $f(x) \in L^2$ and $\hat{F}(k) \in L^2$, where L^2 is the function space of **square-integrable functions**:

$$\|F\|_2 = \left(\int_{-\infty}^{\infty} |f|^2(x) \, dx \right)^{1/2} < \infty. \tag{4.5.3}$$

In those cases where $f(x)$ and $f'(x)$ are continuous everywhere apart from a finite number of points for which $f(x)$ has integrable and bounded discontinuities (such a function $f(x)$ is said to be **piecewise smooth**), it turns out that at each point of discontinuity, call it x_0, the integral given by Eq. (4.5.1) converges to the mean:

$$\lim_{\epsilon \to 0^+} \frac{1}{2} [f(x_0 + \epsilon) + f(x_0 - \epsilon)] .$$

At points where $f(x)$ is continuous, the integral (4.5.1) converges to $f(x)$.

There are other function spaces for which Eqs. (4.5.1)–(4.5.2) hold. If $f(x) \in L^1$, the space of **absolutely integrable functions** (functions f satisfying $\int_{-\infty}^{\infty} |f(x)| dx < \infty$), then $\hat{F}(k)$ tends to zero for $|k| \to \infty$ and belongs to a certain function space of functions decaying at infinity. Conversely, if we start with a suitably decaying function $\hat{F}(k)$, then $f(x) \in L^1$. In a sense, purely from a symmetry point of view, such function spaces may seem less natural than when $f(x), \hat{F}(k)$ are both in L^2 for Eqs. (4.5.1)–(4.5.2) to be valid. Nevertheless, in some $f(x) \in L^1$ and not L^2. Applications sometimes require the use of Fourier transforms in spaces for which no general theory applies, but nevertheless specific results can be attained. It is outside the scope of this text to examine L^p, for $p = 1, 2, \ldots$, function spaces where the general results pertaining to Eqs. (4.5.1)–(4.5.2) can be proven. Interested readers can find such a discussion in various books on complex or Fourier analysis, such as Rudin (1966). Unless otherwise specified, we shall assume that our function $f(x) \in L^1 \cap L^2$; that is, $f(x)$ is both absolutely and square integrable. It follows that

$$|\hat{F}(k)| \leq \|f\|_1 = \int_{-\infty}^{\infty} |f(x)| \, dx$$

and

$$\frac{1}{\sqrt{2\pi}} \|\hat{F}\|_2 = \|f\|_2.$$

The first relationship follows directly. The second is established later in this section. In those cases for which $f(x)$ is piecewise smooth, an elementary though tedious proof of Eqs. (4.5.1)–(4.5.2) can be constructed by suitably breaking up the interval $(-\infty, \infty)$, and using standard results of integration (Titchmarsch, 1948).

Statements analogous to Eqs. (4.5.1)–(4.5.2) hold for functions on finite intervals $(-L, L)$, which may be extended as periodic functions of period $2L$:

$$f(x) = \sum_{n=-\infty}^{\infty} \hat{F}_n \, e^{in\pi x/L}, \qquad (4.5.4)$$

$$\hat{F}_n = \frac{1}{2L} \int_{-L}^{L} f(x) \, e^{-in\pi x/L} \, dx. \qquad (4.5.5)$$

The values $\hat{F}_n$ are called the **Fourier coefficients** of the Fourier series representation of the function $f(x)$ given by Eq. (4.5.4).

A simple illustration of how Fourier transforms may be calculated is afforded by the following.

Example 4.5.1 Let $a > 0$, $b > 0$, and $f(x)$ be given by

$$f(x) = \begin{cases} e^{-ax} & x > 0, \\ e^{bx} & x < 0. \end{cases}$$

Then, from Eq. (4.5.2), we may compute $\hat{F}(k)$ to be

$$\hat{F}(k) = \int_0^{\infty} e^{-ax-ikx} \, dx + \int_{-\infty}^{0} e^{bx-ikx} \, dx$$

$$= \frac{1}{a + ik} + \frac{1}{b - ik}. \qquad (4.5.6)$$

The inversion – i.e., reconstruction of $f(x)$ via Eq. (4.5.1 – is ascertained by calculating the appropriate contour integrals. In this case we use closed semicircles in the upper ($x > 0$) and lower ($x < 0$) half k-planes. The inversion can be done either by combining the two terms in Eq. (4.5.6) or by noting that

$$\frac{1}{2\pi} \int_{-\infty}^{\infty} \frac{e^{ikx}}{a + ik} \, dk = \begin{cases} e^{-ax} & x > 0, \\ 1/2 & x = 0, \\ 0 & x < 0 \end{cases}$$

and

$$\frac{1}{2\pi} \int_{-\infty}^{\infty} \frac{e^{ikx}}{b - ik} \, dk = \begin{cases} 0 & x > 0, \\ 1/2 & x = 0, \\ e^{bx} & x < 0. \end{cases}$$

The values at $x = 0$ take into account both the pole and the principal value contribution at infinity; that is, $\frac{1}{2\pi} \int_{C_R} \frac{dk}{a+ik} = 1/2$ where $k = R\,e^{i\theta}$, for $0 \le \theta \le \pi$, on C_R. Thus Eq. (4.5.1) gives convergence to the mean value at $x = 0$:

$$\frac{1}{2} + \frac{1}{2} = 1 = (\lim_{x \to 0^+} f(x) + \lim_{x \to 0^-} f(x))/2.$$

A function which will lead us to a useful result is the following:

$$\Delta(x; \epsilon) = \begin{cases} \frac{1}{2\epsilon} & |x| < \epsilon, \\ 0 & |x| > \epsilon. \end{cases} \tag{4.5.7}$$

Its Fourier transform is given by

$$\hat{\Delta}(k; \epsilon) = \frac{1}{2\epsilon} \int_{-\epsilon}^{\epsilon} e^{-ikx} \, dx = \frac{\sin k\epsilon}{k\epsilon}. \tag{4.5.8}$$

Certainly $\Delta(x; \epsilon)$ is both absolutely and square integrable; so it is in $L^1 \cap L^2$, and $\hat{\Delta}(k; \epsilon)$ is in L^2. It is natural to ask what happens as $\epsilon \to 0$. The function defined by Eq. (4.5.7) tends, as $\epsilon \to 0$, to a novel "function" called the **Dirac delta function**, denoted by $\delta(x)$, having the following properties:

$$\delta(x) = \lim_{\epsilon \to 0} \Delta(x; \epsilon), \tag{4.5.9}$$

$$\int_{-\infty}^{\infty} \delta(x - x_0) \, dx = \lim_{\epsilon \to 0} \int_{x=x_0-\epsilon}^{x=x_0+\epsilon} \Delta(x - x_0; \epsilon) \, dx = 1, \tag{4.5.10}$$

$$\int_{-\infty}^{\infty} \delta(x - x_0) f(x) \, dx = f(x_0), \tag{4.5.11}$$

where $f(x)$ is continuous. Equations (4.5.10)–(4.5.11) can be ascertained by using the limit definitions (4.5.7) and (4.5.9). The function defined in Eq. (4.5.9) is often called a unit impulse "function"; it has an arbitrarily large value concentrated at the origin, whose integral is unity. The delta function, $\delta(x)$, is not a mathematical function in the conventional sense, as it has an arbitrarily large value at the origin. Nevertheless, there is a rigorous mathematical framework in which these new functions – called distributions – can be analyzed. Interested readers can find such a discussion in, for example, Lighthill (1959). For our purposes the device of the limit process $\epsilon \to 0$ is sufficient. We also note that Eq. (4.5.7) is not the only valid representation of a delta function; for example, others are given by

$$\delta(x) = \lim_{\epsilon \to 0} \left\{ \frac{1}{\sqrt{\pi \epsilon}} e^{-x^2/\epsilon} \right\}, \tag{4.5.12a}$$

$$\delta(x) = \lim_{\epsilon \to 0} \left\{ \frac{\epsilon}{\pi (\epsilon^2 + x^2)} \right\}. \tag{4.5.12b}$$

It should be noted that, formally speaking, the Fourier transform of a delta function is given by

$$\hat{\delta}(k) = \int_{-\infty}^{\infty} \delta(x) e^{-ikx} \, dx = 1. \tag{4.5.13}$$

It does not vanish as $|k| \to \infty$; indeed, it is a constant (namely unity; note that $\delta(x)$ is not L^1). Similarly, it turns out from the theory of distributions (motivated by the inverse Fourier transform), that the following alternative definition of a delta function holds:

$$\delta(x) = \frac{1}{2\pi} \int_{-\infty}^{\infty} \hat{\delta}(k) e^{ikx} \, dk = \frac{1}{2\pi} \int_{-\infty}^{\infty} e^{ikx} \, dk. \tag{4.5.14}$$

Formula (4.5.14) allows us a simple (but formal) way to verify Eqs. (4.5.1)–(4.5.2). Namely, by using Eqs. (4.5.1)–(4.5.2) and assuming that interchanging integrals is valid, we have

$$
\begin{aligned}
f(x) &= \frac{1}{2\pi} \int_{-\infty}^{\infty} dk \, e^{ikx} \left(\int_{-\infty}^{\infty} dx' \, f(x') \, e^{-ikx'} \right) \\
&= \int_{-\infty}^{\infty} dx' \, f(x') \left(\frac{1}{2\pi} \int_{-\infty}^{\infty} dk \, e^{ik(x-x')} \right) \\
&= \int_{-\infty}^{\infty} dx' \, f(x') \delta(x - x'). \tag{4.5.15}
\end{aligned}
$$

In what follows, the Fourier transform of derivatives will be needed. The Fourier transform of a derivative is readily obtained via integration by parts.

$$\hat{F}_1(k) \equiv \int_{-\infty}^{\infty} f'(x) \, e^{-ikx} \, dx = \left[f(x) \, e^{-ikx} \right]_{-\infty}^{\infty} + ik \int_{-\infty}^{\infty} f(x) \, e^{-ikx} \, dx$$

$$= ik \, \hat{F}(k) \tag{4.5.16a}$$

and by repeated integration by parts,

$$\hat{F}_n(k) = \int_{-\infty}^{\infty} f^{(n)}(x) e^{-ikx} \, dx = (ik)^n \hat{F}(k). \tag{4.5.16b}$$

Formulae (4.5.16a,b) will be useful when we examine solutions of differential equations by transform methods.

It is natural to ask what is the Fourier transform of a product. The result is called the **convolution product**; it is *not* the product of the Fourier transforms. We can readily derive this formula. We use two transform pairs: one for a function $f(x)$ (Eqs. (4.5.1)–(4.5.2)) and another for a function $g(x)$, replacing $f(x)$ and $\hat{F}(k)$ in Eqs. (4.5.1)–(4.5.2) by $g(x)$ and $\hat{G}(k)$, respectively. We define the convolution product as follows,

$$(f * g)(x) = \int_{-\infty}^{\infty} f(x - x')g(x')\, dx' = \int_{-\infty}^{\infty} f(x')g(x - x')\, dx';$$

the latter equality follows by renaming variables. We take the Fourier transform of $(f * g)(x)$ and interchange integrals (allowed since both f and g are absolutely integrable) to find

$$\int_{-\infty}^{\infty} (f * g)(x)e^{-ikx}\, dx = \int_{-\infty}^{\infty} dx \left(\int_{-\infty}^{\infty} f(x - x')g(x')\, dx' \right) e^{-ikx}$$

$$= \int_{-\infty}^{\infty} dx'\, e^{-ikx'} g(x') \int_{-\infty}^{\infty} dx\, e^{-ik(x-x')} f(x - x')$$

$$= \hat{F}(k)\hat{G}(k). \tag{4.5.17}$$

Hence, by taking the inverse transform of this result, we get

$$\frac{1}{2\pi} \int_{-\infty}^{\infty} e^{ikx}\, \hat{F}(k)\hat{G}(k)\, dk = \int_{-\infty}^{\infty} g(x')f(x - x')\, dx'$$

$$= \int_{-\infty}^{\infty} g(x - x')f(x')\, dx'. \tag{4.5.18}$$

The latter equality is accomplished by renaming the integration variables. Note that if $f(x) = \delta(x)$, then $\hat{F}(k) = 1$, and Eq. (4.5.18) reduces to the known transform pair for $g(x)$; that is, $g(x) = \frac{1}{2\pi} \int_{-\infty}^{\infty} \hat{G}(k)e^{ikx}\, dk$.

A special case of Eq. (4.5.18) is the so-called **Parseval formula**, obtained by taking $g(x) = \bar{f}(-x)$, where $\bar{f}(x)$ is the complex conjugate of $f(x)$, and evaluating Eq. (4.5.18) at $x = 0$:

$$\int_{-\infty}^{\infty} f(x)\bar{f}(x)\, dx = \frac{1}{2\pi} \int_{-\infty}^{\infty} \hat{F}(k)\hat{G}(k)\, dk.$$

The function $\hat{G}(k)$ is now the Fourier transform of $\bar{f}(-x)$:

$$\hat{G}(k) = \int_{-\infty}^{\infty} e^{-ikx} \bar{f}(-x)\, dx = \int_{-\infty}^{\infty} e^{ikx} \bar{f}(x)\, dx$$

$$= \overline{\left(\int_{-\infty}^{\infty} e^{-ikx} f(x)\, dx \right)} = \overline{\left(\hat{F}(k) \right)}.$$

Hence we have the Parseval formula:

$$\int_{-\infty}^{\infty} |f|^2(x)\, dx = \frac{1}{2\pi} \int_{-\infty}^{\infty} |\hat{F}(k)|^2\, dk. \tag{4.5.19}$$

In some applications, $\int_{-\infty}^{\infty} |f|^2(x)\, dx$ refers to the energy of a signal. Frequently it is the term $\int_{-\infty}^{\infty} |\hat{F}(k)|^2\, dk$ which is really measured (it is sometimes referred to as the **power spectrum**), which then gives the energy as per Eq. (4.5.19).

In those cases where $f(x)$ is an even or an odd function, then the Fourier transform pair reduces to the so-called **cosine transform** or **sine transform** pair. The fact that $f(x)$ is even or odd in x means that $\hat{F}(k)$ will be even or odd in k, and the pair (4.5.1)–(4.5.2) then reduces to statements about semi-infinite functions in the space L^2 on $(0, \infty)$ (i.e., $\int_0^{\infty} |f(x)|^2\, dx < \infty$) or in the space L^1 on $(0, \infty)$ (i.e., $\int_0^{\infty} |f(x)|\, dx < \infty$). For even functions, i.e., when $f(x) = f(-x)$, the following definitions

$$f(x) = \frac{1}{\sqrt{2}} f_c(x), \qquad \hat{F}(k) = \sqrt{2}\hat{F}_c(k) \tag{4.5.20}$$

(or more generally $\hat{F}(k) = a\hat{F}_c(k)$, $f(x) = bf_c(x)$, for $a = 2b$, $b \neq 0$) yield the Fourier cosine transform pair

$$f_c(x) = \frac{2}{\pi} \int_0^{\infty} \hat{F}_c(k) \cos kx\, dk, \tag{4.5.21}$$

$$\hat{F}_c(k) = \int_0^{\infty} f_c(x) \cos kx\, dx. \tag{4.5.22}$$

For odd functions, i.e., when $f(x) = -f(-x)$, the definitions

$$f(x) = \frac{1}{\sqrt{2}} f_s(x), \qquad \hat{F}(k) = -\sqrt{2}i\hat{F}_s(k)$$

yield the Fourier sine transform pair

$$f_s(x) = \frac{2}{\pi} \int_0^{\infty} \hat{F}_s(k) \sin kx\, dk, \tag{4.5.23}$$

$$\hat{F}_s(k) = \int_0^{\infty} f_s(x) \sin kx\, dx. \tag{4.5.24}$$

Obtaining the Fourier sine or cosine transform of a derivative employs integration by parts; for example,

$$\hat{F}_{c,1}(k) = \int_0^{\infty} f'(x) \cos kx\, dk = [f(x) \cos kx]_0^{\infty} + k \int_0^{\infty} f(x) \sin kx\, dk$$

$$= k\hat{F}_s(k) - f(0) \tag{4.5.25}$$

and

$$\hat{F}_{s,1}(k) = \int_0^\infty f'(x) \sin kx\, dk = [f(x) \sin kx]_0^\infty - k \int_0^\infty f(x) \cos kx\, dk$$

$$= -k\hat{F}_c(k), \tag{4.5.26}$$

are formulae for the first derivative. Similar results obtain for higher derivatives.

It turns out to be useful to extend the notion of Fourier transforms. One way to do this is to consider functions which have support only on a semi-interval. We take $f(x) = 0$ on $x < 0$ and replace $f(x)$ by $e^{-cx}f(x)$ with $c > 0$ when $x > 0$. Then Eqs. (4.5.1)–(4.5.2) satisfy, using $\hat{F}(k)$ from Eq. (4.5.2) in Eq. (4.5.1):

$$e^{-cx}f(x) = \frac{1}{2\pi} \int_{-\infty}^\infty dk\, e^{ikx} \left[\int_0^\infty e^{-ikx'} e^{-cx'} f(x')\, dx' \right],$$

hence

$$f(x) = \frac{1}{2\pi} \int_{-\infty}^\infty dk\, e^{(c+ik)x} \left[\int_0^\infty e^{-(c+ik)x'} f(x')\, dx' \right].$$

Within the above integrals we define $s = c + ik$, where c is a fixed real constant, and make the indicated redefinition of the limits of integration to obtain

$$f(x) = \frac{1}{2\pi i} \int_{c-i\infty}^{c+i\infty} ds\, e^{sx} \left(\int_0^\infty e^{-sx'} f(x')\, dx' \right)$$

or, in a form analogous to Eqs. (4.5.1)–(4.5.2),

$$f(x) = \frac{1}{2\pi i} \int_{c-i\infty}^{c+i\infty} \hat{F}(s) e^{sx}\, ds, \tag{4.5.27}$$

$$\hat{F}(s) = \int_0^\infty f(x) e^{-sx}\, dx. \tag{4.5.28}$$

Formulae (4.5.27)–(4.5.28) are referred to as the **Laplace transform** ($\hat{F}(s)$) and the **inverse Laplace transform** of a function ($f(x)$), respectively. The usual function space for $f(x)$ in the Laplace transform (analogous to $L^1 \cap L^2$ for $f(x)$ in Eq. (4.5.1)) consists of those functions satisfying

$$\int_0^\infty e^{-cx} |f(x)|\, dx < \infty. \tag{4.5.29}$$

Note that Re $s = c$ in Eqs. (4.5.27)–(4.5.28). If Eq. (4.5.29) holds for some $c > 0$; then $f(x)$ is said to be of **exponential order**.

The integral (4.5.27) is generally carried out by contour integration. The contour from $c - i\infty$ to $c + i\infty$ is referred to as the **Bromwich contour**, and c is taken to the right of all singularities in order to ensure (4.5.29). Closing the contour to the right will yield $f(x) = 0$ for $x < 0$.

We give two examples.

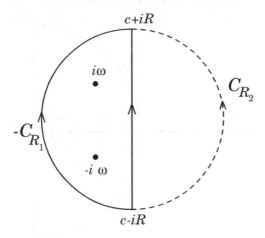

Figure 4.23 The Bromwich contour, Example 4.5.2

Example 4.5.2 Evaluate the inverse Laplace transform of $\hat{F}(s) = \frac{1}{s^2+\omega^2}$, with $\omega > 0$. Consider Figure 4.23 and

$$f(x) = \frac{1}{2\pi i} \int_{c-i\infty}^{c+i\infty} \frac{e^{sx}}{s^2 + w^2} \, ds.$$

For $x < 0$ we close the contour to the right of the Bromwich contour. Because no singularities are enclosed, we have, on C_{R_2},

$$s = c + Re^{i\theta}, \qquad \frac{-\pi}{2} < \theta < \frac{\pi}{2};$$

thus, $f(x) = 0$, for $x < 0$ because $\int_{C_{R_2}} \to 0$ as $R \to \infty$. On the other hand, for $x > 0$, closing to the left, and noting that $\int_{C_{R_1}} \to 0$ as $R \to \infty$ where on C_{R_1} we have $s = c + Re^{i\theta}$, and $\frac{\pi}{2} \le \theta \le \frac{3\pi}{2}$, yields

$$f(x) = \sum_{j=1}^{2} \mathrm{Res}\left(\frac{e^{sx}}{s^2 + w^2}; s_j\right), \qquad s_1 = iw, \quad s_2 = -iw,$$

$$f(x) = \frac{e^{iwx}}{2iw} - \frac{e^{-iwx}}{2iw} = \frac{\sin wx}{w}, \qquad x > 0.$$

Example 4.5.3 Evaluate the inverse Laplace transform of the function

$$\hat{F}(s) = s^{-a}, \qquad 0 < a < 1,$$

where we take the branch cut along the negative real axis (see Figure 4.24).

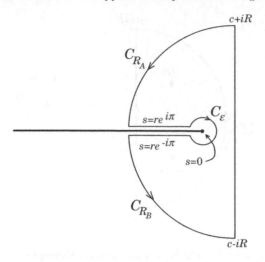

Figure 4.24 Contour for Example 4.5.3

As is always the case, $f(x) = 0$ for $x < 0$ when we close to the right. Closing to the left yields (schematically):

$$f(x) + \left(\int_{C_{R_A}} + \int_{C_{R_B}} + \int_{s=re^{i\pi}} + \int_{s=re^{-i\pi}} - \int_{C_\epsilon} \right) \frac{\hat{F}(s)}{2\pi i} e^{sx} \, ds = 0.$$

The right-hand side vanishes because there are no singularities enclosed by this contour. The integrals $\int_{C_{R_A}}$, $\int_{C_{R_B}}$, and $\int_{C_\epsilon}$ vanish as $R \to \infty$ (because on C_{R_A} we have $s = c + Re^{i\theta}$, $\frac{\pi}{2} < \theta < \pi$; on C_{R_B} we have $s = c + Re^{i\theta}$, $-\pi < \theta < \frac{-\pi}{2}$); and $\epsilon \to 0$ (because on C_ϵ we have $s = \epsilon e^{i\theta}$, $-\pi < \theta < \pi$). On C_ϵ we have

$$\left| \int_{C_\epsilon} \frac{F(s)}{2\pi i} e^{sx} \, ds \right| \leq \int_{-\pi}^{\pi} \epsilon^{1-a} \frac{e^{\epsilon x}}{2\pi} \, d\theta \xrightarrow[\epsilon \to 0]{} 0.$$

Hence,

$$f(x) - \frac{1}{2\pi i} \int_{r=\infty}^{0} r^{-a} e^{-ia\pi} e^{-rx} \, dr - \frac{1}{2\pi i} \int_{r=0}^{\infty} r^{-a} e^{ia\pi} e^{-rx} \, dr = 0$$

or

$$f(x) = \frac{\left(e^{ia\pi} - e^{-ia\pi} \right)}{2\pi i} \int_{0}^{\infty} r^{-a} e^{-rx} \, dr. \qquad (4.5.30)$$

An integral representation of the **gamma function**, or factorial function, is given by

$$\Gamma(z) = \int_{0}^{\infty} u^{z-1} e^{-u} \, du. \qquad (4.5.31)$$

This definition implies that $\Gamma(n) = (n-1)!$ when n is a positive integer: $n = 1, 2, \ldots$. Indeed we have by integration by parts

$$\Gamma(n+1) = [-u^n e^{-u}]_0^\infty + n \int_0^\infty u^{n-1} e^{-u} \, du$$

$$= n\Gamma(n), \tag{4.5.32}$$

and when $n = 1$, Eq. (4.5.31) directly yields $\Gamma(1) = 1$. Equation (4.5.32) is a difference equation, which, when supplemented with the starting condition $\Gamma(1) = 1$, can be solved for all n. So when $n = 1$, Eq. (4.5.32) yields $\Gamma(2) = \Gamma(1) = 1$; when $n = 2$, $\Gamma(3) = 2\Gamma(2) = 2!, \ldots$, and by induction, $\Gamma(n) = (n-1)!$ for positive integer n. We often use Eq. (4.5.31) for general values of z, requiring only that $\mathrm{Re}\, z > 0$ in order for there to be an integrable singularity at $u = 0$. With definition (4.5.31) and rescaling $rx = u$, we have

$$f(x) = \left(\frac{\sin a\pi}{\pi} \right) x^{a-1} \Gamma(1-a).$$

The Laplace transform of a derivative is readily calculated.

$$\hat{F}_1(s) = \int_0^\infty f'(x) e^{-sx} \, dx = [f(x)e^{-sx}]_0^\infty + s \int_0^\infty f(x)e^{-sx} \, dx$$

$$= s\hat{F}(s) - f(0).$$

Integration by parts n times we find that

$$\hat{F}_n(s) = \int_0^\infty f^{(n)}(x)e^{-sx} \, dx$$

$$= s^n \hat{F}(s) - f^{(n-1)}(0) - sf^{(n-2)}(0) - \cdots - s^{n-2} f'(0) - s^{n-1} f(0). \tag{4.5.33}$$

The Laplace transform analog of the convolution result for Fourier transforms (4.5.17) takes the following form. Define

$$h(x) = \int_0^x g(x') f(x - x') \, dx'. \tag{4.5.34}$$

We show that the Laplace transform of h is the product of the Laplace transform of g and f. First,

$$\hat{H}(s) = \int_0^\infty h(x) e^{-sx} \, dx = \int_0^\infty dx \, e^{-sx} \int_0^x g(x') f(x - x') \, dx'.$$

By interchanging integrals, we have that

$$\hat{H}(s) = \int_0^\infty dx' \, g(x') \int_{x'}^\infty dx \, e^{-sx} \, f(x - x')$$

$$= \int_0^\infty dx' \, g(x') \, e^{-sx'} \int_{x'}^\infty dx \, e^{-s(x-x')} \, f(x - x')$$

$$= \int_0^\infty dx' \, g(x') \, e^{-sx'} \int_0^\infty du \, e^{-su} \, f(u),$$

hence

$$\hat{H}(s) = \hat{G}(s)\hat{F}(s). \tag{4.5.35}$$

The convolution formulae (4.5.34–4.5.35) can be used in a variety of ways. We note that if $\hat{F}(s) = 1/s$, then Eq. (4.5.27) implies $f(x) = 1$. We use this in the following example.

Example 4.5.4   Evaluate $h(x)$ where the Laplace transform of $h(x)$ is given by

$$\hat{H}(s) = \frac{1}{s \, (s^2 + 1)}.$$

We have two functions:

$$\hat{F}(s) = \frac{1}{s}, \qquad \hat{G}(s) = \frac{1}{s^2 + 1}.$$

Using the result of Example 4.5.2 gives $g(x) = \sin x$, hence Eq. (4.5.35) implies that

$$h(x) = \int_0^x \sin x' \, dx' = 1 - \cos x.$$

This may also be found by the partial fractions decomposition

$$\hat{H}(s) = \frac{1}{s} - \frac{s}{s^2 + 1}$$

and noting that $\hat{F}(s) = \frac{s}{s^2+w^2}$ has, as its Laplace transform, $\hat{f}(x) = \cos wx$.

4.5.1 Problems for Section 4.5

1. Obtain the Fourier transforms of the following functions:

(a) $e^{-|x|}$;   (b) $\dfrac{1}{x^2 + a^2}$, $a^2 > 0$;

(c) $\dfrac{1}{(x^2 + a^2)^2}$, $a^2 > 0$;   (d) $\dfrac{\sin ax}{(x + b)^2 + c^2}$, $a, b, c > 0$.

2. Obtain the inverse Fourier transform of the following functions:

(a) $\dfrac{1}{k^2 + w^2}$, $w^2 > 0$;   (b) $\dfrac{1}{(k^2 + w^2)^2}$, $w^2 > 0$.

Note the duality between direct and inverse Fourier transforms.

3. Show that the Fourier transform of the "Gaussian" $f(x) = \exp\left(-\dfrac{(x-x_0)^2}{a^2}\right)$, for x_0, a real, is also a Gaussian:

$$\hat{F}(k) = a\sqrt{\pi}\, e^{-\left(\frac{ka}{2}\right)^2} e^{-ikx_0}.$$

4. Obtain the Fourier transform of the following functions, and thereby show that the Fourier transforms of hyperbolic secant functions are also related to hyperbolic secant functions.

(a) $\text{sech}\,[a(x - x_0)]\, e^{iwx}$,   for real a, x_0, w;

(b) $\text{sech}^2\,[a(x - x_0)]$,   for real a, x_0.

5. (a) Obtain the Fourier transform of

$$f(x) = \dfrac{\sin wx}{x},   w > 0.$$

(b) Show that $f(x)$ in part (a) is not L_1; that is $\displaystyle\int_{-\infty}^{\infty} |f(x)|\, dx$ does not exist. Despite this fact, we can obtain the Fourier transform; so $f(x) \in L_1$ is a sufficient condition, but is not necessary, for the Fourier transform to exist.

6. Suppose we are given the differential equation

$$\dfrac{d^2u}{dx^2} - w^2 u = -f(x),$$

with $u(x = \pm\infty) = 0$, $w > 0$.

(a) Take the Fourier transform of this equation to find (using Eq. (4.5.16ab))

$$\hat{U}(k) = \hat{F}(k)/\left(k^2 + w^2\right),$$

where $\hat{U}(k)$ and $\hat{F}(k)$ are the Fourier transform of $u(x)$ and $f(x)$, respectively.

(b) Use the convolution product (4.5.18) to deduce that

$$u(x) = \dfrac{1}{2w}\int_{-\infty}^{\infty} e^{-w|x-\zeta|} f(\zeta)\, d\zeta$$

and thereby obtain the solution of the differential equation.

7. Obtain the Fourier sine transform of the following functions:

$$\text{(a) } e^{-\omega x}, \quad \omega > 0; \qquad \text{(b) } \frac{x}{x^2 + 1}; \qquad \text{(c) } \frac{\sin \omega x}{x^2 + 1}, \quad \omega > 0.$$

8. Obtain the Fourier cosine transform of the following functions:

$$\text{(a) } e^{-\omega x}, \quad \omega > 0; \qquad \text{(b) } \frac{1}{x^2 + 1}; \qquad \text{(c) } \frac{\cos \omega x}{x^2 + 1}, \quad \omega > 0.$$

9. (a) Assuming $u(\infty) = 0$, establish that

$$\int_0^\infty \frac{d^2 u}{dx^2} \sin kx \, dx \, k u(0) - k^2 \hat{U}_s(k),$$

where $\hat{U}_s(k) = \int_0^\infty u(x) \sin kx \, dx$ is the sine transform of $u(x)$ (Eq. (4.5.24)).

(b) Use this result to show that taking the Fourier sine transform of

$$\frac{d^2 u}{dx^2} - \omega^2 u = -f(x),$$

with $u(0) = u_0$, $u(\infty) = 0$, and $\omega > 0$, yields for the Fourier sine transform of $u(x)$,

$$\hat{U}_s(k) = \frac{u_0 k + \hat{F}_s(k)}{k^2 + \omega^2},$$

where $\hat{F}_s(k)$ is the Fourier sine transform of $f(x)$.

(c) Use the analogue of the convolution product for the Fourier sine transform:

$$\frac{1}{2} \int_0^\infty [f(|x - \zeta|) - f(x + \zeta)] g(\zeta) \, d\zeta = \frac{2}{\pi} \int_0^\infty \sin kx \hat{F}_c(k) \hat{G}_s(k) \, dk$$

(where $\hat{F}_c(k)$ is the Fourier cosine transform of $f(x)$ and $\hat{G}_s(k)$, is the Fourier sine transform of $g(x)$) to show that the solution of the differential equation is given by

$$u(x) = u_0 e^{-\omega x} + \frac{1}{2\omega} \int_0^\infty \left(e^{-\omega |x - \zeta|} - e^{-\omega (x + \zeta)} \right) f(\zeta) \, d\zeta.$$

(The convolution product for the sine transform can be deduced from the usual convolution product (4.5.18) by assuming in the latter formula that $f(x)$ is even and $g(x)$ is an odd function of x.)

10. (a) Assuming $u(\infty) = 0$ establish that

$$\int_0^\infty \frac{d^2 u}{dx^2} \cos kx \, dx = -\frac{du}{dx}(0) - k^2 \hat{U}_c(k),$$

where $\hat{U}_c(k) = \int_0^\infty u(x) \cos kx \, dx$ is the cosine transform of $u(x)$.

(b) Use this result to show that by taking the Fourier cosine transform of

$$\frac{d^2u}{dx^2} - \omega^2 u = -f(x), \qquad \text{with } \frac{du}{dx}(0) = u_0', \qquad u(\infty) = 0, \quad \omega > 0,$$

yields for the Fourier cosine transform of $u(x)$,

$$\hat{U}_c(k) = \frac{\hat{F}_c(k) - u_0'}{k^2 + \omega^2},$$

where $\hat{F}_c(k)$ is the Fourier cosine transform of $f(x)$.

(c) Use the analogue of the convolution product of the Fourier cosine transform,

$$\frac{1}{2} \int_0^\infty (f(|x - \zeta|) + f(x + \zeta)) g(\zeta) \, d\zeta = \frac{2}{\pi} \int_0^\infty \cos kx \hat{F}_c(k) \hat{G}_c(k) \, dk,$$

to show that the solution of the differential equation is given by

$$u(x) = -\frac{u_0'}{\omega} e^{-\omega x} + \frac{1}{2\omega} \int_0^\infty \left(e^{-\omega|x - \zeta|} + e^{-\omega|x + \zeta|} \right) f(\zeta) \, d\zeta.$$

(The convolution product for the cosine transform can be deduced from the usual convolution product (4.5.18) by assuming in the latter formula that $f(x)$ and $g(x)$ are even functions of x.)

11. Obtain the inverse Laplace transforms of the following functions, assuming ω, $\omega_1, \omega_2 > 0$:

(a) $\dfrac{s}{s^2 + \omega^2}$;   (b) $\dfrac{1}{(s + \omega)^2}$;   (c) $\dfrac{1}{(s + \omega)^n}$;

(d) $\dfrac{s}{(s + \omega)^n}$;   (e) $\dfrac{1}{(s + \omega_1)(s + \omega_2)}$;   (f) $\dfrac{1}{s^2(s^2 + \omega^2)}$;

(g) $\dfrac{1}{(s + \omega_1)^2 + \omega_2^2}$;   (h) $\dfrac{1}{(s^2 - \omega^2)^2}$.

12. Show explicitly that the Laplace transform of the second derivative of a function of x satisfies

$$\int_0^\infty f''(x) e^{-sx} \, dx = s^2 \hat{F}(s) - s f(0) - f'(0).$$

13. Establish the following relationships, where we use the notation $\mathcal{L}(f(x)) \equiv \hat{F}(s)$.

(a)

$$\mathcal{L}\left(e^{ax} f(x)\right) = \hat{F}(s - a).$$

(b)

$$\mathcal{L}\left(f(x-a)H(x-a)\right) = e^{-as}\hat{F}(s),$$

where

$$H(x) = \begin{cases} 1 & x \geq 0, \\ 0 & x < 0. \end{cases}$$

(c) Use the convolution product formula for Laplace transforms to show that the inverse Laplace transform of

$$\hat{H}(s) = \frac{1}{s^2\left(s^2 + \omega^2\right)}, \qquad \omega > 0$$

satisfies

$$h(x) = \frac{1}{\omega} \int_0^x x' \sin\omega(x - x')\,dx' = \frac{x}{\omega^2} - \frac{\sin\omega x}{\omega^3}.$$

Verify this result by using partial fractions.

14. (a) Show that the inverse Laplace transform of $\hat{F}(s) = e^{-as^{1/2}}/s$, with $a > 0$, is given by

$$f(x) = 1 - \frac{1}{\pi} \int_0^\infty \frac{\sin(ar^{1/2})}{r} e^{-rx}\,dr.$$

Note that the integral converges at $r = 0$.

(b) Use the definition of the error function integral,

$$\mathrm{erf}\,x = \frac{2}{\sqrt{\pi}} \int_0^x e^{-r^2}\,dr,$$

to show that an alternative form for $f(x)$ is

$$f(x) = 1 - \mathrm{erf}\left(\frac{a}{2\sqrt{x}}\right).$$

(c) Show that the inverse Laplace transform of $\hat{F}(s) = e^{-as^{1/2}}/s^{1/2}$, for $a > 0$, is given by

$$f(x) = \frac{1}{\pi} \int_0^\infty \frac{\cos(ar^{1/2})}{r^{1/2}} e^{-rx}\,dr,$$

or the equivalent forms

$$f(x) = \frac{2}{\pi\sqrt{x}} \int_0^\infty \cos\frac{au}{\sqrt{x}} e^{-u^2}\,du = \frac{1}{\sqrt{\pi x}} e^{-a^2/4x}.$$

Verify this result by taking the derivative with respect to a in the formula of part (a).

(d) Following the procedure of part (c) show that the inverse Laplace transform of $\hat{F}(s) = e^{-as^{1/2}}$ is given by

$$f(x) = \frac{a}{2\sqrt{\pi}x^{3/2}} e^{-a^2/4x}.$$

15. Show that the inverse Laplace transform of the function $\hat{F}(s) = \dfrac{1}{\sqrt{s^2 + \omega^2}}$, for $\omega > 0$, is given by

$$f(x) = \frac{1}{\pi} \int_{-\omega}^{\omega} \frac{e^{ixr}}{\sqrt{\omega^2 - r^2}} \, dr = \frac{2}{\pi} \int_0^1 \frac{\cos(\omega x \rho)}{\sqrt{1 - \rho^2}} \, d\rho.$$

(The latter integral is a representation of $J_0(\omega x)$, the Bessel function of order zero.) Hint: deform the contour around the branch points $s = \pm i\omega$, then show that the large contour at infinity and small contours encircling $\pm i\omega$ are vanishingly small. It is convenient to use the polar representations $s + i\omega_1 = r_1 e^{i\theta_1}$ and $s - i\omega_2 = r_2 e^{i\theta_2}$, where $-\dfrac{3\pi}{2} < \theta_i \le \dfrac{\pi}{2}$ for $i = 1, 2$, and $\left(s^2 + \omega^2\right)^{1/2} = \sqrt{r_1 r_2} \, e^{i(\theta_1 + \theta_2)/2}$. The contributions on both sides of the cut add to give the result.

16. Show that the inverse Laplace transform of the function $\hat{F}(s) = \dfrac{\log s}{s^2 + \omega^2}$, for $\omega > 0$, is given by

$$f(x) = \frac{\pi}{2\omega} \cos \omega x - \int_0^\infty \frac{e^{-rx}}{r^2 + \omega^2} \, dr \qquad \text{for } x > 0.$$

Hint: Choose the branch $s = re^{i\theta}$, for $-\pi \le \theta < \pi$. Show that the contour at infinity and around the branch point $s = 0$ are vanishingly small. There are two contributions along the branch cut which add to give the second (integral) term: the first is due to the poles at $s = \pm i\omega$.

17. (a) Show that the inverse Laplace transform of $\hat{F}_1(s) = \log s$ is given by $f_1(x) = -1/x$.

(b) Do the same for $\hat{F}_2(s) = \log(s + 1)$ to obtain $f_2(x) = -\dfrac{e^{-x}}{x}$.

(c) Find the inverse Laplace transform $\hat{F}(s) = \log\left(\dfrac{s+1}{s}\right)$ to obtain $f(x) = \dfrac{1 - e^{-x}}{x}$, by subtracting the results of parts (a) and (b).

(d) Show that we can get this result directly by encircling both the $s = 0$ and $s = -1$ branch points and using the polar representations $s + 1 = r_1 e^{i\theta_1}$, $s = r_2 e^{i\theta_2}$, with $-\pi \le \theta_i < \pi$, $i = 1, 2$.

18. Establish the following results by formally inverting the Laplace transform.

$$\hat{F}(s) = \frac{1}{s} \frac{1 - e^{-\ell s}}{1 + e^{-\ell s}}, \quad \ell > 0,$$

$$f(x) = \sum_{n=1,3,5,\ldots} \left(\frac{4}{n\pi}\right) \sin \frac{n\pi x}{\ell}.$$

Note that there are an infinite number of poles present in $\hat{F}(s)$; consequently a straightforward continuous limit as $R \to \infty$ on a large semicircle will pass through one of these poles. Consider a large semi-circle C_{R_N}, where R_N encloses N poles (e.g. $R_N = \frac{\pi i}{\ell}(N + \frac{1}{2})$) and show that as $N \to \infty$, $R_N \to \infty$, and the integral along C_{R_N} will vanish. Choosing appropriate sequences such as those in this example, the inverse Laplace transform containing an infinite number of poles can be calculated.

19. Establish the following result by formally inverting the Laplace transform:

$$\hat{F}(s) = \frac{1}{s} \frac{\sinh(sy)}{\sinh(s\ell)}, \quad \ell > 0,$$

$$f(x) = \frac{y}{\ell} + \sum_{n=1}^{\infty} \frac{2(-1)^n}{n\pi} \sin\left(\frac{n\pi y}{\ell}\right) \cos\left(\frac{n\pi x}{\ell}\right).$$

See the remark at the end of Problem 18, which explains how to show how the inverse Laplace transform can be proven to be valid in a situation such as this where there are an infinite number of poles.

4.6 Applications of Transforms to Differential Equations

Fourier and Laplace transforms are particularly valuable for solving differential equations in infinite and semi-infinite domains. In this section we shall describe some typical examples. The discussion is not intended to be complete. The aim of this section is to elucidate the transform technique, not to detail theoretical aspects regarding differential equations. The reader only needs basic training in the calculus of several variables to be able to follow the analysis. First we give an elementary example from ordinary differential equations (ODEs). Then we will use various classical partial differential equations (PDEs) as vehicles to illustrate the associated methodology. Herein we will consider well-posed problems that will yield unique solutions. More general PDEs and the notion of well-posedness are investigated in considerable detail in courses on PDEs.

Example 4.6.1 Solve the following ODE

$$\frac{d^2 y}{dx^2} + \omega^2 y = \alpha, \quad \text{with } y(x = 0) = \frac{dy}{dx}(x = 0) = 0, \quad \text{constant } \omega, \alpha.$$

Using the methods described in the previous section we take the Laplace transform of the equation. This yields

$$s^2 \widehat{Y}(s) - sy(0) - y'(0) + \omega^2 \widehat{Y}(s) = \frac{\alpha}{s},$$

where $\widehat{Y}(s)$ denotes the Laplace transform of $y(x)$. Using the given initial conditions we find

$$\widehat{Y}(s) = \frac{\alpha}{s(s^2 + \omega^2)} = \frac{\alpha}{\omega^2} \left(\frac{1}{s} - \frac{s}{s^2 + \omega^2} \right).$$

Then, carrying out the inverse Laplace transform, yields

$$y(x) = \frac{\alpha}{\omega^2} (1 - \cos \omega x),$$

which satisfies the given ODE and initial conditions.

Example 4.6.2 Steady state heat flow in a semi-infinite domain obeys Laplace's equation. Solve for the bounded solution of Laplace's equation,

$$\frac{\partial^2 \phi(x, y)}{\partial x^2} + \frac{\partial^2 \phi(x, y)}{\partial y^2} = 0, \tag{4.6.1}$$

in the region $-\infty < x < \infty$, $y > 0$, where on $y = 0$ we are given $\phi(x, 0) = h(x)$ real, with $h(x) \in L^1 \cap L^2$ (i.e., $\int_{-\infty}^{\infty} |h(x)| \, dx < \infty$ and $\int_{-\infty}^{\infty} |h(x)|^2 \, dx < \infty$).

This example will allow us to solve (4.6.1) by Fourier transforms. Denoting the Fourier transform in x of $\phi(x, y)$ by $\widehat{\Phi}(k, y)$:

$$\widehat{\Phi}(k, y) = \int_{-\infty}^{\infty} e^{-ikx} \phi(x, y) \, dx,$$

taking the Fourier transform of Eq. (4.6.1), and using Eqs. (4.5.16a,b) for the Fourier transform of derivatives (assuming the validity of interchanging y-derivatives and integrating over k, which can be verified *a posteriori*), we have

$$\frac{\partial^2 \widehat{\Phi}}{\partial y^2} - k^2 \widehat{\Phi} = 0. \tag{4.6.2}$$

Hence

$$\widehat{\Phi}(k, y) = A(k)e^{ky} + B(k)e^{-ky},$$

where $A(k)$ and $B(k)$ are arbitrary functions of k, to be specified by the boundary conditions. We require that $\widehat{\Phi}(k, y)$ be bounded for all $y > 0$. In order that $\widehat{\Phi}(k, y)$ yield a bounded function $\phi(x, y)$, we need

$$\widehat{\Phi}(k, y) = C(k)e^{-|k|y}. \tag{4.6.3}$$

Denoting the Fourier transform of $\phi(x, 0) = h(x)$ by $\widehat{H}(k)$ fixes $C(k) = \widehat{H}(k)$, so that

$$\widehat{\Phi}(k, y) = \widehat{H}(k)e^{-|k|y}. \tag{4.6.4}$$

From Eq. (4.5.1) by direct integration (contour integration is not necessary) we find that $\widehat{F}(k, y) = e^{-|k|y}$ is the Fourier transform of $f(x, y) = \frac{1}{\pi}\frac{y}{x^2+y^2}$; thus from the convolution formula Eq. (4.5.17) the solution to Eq. (4.6.1) is given by

$$\phi(x, y) = \frac{1}{\pi} \int_{-\infty}^{\infty} \frac{y\, h(x')}{(x - x')^2 + y^2}\, dx'. \tag{4.6.5}$$

If $h(x)$ were taken to be a Dirac delta function concentrated at $x = \zeta$, i.e., $h(x) = h_s(x - \zeta) = \delta(x - \zeta)$, then $\widehat{H}(k) = e^{-ik\zeta}$, and from Eq. (4.6.4) directly (or Eq. (4.6.5)) a special solution to Eq. (4.6.1), $\phi_s(x, y)$ is

$$\phi_s(x, y) = G(x - \zeta, y) = \frac{1}{\pi}\left(\frac{y}{(x - \zeta)^2 + y^2}\right). \tag{4.6.6}$$

The function $G(x-\zeta, y)$ is called a **Green's function**; it is a fundamental solution to Laplace's equation in this region. Green's functions have the property of solving a given equation with delta function inhomogeneity. From the boundary values $h_s(x - \zeta) = \delta(x - \zeta)$ we may construct arbitrary initial values

$$\phi(x, 0) = \int_{-\infty}^{\infty} h(\zeta)\, \delta(x - \zeta)\, d\zeta = h(x) \tag{4.6.7a}$$

and, because Laplace's equation is linear, we have by superposition that the general solution satisfies

$$\phi(x, y) = \int_{-\infty}^{\infty} h(\zeta)\, G(x - \zeta, y)\, d\zeta, \tag{4.6.7b}$$

which is Eq. (4.6.5), noting that ζ or x' are dummy integration variables. In many applications it is sufficient to obtain the Green's function of the underlying differential equation.

The formula (4.6.5) is sometimes referred to as the **Poisson formula for a half plane**. Although we derived it via transform methods, it is worth noting that a pair of such formulae can be derived from Cauchy's integral formula. We describe this alternative method now.

Let $f(z)$ be analytic on the real axis and in the upper half plane, and assume $f(\zeta) \to 0$ uniformly as $\zeta \to \infty$. Using a large closed semicircular contour such as that depicted in Figure 4.3 we have

$$f(z) = \frac{1}{2\pi i} \oint_C \frac{f(\zeta)}{\zeta - z} \, d\zeta,$$

$$0 = \frac{1}{2\pi i} \oint_C \frac{f(\zeta)}{\zeta - \bar{z}} \, d\zeta,$$

where Im $z > 0$ (in the second formula there is no singularity because the contour closes in the upper half plane and $\zeta = \bar{z}$ in the lower half plane). Adding and subtracting yields

$$f(z) = \frac{1}{2\pi i} \oint_C f(\zeta) \left(\frac{1}{\zeta - z} \pm \frac{1}{\zeta - \bar{z}} \right) d\zeta.$$

The semicircular portion of the contour C_R vanishes as $R \to \infty$, implying the following on Im $\zeta = 0$ for the plus and minus parts of the above integral, respectively, where we write writing $z = x + iy$ and $\zeta = x' + iy'$:

$$f(x, y) = \frac{1}{\pi i} \int_{-\infty}^{\infty} f(x', y' = 0) \left(\frac{x' - x}{(x - x')^2 + y^2} \right) dx',$$

$$f(x, y) = \frac{1}{\pi} \int_{-\infty}^{\infty} f(x', y' = 0) \left(\frac{y}{(x - x')^2 + y^2} \right) dx'.$$

Putting

$$f(z) = f(x, y) = u(x, y) + i v(x, y), \qquad \text{Re } f(x, y = 0) = u(x, 0) = h(x),$$

and taking the imaginary part of the first and the real part of the second of the above formulae, yields the conjugate Poisson formulae for a half plane:

$$v(x, y) = \frac{-1}{\pi} \int_{-\infty}^{\infty} h(x') \left(\frac{x' - x}{(x - x')^2 + y^2} \right) dx',$$

$$u(x, y) = \frac{1}{\pi} \int_{-\infty}^{\infty} h(x') \left(\frac{y}{(x - x')^2 + y^2} \right) dx'.$$

Identifying $u(x, y)$ as $\phi(x, y)$, we see that the harmonic function $u(x, y)$ (because $f(z)$ is analytic its real and imaginary parts satisfy Laplace's equation) is given by the same formula as Eq. (4.6.5). Moreover, we note that the imaginary part of $f(z)$, namely $v(x, y)$, is determined by the real part of $f(z)$ on the boundary. We see that we cannot arbitrarily prescribe both the real and imaginary parts of $f(z)$ on the boundary. These formulae are valid for a half plane. Similar formulae can be obtained by this method for a circle.

Laplace's equation, (4.6.1), is typical of a steady-state situation, for example, as mentioned earlier, steady state heat flow in a uniform metal plate. If we have time-dependent heat flow, the diffusion equation

$$\frac{\partial \phi}{\partial t} = k \nabla^2 \phi \tag{4.6.8}$$

is a relevant equation where k is the diffusion coefficient. In Eq. (4.6.8), ∇^2 is the Laplacian operator, which in two dimensions is given by $\nabla^2 = \frac{\partial^2}{\partial x^2} + \frac{\partial^2}{\partial y^2}$. In one dimension, taking $k = 1$ for convenience, we have the following initial value problem:

$$\frac{\partial \phi(x,t)}{\partial t} = \frac{\partial^2 \phi(x,t)}{\partial x^2}. \tag{4.6.9}$$

The Green's function for the problem on the line $-\infty < x < \infty$ is obtained by solving Eq. (4.6.9) subject to

$$\phi(x,0) = \delta(x - \zeta).$$

Example 4.6.3   Solve for the Green's function of Eq. (4.6.9). Define

$$\widehat{\Phi}(k,t) = \int_{-\infty}^{\infty} e^{-ikx} \phi(x,t)\, dx$$

whereupon the Fourier transform of Eq. (4.6.9) satisfies

$$\frac{\partial \widehat{\Phi}(k,t)}{\partial t} = -k^2 \widehat{\Phi}(k,t), \tag{4.6.10}$$

hence

$$\widehat{\Phi}(k,t) = \widehat{\Phi}(k,0) e^{-k^2 t} = e^{-ik\zeta - k^2 t}, \tag{4.6.11}$$

where $\widehat{\Phi}(k,0) = e^{-ik\zeta}$ is the Fourier transform of $\phi(x,0) = \delta(x - \zeta)$. Thus, by the inverse Fourier transform, and denoting by $G(x - \zeta, t)$ the inverse transform of (4.6.11), gives

$$G(x - \zeta, t) = \frac{1}{2\pi} \int_{-\infty}^{\infty} e^{ik(x-\zeta) - k^2 t}\, dk = e^{-(x-\zeta)^2/4t} \cdot \frac{1}{2\pi} \int_{-\infty}^{\infty} e^{-\left(k - i\frac{x-\zeta}{2t}\right)^2 t}\, dk$$

$$= \frac{e^{-\frac{(x-\zeta)^2}{4t}}}{2\sqrt{\pi t}}, \tag{4.6.12}$$

where we use $\int_{-\infty}^{\infty} e^{-u^2}\, du = \sqrt{\pi}$. Arbitrary initial values are included by again observing that

$$\phi(x,0) = h(x) = \int_{-\infty}^{\infty} h(\zeta)\delta(x - \zeta)\, d\zeta,$$

which implies

$$\phi(x,t) = \int_{-\infty}^{\infty} G(x - \zeta, t)h(\zeta)\, d\zeta = \frac{1}{2\sqrt{\pi t}} \int_{-\infty}^{\infty} h(\zeta)\, e^{-\frac{(x-\zeta)^2}{4t}}\, d\zeta. \tag{4.6.13}$$

The above solution to Eq. (4.6.9) could also be obtained by using Laplace trans-forms. It is instructive to show how the method proceeds in this case. We begin by introducing the Laplace transform of $\phi(x, t)$ with respect to t:

$$\widehat{\Phi}(x, s) = \int_0^\infty e^{-st} \phi(x, t) \, dt. \tag{4.6.14}$$

Taking the Laplace transform in t of Eq. (4.6.8), with $\phi(x, 0) = \delta(x - \zeta)$, yields

$$\frac{\partial^2 \widehat{\Phi}}{\partial x^2}(x, s) - s\widehat{\Phi}(x, s) = -\delta(x - \zeta). \tag{4.6.15}$$

Hence the Laplace transform of the Green's function to Eq. (4.6.9) satisfies Eq. (4.6.15). We remark that generally speaking, any function $\hat{G}(x - \zeta)$ satisfying

$$L\,\hat{G}(x - \zeta) = -\delta(x - \zeta),$$

where L is a linear differential operator, is referred to as a Green's function. The general solution corresponding to $\phi(x, 0) = h(x)$ is obtained by superposition: $\phi(x, t) = \int_{-\infty}^\infty G(x - \zeta)h(\zeta)\,d\zeta$. Equation (4.6.15) is solved by first finding the bounded homogeneous solutions on $-\infty < x < \infty$, for $(x - \zeta) > 0$ and $(x - \zeta) < 0$:

$$\begin{aligned} \widehat{\Phi}_+(x - \zeta, s) &= A(s)e^{-s^{1/2}(x-\zeta)} &\quad \text{for } x - \zeta > 0; \\ \widehat{\Phi}_-(x - \zeta, s) &= B(s)e^{s^{1/2}(x-\zeta)} &\quad \text{for } x - \zeta < 0 \end{aligned} \right\} \tag{4.6.16}$$

where we take $s^{1/2}$ to have a branch cut on the negative real axis; that is, $s = re^{i\theta}$, $-\pi \le \theta < \pi$. This will allow us to readily invert the Laplace transform (Re $s > 0$).

The coefficients $A(s)$ and $B(s)$ in Eq. (4.6.16) are found by:
(a) requiring continuity of $\widehat{\Phi}(x - \zeta, s)$ at $x = \zeta$; and by
(b) integrating Eq. (4.6.15) from $x = \zeta - \epsilon$, to $x = \zeta + \epsilon$, and taking the limit as $\epsilon \to 0^+$.

This yields a jump condition on $\dfrac{\partial \widehat{\Phi}}{\partial x}(x - \zeta, s)$:

$$\left[\frac{\partial \widehat{\Phi}}{\partial x}(x - \zeta, s) \right]_{x-\zeta=0^-}^{x-\zeta=0^+} = -1. \tag{4.6.17}$$

Continuity yields $A(s) = B(s)$, and Eq. (4.6.17) gives

$$-s^{1/2}A(s) - s^{1/2}B(s) = -1, \tag{4.6.18a}$$

hence

$$A(s) = B(s) = \frac{1}{2s^{1/2}}. \tag{4.6.18b}$$

Using Eq. (4.6.16), $\widehat{\Phi}(x - \zeta, s)$ is written in the compact form:

$$\widehat{\Phi}(x - \zeta, s) = \frac{e^{-s^{1/2}|x-\zeta|}}{2s^{1/2}}. \tag{4.6.19}$$

The solution $\phi(x, t)$ is found from the inverse Laplace transform:

$$\phi(x, t) = \frac{1}{2\pi i} \int_{c-i\infty}^{c+i\infty} \frac{e^{-s^{1/2}|x-\zeta|} e^{st}}{2s^{1/2}} \, ds \tag{4.6.20}$$

for $c > 0$. To evaluate Eq. (4.6.20), we employ the same "keyhole" contour as in Example 4.5.3 (see Figure 4.24). There are no singularities enclosed, and the contours C_R and C_ϵ at infinity and at the origin vanish in the limit $R \to \infty$, $\epsilon \to 0$ respectively. We only obtain contributions along the top and bottom of the branch cut to find

$$\phi(x, t) = \frac{-1}{2\pi i} \int_{\infty}^{0} \frac{e^{-ir^{1/2}|x-\zeta|} e^{-rt}}{2r^{1/2}e^{i\pi/2}} e^{i\pi} \, dr$$

$$+ \frac{-1}{2\pi i} \int_{0}^{\infty} \frac{e^{ir^{1/2}|x-\zeta|} e^{-rt}}{2r^{1/2}e^{-i\pi/2}} e^{-i\pi} \, dr. \tag{4.6.21}$$

In the second integral we put $r^{1/2} = u$; in the first we put $r^{1/2} = u$ and then take $u \to -u$, whereupon we find the same answer as before (see Eq. (4.6.12)):

$$\phi(x, t) = \frac{1}{2\pi} \int_{-\infty}^{\infty} e^{-u^2 t + iu|x-\zeta|} \, du = \frac{1}{2\pi} \int_{-\infty}^{\infty} e^{-\left(u - i\frac{|x-\zeta|}{2t}\right)^2 t} e^{-\frac{(x-\zeta)^2}{4t}} \, du$$

$$= \frac{e^{-(x-\zeta)^2/4t}}{2\sqrt{\pi t}}. \tag{4.6.22}$$

The Laplace transform method can also be applied to problems in which the spatial variable is on the semi-infinite domain. However, rather than use Laplace transforms, for variety and illustration, we show below how the sine transform can be used on Eq. (4.6.9) with the following boundary conditions:

$$\phi(x, 0) = 0, \qquad \phi(x = 0, t) = h(t), \qquad \lim_{x \to \infty} \frac{\partial \phi}{\partial x}(x, t) = 0. \tag{4.6.23}$$

Define, following Section 4.5,

$$\phi(x, t) = \frac{2}{\pi} \int_{0}^{\infty} \widehat{\Phi}_s(k, t) \sin kx \, dk, \tag{4.6.24a}$$

$$\widehat{\Phi}_s(k, t) = \int_{0}^{\infty} \phi(x, t) \sin kx \, dx. \tag{4.6.24b}$$

We now operate on Eq. (4.6.9) with the integral $\int_0^\infty dx \sin kx$, and via integration by parts, find

$$\int_0^\infty \frac{\partial^2 \phi}{\partial x^2} \sin kx \, dx = \left[\frac{\partial \phi}{\partial x}(x,t) \sin kx \right]_{x=0}^\infty - k \int_0^\infty \frac{\partial \phi}{\partial x} \cos kx \, dx$$

$$= k\phi(0,t) - k^2 \widehat{\Phi}_s(k,t), \tag{4.6.25}$$

whereupon the transformed version of Eq. (4.6.9) is

$$\frac{\partial \widehat{\Phi}_s}{\partial t}(k,t) + k^2 \widehat{\Phi}_s(k,t) = k \, h(t). \tag{4.6.26}$$

The solution of Eq. (4.6.26) with $\phi(x,0) = 0$ is given by

$$\widehat{\Phi}_s(k,t) = \int_0^t h(t') k \, e^{-k^2(t-t')} \, dt'. \tag{4.6.27}$$

If $\phi(x,0)$ were nonzero, then Eq. (4.6.27) would have another term. For simplicity we only consider the case $\phi(x,0) = 0$. Therefore

$$\phi(x,t) = \frac{2}{\pi} \int_0^\infty dk \sin kx \left\{ \int_0^t h(t') e^{-k^2(t-t')} k \, dt' \right\}. \tag{4.6.28}$$

By integration (use $\sin kx = (e^{ikx} - e^{-ikx})/(2i)$ and integrate by parts to obtain integrals such as those in (4.6.12)) we can show that

$$J(x,t-t') = \int_0^\infty k \, e^{-k^2(t-t')} \sin kx \, dk = \frac{\sqrt{\pi} \, x e^{-x^2/4(t-t')}}{4(t-t')^{3/2}}, \tag{4.6.29}$$

hence by interchanging integrals in Eq. (4.6.28) we have

$$\phi(x,t) = \frac{2}{\sqrt{\pi}} \int_0^t h(t') J(x,t-t') \, dt'.$$

When $h(t) = 1$, if we call $\eta = \frac{x}{2(t-t')^{1/2}}$, then $d\eta = \frac{x}{4(t-t')^{3/2}} \, dt'$, and we have

$$\phi(x,t) = \frac{2}{\sqrt{\pi}} \int_{\frac{x}{2\sqrt{t}}}^\infty e^{-\eta^2} \, d\eta \equiv \text{erfc}\left(\frac{x}{2\sqrt{t}} \right). \tag{4.6.30}$$

We note that erfc(x) is a well-known function, called the **complementary error function**: erfc$(x) \equiv 1 - \text{erf}(x)$, where erf$(x) \equiv \frac{2}{\sqrt{\pi}} \int_0^x e^{-y^2} \, dy$.

It should be mentioned that the Fourier sine transform applies to problems such as Eq. (4.6.23) with fixed conditions on ϕ at the origin. Such solutions can be extended to the interval $-\infty < x < \infty$ where the initial values $\phi(x,0)$ are themselves extended as an odd function on $(-\infty, 0)$. However, if we should replace $\phi(x = 0, t) = h(t)$ by a derivative condition at the origin, say $\frac{\partial \phi}{\partial x}(x = 0, t) = h(t)$, then the appropriate transform to use is a cosine transform.

Figure 4.25 Indented contour C

Another type of partial differential equation which is encountered frequently in applications is the wave equation

$$\frac{\partial^2 \phi}{\partial x^2} - \frac{1}{c^2}\frac{\partial^2 \phi}{\partial t^2} = F(x,t), \qquad (4.6.31)$$

where the constant c, $c > 0$, is the speed of propagation of the unforced wave. The wave equation governs vibrations of many types of continuous media with external forcing $F(x,t)$. If $F(x,t)$ vibrates periodically in time with constant frequency $\omega > 0$, say, $F(x,t) = f(x)e^{i\omega t}$, then it is natural to look for special solutions to Eq. (4.6.31) of the form $\phi(x,t) = \Phi(x)e^{i\omega t}$. Then $\Phi(x)$ satisfies

$$\frac{\partial^2 \Phi}{\partial x^2} + \left(\frac{\omega}{c}\right)^2 \Phi = f(x). \qquad (4.6.32)$$

A real solution to Eq. (4.6.32) is obtained by taking the real part; this would correspond to forcing of $\phi(x,t) = \phi(x)\cos\omega t$. If we simply look for a Fourier transform solution of Eq. (4.6.32) we arrive at

$$\Phi(x) = \frac{-1}{2\pi}\int_C \frac{\hat{F}(k)}{k^2 - (\omega/c)^2}e^{ikx}\,dk, \qquad (4.6.33)$$

where $\hat{F}(k)$ is the Fourier transform of $f(x)$. Unfortunately, for the standard contour C, k real, $-\infty < k < \infty$, Eq. (4.6.33) is not well defined, because there are singularities in the denominator of the integrand in Eq. (4.6.33) when $k = \pm\omega/c$. Without further specification the problem is not well posed. The standard acceptable solution is found by specifying a contour C that is indented below $k = -\omega/c$ and above $k = +\omega/c$ (see Figure 4.25); this removes the singularities in the denominator.

This choice of contour turns out to yield solutions with outgoing waves at large distances from the source $F(x,t)$. A choice of contour reflects an imposed boundary condition. In this problem it is well known and is referred to as the Sommerfeld Radiation Condition. An outgoing wave has the form $e^{i\omega(t-|x|/c)}$ (as t increases, $|x|$ increases for a given choice of phase, i.e., on a fixed point on a wave crest). An incoming wave has the form: $e^{i\omega(t+|x|/c)}$. Using the Fourier representation $\hat{F}(k) = \int_{-\infty}^{\infty} f(\zeta)e^{-i\zeta k}\,d\zeta$ in Eq. (4.6.33) and interchanging integrals, we can write the function in the form

$$\Phi(x) = \int_{-\infty}^{\infty} f(\zeta)H(x - \zeta, \omega/c)\,d\zeta, \qquad (4.6.34a)$$

where

$$H(x - \zeta, \omega/c) = \frac{-1}{2\pi} \int_C \frac{e^{ik(x-\zeta)}}{k^2 - (\omega/c)^2} \, dk \qquad (4.6.34b)$$

and the contour C is specified as in Figure 4.25. Contour integration of Eq. (4.6.34b) yields

$$H(x - \zeta, \omega/c) = \frac{i \, e^{-i|x-\zeta|(\omega/c)}}{2(\omega/c)}. \qquad (4.6.34c)$$

At large distances from the source, $|x| \to \infty$, we have outgoing waves for the solution $\phi(x, t)$:

$$\phi(x, t) = \text{Re} \left\{ \frac{i}{2(\omega/c)} \int_{-\infty}^{\infty} f(\zeta) e^{i\omega(t - |x-\zeta|/c)} \, d\zeta \right\}. \qquad (4.6.34d)$$

Thus, for example, if $f(\zeta)$ is a point source: $f(\zeta) = \delta(\zeta - x_0)$ where $\delta(\zeta - x_0)$ is a Dirac delta function concentrated at x_0, then (4.6.34d) yields

$$\phi(x, t) = -\frac{1}{2(\omega/c)} \sin \omega(t - |x - x_0|/c). \qquad (4.6.34e)$$

An alternative method for finding this result is to add a damping mechanism to the original equation. Namely, if we add the term $-\epsilon \frac{\partial \phi}{\partial t}$ to the left-hand side of Eq. (4.6.31), then Eq. (4.6.33) is modified by adding the term $i\epsilon\omega$ to the denominator of the integrand. This has the desired effect of moving the poles off the real axis ($k_1 = -\omega/c + i\epsilon\alpha$, $k_2 = +\omega/c - i\epsilon\alpha$, where α =constant) in the same manner as indicated by Figure 4.25. By using Fourier transforms, and then taking the limit $\epsilon \to 0$ (small damping), the above results could have been obtained.

In practice, wave propagation problems such as

$$\frac{\partial u}{\partial t} + \frac{\partial^3 u}{\partial x^3} = 0, \qquad -\infty < x < \infty, \qquad u(x, 0) = f(x), \qquad (4.6.35)$$

where again $f(x) \in L^1 \cap L^2$, are solved by Fourier transforms. The function $u(x, t)$ typically represents the small amplitude vibrations of a continuous medium such as water waves. One looks for a solution to Eq. (4.6.35) of the form

$$u(x, t) = \frac{1}{2\pi} \int_{-\infty}^{\infty} b(k, t) e^{ikx} \, dk. \qquad (4.6.36)$$

Taking the Fourier transform of (4.6.35) and using (4.5.16a,b) or alternatively, substitution of Eq. (4.6.36) into Eq. (4.6.35) – assuming interchanges of derivative and integrand are valid (a fact which can be shown to follow from rapid enough decay of $f(x)$ at infinity; that is, $f \in L^1 \cap L^2$) – yields

$$\frac{\partial b}{\partial t} - ik^3 b = 0 \qquad (4.6.37a)$$

hence

$$b(k,t) = b(k,0)e^{ik^3t}, \qquad (4.6.37b)$$

where

$$b(k,0) = \int_{-\infty}^{\infty} f(x)e^{-ikx}\, dx. \qquad (4.6.37c)$$

The solution (4.6.36) can be viewed as a superposition of waves of the form $e^{ikx-i\omega(k)t}$, $\omega(k) = -k^3$. The function $\omega(k)$ is referred to as the **dispersion relation**. The above integral representation, for general $f(x)$, is the "best" one can do, because we cannot evaluate it in closed form. However, as $t \to \infty$, the integral can be approximated by asymptotic methods as discussed in Chapter 6 of Ablowitz and Fokas (2003); that is, the methods of stationary phase and steepest descents. Suffice it to say that the solution $u(x,t) \to 0$ as $t \to \infty$ (the initial values are said to disperse as $t \to \infty$) and the major contribution to the integral is found near the location where $\omega'(k) = x/t$; that is, $x/t = -3k^2$ in the integrand (where the phase $\Psi = kx - \omega(k)t$ is stationary: $\frac{\partial\Psi}{\partial k} = 0$). The quantity $\omega'(k)$ is called the **group velocity**, and it represents the speed of a packet of waves centered around wave number k. Using asymptotic methods for $x/t < 0$ and as $t \to \infty$, we can show that $u(x,t)$ has the following approximate form:

$$\left. \begin{array}{c} u(x,t) \approx \dfrac{c}{\sqrt{t}}\left(\displaystyle\sum_{i=1}^{2} \dfrac{b(k_i)}{\sqrt{|k_i|}}\, e^{i(k_i x + k_i^3 t + \phi_i)}\right), \\[1em] k_1 = \sqrt{-x/3t}, \quad k_2 = -\sqrt{-x/3t}, \quad \text{with } c,\, \phi_i \text{ constant} \end{array} \right\} . \qquad (4.6.38)$$

When $x/t > 0$ the solution decays exponentially. As $x/t \to 0$, Eq. (4.6.38) may be rearranged and put into the following self-similar form

$$u(x,t) \approx \frac{d}{(3t)^{1/3}} A\left(x/(3t)^{1/3}\right), \qquad (4.6.39)$$

where d is constant and $A(\eta)$ satisfies (by substitution of (4.6.39) into (4.6.35))

$$A_{\eta\eta\eta} - \eta A_\eta - A = 0$$

or

$$A_{\eta\eta} - \eta A = 0, \qquad (4.6.40a)$$

with the boundary condition $A \to 0$ as $\eta \to \infty$. Equation (4.6.40a) is called Airy's equation. The integral representation of the solution to Airy's equation with $A \to 0$, $\eta \to \infty$, is given by

$$A(\eta) = \frac{1}{2\pi} \int_{-\infty}^{\infty} e^{i(k\eta + k^3/3)}\, dk; \qquad (4.6.40b)$$

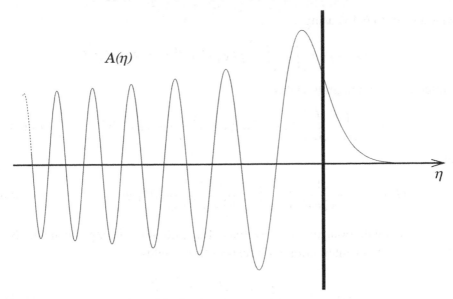

$A(\eta)$

η

Figure 4.26 The Airy function

see also the end of this section, Eq. (4.6.57). Its wave form is depicted in Figure 4.26. The function $A(\eta)$ acts like "matching" or "turning" of solutions from one type of behavior to another; i.e., from exponential decay as $\eta \to +\infty$ to oscillation as $\eta \to -\infty$ (see also Section 6.7 of Ablowitz and Fokas, 2003).

Sometimes there is a need to use multiple transforms. For example, consider finding the solution to the following problem:

$$\frac{\partial^2 \phi}{\partial x^2} + \frac{\partial^2 \phi}{\partial y^2} - m^2 \phi = f(x, y), \qquad \phi(x, y) \to 0 \text{ as } x^2 + y^2 \to \infty. \qquad (4.6.41)$$

A simple transform in x satisfies

$$\phi(x, y) = \frac{1}{2\pi} \int_{-\infty}^{\infty} \Phi(k_1, y) e^{ik_1 x} \, dk_1.$$

We can take another transform in y to obtain

$$\phi(x, y) = \frac{1}{(2\pi)^2} \int_{-\infty}^{\infty} \int_{-\infty}^{\infty} \hat{\Phi}(k_1, k_2) e^{ik_1 x + ik_2 y} \, dk_1 \, dk_2. \qquad (4.6.42)$$

Using a similar formula for $f(x, y)$ in terms of its transform $\hat{F}(k_1, k_2)$, we find, by substitution into Eq. (4.6.41),

$$\phi(x, y) = \frac{-1}{(2\pi)^2} \int \int \frac{\hat{F}(k_1, k_2) e^{ik_1 x + ik_2 y}}{k_1^2 + k_2^2 + m^2} \, dk_1 \, dk_2. \qquad (4.6.43)$$

Rewriting Eq. (4.6.43) using

$$\hat{F}(k_1, k_2) = \int_{-\infty}^{\infty} \int_{-\infty}^{\infty} f(x', y') e^{-ik_1 x' - ik_2 y'} \, dx' \, dy'$$

and interchanging integrals yields

$$\phi(x, y) = \int_{-\infty}^{\infty} \int_{-\infty}^{\infty} f(x', y') G(x - x', y - y') \, dx' \, dy', \qquad (4.6.44a)$$

where

$$G(x, y) = -\frac{1}{(2\pi)^2} \int_{-\infty}^{\infty} \int_{-\infty}^{\infty} \frac{e^{ik_1 x + ik_2 y}}{k_1^2 + k_2^2 + m^2} dk_1 \, dk_2. \qquad (4.6.44b)$$

By clever manipulation, one can evaluate Eq. (4.6.44b). Using the methods of Section 4.3, contour integration with respect to k_1 yields

$$G(x, y) = -\frac{1}{4\pi} \int_{-\infty}^{\infty} \frac{e^{ik_2 y - \sqrt{k_2^2 + m^2} |x|}}{\sqrt{k_2^2 + m^2}} \, dk_2. \qquad (4.6.45a)$$

Thus, for $x \neq 0$,

$$\frac{\partial G}{\partial x}(x, y) = \frac{\text{sgn}(x)}{4\pi} \int_{-\infty}^{\infty} e^{ik_2 y - \sqrt{k_2^2 + m^2} |x|} \, dk_2, \qquad (4.6.45b)$$

where sgn $x = 1$ for $x > 0$, and sgn $x = -1$ for $x < 0$. Equation (4.6.45b) takes on an elementary form for $m = 0$ (i.e., $\sqrt{k_2^2} = |k_2|$):

$$\frac{\partial G}{\partial x}(x, y) = \frac{x}{2\pi(x^2 + y^2)}$$

and we have

$$G(x, y) = \frac{1}{4\pi} \log(x^2 + y^2). \qquad (4.6.46)$$

The constant of integration is immaterial, because to have a vanishing solution $\phi(x, y)$ as $x^2 + y^2 \rightarrow \infty$, Eq. (4.6.44a) necessarily requires that $\int_{-\infty}^{\infty} \int_{-\infty}^{\infty} f(x, y) \, dx \, dy = 0$, which follows directly from Eq. (4.6.41) by integration when $m = 0$. Note that Eq. (4.6.43) implies that when $m = 0$, for the integral to be well defined, $\hat{F}(k_1 = 0, k_2 = 0) = 0$, which in turn implies the need for the vanishing of the double integral of $f(x, y)$. Finally, if $m \neq 0$, we only remark that Eq. (4.6.44b) or Eq. (4.6.45a) is transformable to an integral representation of a modified Bessel function of order zero:

$$G(x, y) = -\frac{1}{2\pi} K_0 \left(m \sqrt{x^2 + y^2} \right). \qquad (4.6.47)$$

Interested readers can find contour integral representations of Bessel functions in many books on special functions.

Frequently in the study of differential equations, integral representations can be found for the solution. Integral representations supplement series methods discussed in Chapter 3 can provide an alternative representation of a class of solutions. We give one example in what follows. Consider Airy's equation in the form (see also Eq. (4.6.40a)

$$\frac{d^2y}{dz^2} - zy = 0 \tag{4.6.48}$$

and look for an integral representation of the form

$$y(z) = \int_C e^{z\zeta} v(\zeta)\, d\zeta \tag{4.6.49}$$

where the contour C and the function $v(\zeta)$ are to be determined. Formula (4.6.49) is frequently referred to as a generalized Laplace transform, and the method as the generalized Laplace transform method. (Here C is generally not the Bromwich Contour.) Equation (4.6.49) is a special case of the more general integral representation $\int_C K(z, \zeta)v(\zeta)\, d\zeta$. Substitution of Eq. (4.6.49) into Eq. (4.6.48), and assuming the interchange of differentiation and integration (which is verified *a posteriori*) gives

$$\int_C \left(\zeta^2 - z \right) v(\zeta)e^{z\zeta}\, d\zeta = 0. \tag{4.6.50}$$

Using

$$z e^{z\zeta} v = \frac{d}{d\zeta}\left(e^{z\zeta} v \right) - e^{z\zeta}\frac{dv}{d\zeta},$$

rearranging and integrating, yields

$$-\left[e^{z\zeta} v(\zeta) \right]_C + \int_C \left(\zeta^2 v + \frac{dv}{d\zeta} \right) e^{z\zeta}\, d\zeta = 0, \tag{4.6.51}$$

where the term in brackets, $[\cdot]_C$, stands for evaluation at the endpoints of the contour. The essence of the method is to choose C and $v(\zeta)$ such that both terms in Eq. (4.6.51) vanish. Taking

$$\frac{dv}{d\zeta} + \zeta^2 v = 0 \tag{4.6.52}$$

implies that

$$v(\zeta) = Ae^{-\zeta^3/3}, \qquad \text{for a constant } A. \tag{4.6.53}$$

Thinking of an infinite contour, and calling $\zeta = R\,e^{i\theta}$, we see that the dominant term as $R \to \infty$ in $[\cdot]_C$ is due to $v(\zeta)$, which in magnitude is given by

$$|v(\zeta)| = |A|e^{-R^3(\cos 3\theta)/3}. \tag{4.6.54}$$

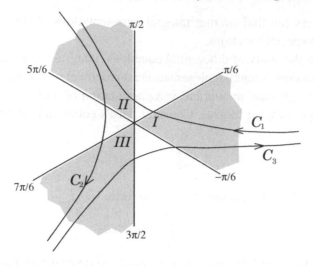

Figure 4.27 Three standard contours

This contribution will vanish for large values of ζ when $\cos 3\theta > 0$; that is, for

$$-\frac{\pi}{2} + 2n\pi < 3\theta < \frac{\pi}{2} + 2n\pi, \qquad n = 0, 1, 2. \qquad (4.6.55)$$

Thus, we have three regions in which there is decay:

$$\left.\begin{array}{cc} -\dfrac{\pi}{6} < \theta < \dfrac{\pi}{6} & \text{(I)} \\[2mm] \dfrac{\pi}{2} < \theta < \dfrac{5\pi}{6} & \text{(II)} \\[2mm] \dfrac{7\pi}{6} < \theta < \dfrac{3\pi}{2} & \text{(III)} \end{array}\right\}. \qquad (4.6.56)$$

There are three standard contours $C_i, i = 1, 2, 3$, in which the integrated term $[.]_C$ vanishes as $R \to \infty$, depicted in Figure 4.27. The shaded region refers to regions of decay of $v(\zeta)$.

The three solutions, of which only two are linearly independent (because the equation is of second order), are denoted by

$$y_i(z) = \alpha_i \int_{C_i} e^{z\zeta - \zeta^3/3} \, d\zeta, \qquad (4.6.57)$$

where α_i is a convenient normalizing factor, $i = 1, 2, 3$. In order not to have a trivial solution, one must take the contour C between any two of the decaying regions. If we change variables to $\zeta = ik$, then the solution corresponding to $i = 2$, namely $y_2(z)$, is proportional to the Airy function solution $A(\eta)$ discussed earlier (see Eq. (4.6.40b)).

Finally, we remark that this method applies to linear differential equations with coefficients depending linearly on the independent variable. Generalizations to other kernels $K(z, \zeta)$ (mentioned after Eq. (4.6.49)) can be made and are employed to solve other linear differential equations involving Bessel functions, Legendre functions, etc. The interested reader may wish to consult a reference such as Jeffreys and Jeffreys (1962).

It is interesting that whereas the use of contours in the complex plane, like the contours in (4.6.49), have been employed for a long time, the use of such contours for solving PDEs was developed only recently (Fokas, 2006).

4.6.1 Problems for Section 4.6

1. Use Laplace transform methods to solve the ODE

$$L\frac{dy}{dt} + Ry = f(t), \qquad y(0) = y_0, \quad \text{for constant } L, R > 0.$$

   (a) Let $f(t) = \sin \omega_0 t$, $\omega_0 > 0$, so that the Laplace transform of $f(t)$ is $\hat{F}(s) = \omega_0/(s^2 + \omega_0^2)$. Find

   $$y(t) = y_0\, e^{-\frac{R}{L}t} + \frac{\omega_0}{L\left(\frac{R^2}{L^2} + \omega_0^2\right)}\, e^{-\frac{R}{L}t}$$

   $$+ \frac{\omega_0 R}{L^2}\, \frac{\sin \omega_0 t}{\left(\left(\frac{R}{L}\right)^2 + \omega_0^2\right)} - \frac{\omega_0}{L}\, \frac{\cos \omega_0 t}{\left(\left(\frac{R}{L}\right)^2 + \omega_0^2\right)}.$$

   (b) Suppose $f(t)$ is an arbitrary continuous function which possesses a Laplace transform. Use the convolution product for Laplace transforms (Section 4.5) to find

   $$y(t) = y_0\, e^{-\frac{R}{L}t} + \frac{1}{L}\int_0^t f(t')e^{-\frac{R}{L}(t-t')}\, dt'.$$

   (c) Let $f(t) = \sin \omega_0 t$ in (b) to obtain the result of part (a), and thereby verify your answer.

   This is an example of an "L, R circuit" with impressed voltage $f(t)$ arising in basic electric circuit theory.

2. Use Laplace transform methods to solve the ODE

$$\frac{d^2 y}{dt^2} - k^2 y = f(t), \qquad k > 0, \ y(0) = y_0, \ y'(0) = y_0'.$$

   (a) Let $f(t) = e^{-k_0 t}$, $k_0 \neq k$, $k_0 > 0$, so that the Laplace transform of $f(t)$ is $\hat{F}(s) = \dfrac{1}{s + k_0}$. Find

   $$y(t) = y_0 \cosh kt + \frac{y_0'}{k} \sinh kt + \frac{e^{-k_0 t}}{k_0^2 - k^2} - \frac{\cosh kt}{k_0^2 - k^2} + \frac{k_0/k}{k_0^2 - k^2} \sinh kt.$$

(b) Suppose $f(t)$ is an arbitrary continuous function which possesses a Laplace transform. Use the convolution product for Laplace transforms (Section 4.5) to find

$$y(t) = y_0 \cosh kt + \frac{y_0'}{k} \sinh kt + \int_0^t f(t') \frac{\sinh k(t - t')}{k} \, dt'.$$

(c) Set $f(t) = e^{-k_0 t}$ in (b) to obtain the result in part (a). What happens when $k_0 = k$?

3. Consider the differential equation

$$\frac{d^3 y}{dt^3} + \omega_0{}^3 y = f(t), \qquad \omega_0 > 0, \ y(0) = y'(0) = y''(0) = 0.$$

(a) Assuming that $f(t)$ has a Laplace transform $\widehat{F}(s)$, show that the Laplace transform of the solution, $\widehat{Y}(s)$, satisfies

$$\widehat{Y}(s) = \frac{\widehat{F}(s)}{s^3 + \omega_0{}^3}.$$

(b) Deduce that the inverse Laplace transform of $\dfrac{1}{s^3 + \omega_0{}^3}$ is given by

$$h(t) = \frac{e^{-\omega_0 t}}{3\omega_0{}^2} - \frac{2}{3\omega_0{}^2} e^{\omega_0 t/2} \cos\left(\frac{\omega_0}{2}\sqrt{3}t + \frac{\pi}{3}\right),$$

and show that

$$y(t) = \int_0^t h(t') f(t - t') \, dt'$$

by using the convolution product for Laplace transforms.

4. Let us consider Laplace's equation:

$$\frac{\partial^2 \phi}{\partial x^2} + \frac{\partial^2 \phi}{\partial y^2} = 0,$$

for $-\infty < x < \infty$ and $y > 0$, with the boundary conditions $\dfrac{\partial \phi}{\partial y}(x, 0) = h(x)$ and $\phi(x, y) \to 0$ as $x^2 + y^2 \to \infty$. Find the Fourier transform solution. Is there a constraint on the data $h(x)$ for a solution to exist? If so, can this be explained another way?

5. Consider the linear "free" Schrödinger equation (without a potential),

$$i\frac{\partial u}{\partial t} + \frac{\partial^2 u}{\partial x^2} = 0, \qquad \text{with } u(x, 0) = f(x).$$

(a) Solve this problem using Fourier transforms, by obtaining the Green's function in closed form, and using superposition. Recall that

$$\int_{-\infty}^{\infty} e^{iu^2} du = e^{i\pi/4} \sqrt{\pi}.$$

(b) Obtain the above solution using Laplace transforms.

6. Consider the heat equation,

$$\frac{\partial \phi}{\partial t} = \frac{\partial^2 \phi}{\partial x^2}, \qquad 0 < x < \infty, t > 0,$$

with the following initial and boundary conditions:

$$\phi(x, 0) = 0, \qquad \frac{\partial \phi}{\partial x}(x = 0, t) = g(t), \qquad \lim_{x \to \infty} \phi(x, t) = \lim_{x \to \infty} \frac{\partial \phi}{\partial x} = 0.$$

(a) Solve this problem using Fourier cosine transforms.

(b) Solve this problem using Laplace transforms.

(c) Show that the representations of (a) and (b) are equivalent.

7. Consider the wave equation (with wave speed being unity),

$$\frac{\partial^2 \phi}{\partial t^2} - \frac{\partial^2 \phi}{\partial x^2} = 0, \qquad 0 < x < \ell, t > 0$$

and the boundary conditions

$$\phi(x, t = 0) = 0, \qquad \frac{\partial \phi}{\partial t}(x, t = 0) = 0, \qquad \phi(x = 0, t) = 0, \qquad \phi(x = \ell, t) = 1.$$

(a) Obtain the Laplace transform of the solution $\widehat{\Phi}(x, s)$,

$$\widehat{\Phi}(x, s) = \frac{\sinh sx}{s \sinh s\ell}.$$

(b) Obtain the solution $\phi(x, t)$ by inverting the Laplace transform to find

$$\phi(x, t) = \frac{x}{\ell} + \sum_{n=1}^{\infty} \frac{2(-1)^n}{n\pi} \sin\left(\frac{n\pi x}{\ell}\right) \cos\left(\frac{n\pi t}{\ell}\right);$$

see also Problem 19, Section 4.5.

8. Consider the wave equation

$$\frac{\partial^2 \phi}{\partial t^2} - \frac{\partial^2 \phi}{\partial x^2} = 0, \qquad 0 < x < \ell, t > 0$$

and the boundary conditions

$$\phi(x, t = 0) = 0, \qquad \frac{\partial \phi}{\partial t}(x, t = 0) = 0,$$

$$\phi(x = 0, t) = 0, \qquad \phi(x = \ell, t) = f(t).$$

(a) Show that the Laplace transform of the solution is given by

$$\widehat{\Phi}(x, s) = \frac{\widehat{F}(s)\sinh sx}{\sinh s\ell},$$

where $\widehat{F}(s)$ is the Laplace transform of $f(t)$.

(b) Denote the solution of the problem when $f(t) = 1$ (so that $\widehat{F}(s) = 1/s$) by $\phi_s(x, t)$. Show that the general solution is given by

$$\phi(x, t) = \int_0^t \frac{\partial \phi_s}{\partial t'}(x, t') f(t - t') \, dt'.$$

9. Use multiple Fourier transforms to solve

$$\frac{\partial \phi}{\partial t} - \left(\frac{\partial^2 \phi}{\partial x^2} + \frac{\partial^2 \phi}{\partial y^2}\right) = 0$$

on the infinite domain $-\infty < x < \infty$, $-\infty < y < \infty$, $t > 0$, with $\phi(x, y) \to 0$ as $x^2 + y^2 \to \infty$, and $\phi(x, y, 0) = f(x, y)$. How does the solution simplify if $f(x, y)$ is a function of $x^2 + y^2$? What is the Green's function in this case?

10. Consider the forced heat equation,

$$\frac{\partial \phi}{\partial t} - \frac{\partial^2 \phi}{\partial x^2} = f(x, t), \qquad \phi(x, 0) = g(x)$$

on $-\infty < x < \infty$, $t > 0$, with $\phi, g, f \to 0$ as $|x| \to \infty$.

(a) Use Fourier transforms to solve this equation. How does the solution compare with the case $f = 0$?

(b) Use Laplace transforms to solve this equation. How does the method compare with that described in this section for the case $f = 0$?

11. Given the ODE

$$zy'' + (2r + 1)y' + zy = 0,$$

look for a contour representation of the form $y = \int_C e^{z\zeta} v(\zeta) d\zeta$.

(a) Show that if C is a *closed* contour and $v(\zeta)$ is single valued on this contour, then it follows that $v(\zeta) = A(\zeta^2 + 1)^{r-1/2}$.

(b) Show that if $y = z^{-s}w$, then, when $s = r$, w satisfies Bessel's equation, $z^2 w'' + zw' + (z^2 - r^2)w = 0$, and a contour representation of the solution is given by

$$w = Az^r \oint_C e^{z\zeta} (\zeta^2 + 1)^{r-1/2} \, d\zeta.$$

Note that if $r = n + 1/2$ for integer n, then this representation yields the trivial solution. We take the branch cut to be inside the circle C when $(r - 1/2)$ is not an integer.

12. The hypergeometric equation,

$$zy'' + (a - z)y' - by = 0,$$

has a contour integral representation of the form $y = \int_C e^{z\zeta} v(\zeta)\, d\zeta$.

(a) Show that one solution is given by

$$y = \int_0^1 e^{z\zeta} \zeta^{b-1} (1 - \zeta)^{a-b-1}\, d\zeta,$$

where Re $b > 0$ and Re $(a - b) > 0$.

(b) Letting $b = 1$, $a = 2$, show that this solution is $y = \dfrac{e^z - 1}{z}$, and verify that it satisfies the equation.

(c) Show that a second solution, $y_2 = v y_1$ (where the first solution is denoted by y_1) obeys

$$z y_1 v'' + (2 z y_1' + (a - z) y_1)\, v' = 0.$$

Integrate this equation to find v, and thereby obtain a formal representation for y_2. What can be said about the analytic behavior of y_2 near $z = 0$?

13. Suppose we are given the following damped wave equation:

$$\frac{\partial^2 \phi}{\partial x^2} - \frac{1}{c^2}\frac{\partial^2 \phi}{\partial t^2} - \epsilon \frac{\partial \phi}{\partial t} = e^{i\omega t} \delta(x - \zeta), \qquad \omega, \epsilon > 0.$$

(a) Show that $\psi(x)$, where $\phi(x, t) = e^{i\omega t} \psi(x)$, satisfies

$$\psi'' + \left(\left(\frac{\omega}{c}\right)^2 - i\omega\epsilon \right) \psi = \delta(x - \zeta).$$

(b) Show that $\widehat{\Psi}(k)$, the Fourier transform of $\psi(x)$, is given by

$$\widehat{\Psi}(k) = \frac{-e^{-ik\zeta}}{k^2 - \left(\frac{\omega}{c}\right)^2 + i\omega\epsilon}.$$

(c) Invert $\widehat{\Psi}(k)$ to obtain $\psi(x)$, and in particular show that as $\epsilon \to 0^+$ we have

$$\psi(x) = \frac{i e^{-i\frac{\omega}{c}|x-\zeta|}}{2\left(\frac{\omega}{c}\right)},$$

and that this has the effect of deforming the contour as described in Figure 4.25.

14. In this problem we obtain the Green's function of Laplace's equation in the upper half plane, $-\infty < x < \infty, 0 < y < \infty$, by solving

$$\frac{\partial^2 G}{\partial x^2} + \frac{\partial^2 G}{\partial y^2} = \delta(x - \zeta)\delta(y - \eta),$$

$$G(x, y = 0) = 0, \quad G(x, y) \to 0 \text{ as } r^2 = x^2 + y^2 \to \infty.$$

(a) Take the Fourier transform of the equation with respect to x, and show that the Fourier transform $\hat{G}(k, y) = \int_{-\infty}^{\infty} G(x, y)e^{-ikx} \, dk$, satisfies

$$\frac{\partial^2 \hat{G}}{\partial y^2} - k^2 \hat{G} = \delta(y - \eta)e^{-ik\zeta} \qquad \text{with } \hat{G}(k, y = 0) = 0.$$

(b) Take the Fourier sine transform of $\hat{G}(k, y)$ with respect to y and show, for

$$\hat{G}_s(k, l) = \int_0^\infty \hat{G}(k, y) \sin ly \, dy,$$

that it satisfies

$$\hat{G}_s(k, l) = -\frac{\sin l\eta e^{-ik\zeta}}{l^2 + k^2}.$$

(c) Invert this expression with respect to k and find

$$G(x, l) = -\frac{e^{-l|x-\zeta|} \sin l\eta}{2l},$$

whereupon show

$$G(x, y) = -\frac{1}{\pi} \int_0^\infty \frac{e^{-l|x-\zeta|} \sin l\eta \sin ly}{l} \, dl.$$

(d) Evaluate $G(x, y)$ to show

$$G(x, y) = \frac{1}{4\pi} \log \left(\frac{(x - \zeta)^2 + (y - \eta)^2}{(x - \zeta)^2 + (y + \eta)^2} \right).$$

Hint: note that taking the derivative with respect to y of $G(x, y)$ in part (c), yields an integral for $\dfrac{\partial G}{\partial y}$ that is elementary. Then one can integrate this result using $G(x, y = 0) = 0$, to obtain $G(x, y)$.

5

Conformal Mappings and Applications

5.1 Introduction

A large number of problems arising in fluid mechanics, electrostatics, heat conduction, and many other physical situations, can be mathematically formulated in terms of Laplace's equation (see also the discussion in Section 2.1). That is, all these physical problems reduce to solving the equation

$$\Phi_{xx} + \Phi_{yy} = 0 \qquad (5.1.1)$$

in a certain region D of the z-plane. The function $\Phi(x, y)$, in addition to satisfying equation Eq. (5.1.1), also satisfies certain boundary conditions on the boundary C of the region D. Recalling that the real and the imaginary parts of an analytic function satisfy Eq. (5.1.1), it follows that solving the above problem reduces to finding a function which is analytic in D and that satisfies certain boundary conditions on C. It turns out that the solution of this problem can be greatly simplified if the region D is either the upper half of the z-plane or the unit disk. This suggests that instead of solving Eq. (5.1.1) in D, one should first perform a change of variables from the complex variable z to the complex variable $w = f(z)$, such that the region D of the z-plane is mapped to the upper half plane of the w-plane. Generally speaking, such transformations are called conformal, and their study is the main content of this chapter.

General properties of conformal transformations are studied in Sections 5.2 and 5.3. In Section 5.3 a number of theorems are stated, which are quite natural and motivated by heuristic considerations. The rigorous proofs are deferred to Section *5.5, which deals with more theoretical issues. We have denoted Section *5.5 as an optional (more difficult) section. In Section 5.4 a number of basic physical applications of conformal mapping are discussed, including problems from ideal fluid flow, steady state heat conduction, and electrostatics. Physical applications which require more advanced methods of conformal mapping are also included in later sections.

According to a celebrated theorem first discussed by Riemann, if D is a simply connected region D which is not the entire complex z-plane, then there exists an analytic function $f(z)$ such that $w = f(z)$ transforms D onto the upper half w-plane. Unfortunately, this theorem does not provide a constructive approach for finding $f(z)$. However, for certain simple domains, such as domains bounded by polygons, it is possible to find an explicit formula (in terms of quadratures) for $f(z)$. Transformations of polygonal domains to the upper half plane are called Schwarz–Christoffel transformations and are studied in Section 5.6. A classically important case is the transformation of a rectangle to the upper half plane, which leads to elliptic integrals and elliptic functions. An important class of conformal transformations, called bilinear transformations, is the subject of Section 5.7. Another interesting class of transformations involves a "circular polygon" (i.e., a polygon whose sides are circular arcs), which is studied in Section 5.8. The case of a circular triangle is discussed in some detail and relevant classes of functions such as Schwarzian functions and elliptic modular functions arise naturally. Some further interesting mathematical problems related to conformal transformations are discussed in Section 5.9.

5.2 Conformal Transformations

Let C be a curve in the complex z-plane. Let $w = f(z)$, where $f(z)$ is some analytic function of z, define a change of variables from the complex variable z to the complex variable w. Under this transformation, the curve C is mapped to some curve C^* in the complex w-plane. The precise form of C^* will depend on the precise form of C. However, there exists a geometrical property of C^* that is independent of the particular choice of C: let z_0 be a point of the curve C, and assume that $f'(z_0) \neq 0$; under the transformation $w = f(z)$ the tangent to the curve C at the point z_0 is rotated counterclockwise by $\arg f'(z_0)$ (see Figure 5.1), $w_0 = f(z_0)$.

Before proving this statement, let us first consider the particular case that the transformation $f(z)$ is linear; that is, $f(z) = az + b$, $a, b \in \mathbb{C}$, and the curve C is a straight ray going through the origin. The mathematical description of such a curve is given by $z(s) = se^{i\varphi}$, where φ is constant, and the notation $z(s)$ indicates that for points on this curve, z is a function of s only. Under the transformation $w = f(z)$, this curve is mapped to $w(s) = az(s) + b = |a|s \exp[i(\varphi + \arg(a))] + b$, that is, a ray through z_0 in the z-plane is rotated by $\arg(a) = \arg(f'(z))$ in the w-plane, see Figure 5.2.

Let us now consider the general case. Points on a continuous curve C are characterized by the fact that their x and y coordinates are related. It turns out that, rather than describing this relationship directly, it is more convenient to describe it parametrically through the equations $x = x(s)$, $y = y(s)$, where $x(s)$ and $y(s)$ are

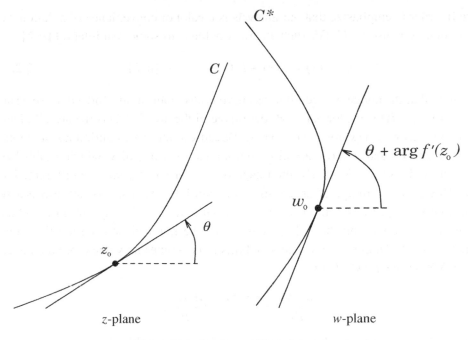

Figure 5.1 Conformal Transformation

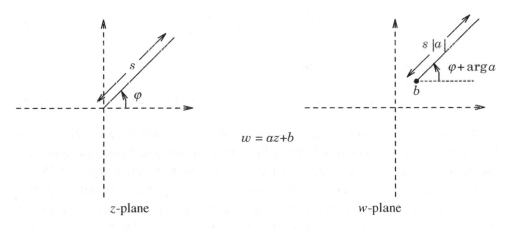

Figure 5.2 Ray rotated by $\arg(a)$

real differentiable functions of the real parameter s. For example, for the straight ray of Figure 5.2, $x = s \cos \varphi$, $y = s \sin \varphi$; for a circle with center at the origin and radius R, $x = R \cos s$, $y = R \sin s$, etc. More generally, the mathematical description of a curve C can be given by $z(s) = x(s) + iy(s)$. Suppose that $f(z)$ is analytic for z in some domain of the complex z-plane denoted by D. Our considerations are applicable to that part of C that is contained in D. We shall refer to this part as an

arc in order to emphasize that our analysis is local. For convenience of notation we shall denote it also by C. For such an arc, s belongs to some real interval $[a, b]$.

$$C : z(s) = x(s) + iy(s), \qquad s \in [a, b]. \tag{5.2.1}$$

We note that the image of a continuous curve is also continuous. Indeed, if we write $w = u(x, y) + iv(x, y)$, for $u, v \in \mathbb{R}$, the image of the arc (5.2.1) is the arc C^* given by $w(s) = u(x(s), y(s)) + iv(x(s), y(s))$. Because x and y are continuous functions of s, it follows that u and v are also continuous functions of s, which establishes the continuity of C^*. Similarly, the image of a differentiable arc is a differentiable arc. However, the image of an arc that does not intersect itself is not necessarily non-intersecting. In fact, if $f(z_1) = f(z_2)$, with $z_1, z_2 \in D$, any non-intersecting continuous arc passing through z_1 and z_2 will be mapped onto an arc that does intersect itself. Of course, one can avoid this if $f(z)$ takes no value more than once in D. We define $dz(s)/ds$ by

$$\frac{dz(s)}{ds} = \frac{dx(s)}{ds} + i\frac{dy(s)}{ds}.$$

Let $f(z)$ be analytic in a domain containing the open neighborhood of $z_0 \equiv z(s_0)$. The image of C is $w(s) = f(z(s))$; thus, by the chain rule

$$\left.\frac{dw(s)}{ds}\right|_{s=s_0} = f'(z_0)\left.\frac{dz(s)}{ds}\right|_{s=s_0}. \tag{5.2.2}$$

If $f'(z_0) \neq 0$ and $z'(s_0) \neq 0$, it follows that $w'(s_0) \neq 0$ and

$$\arg(w'(s_0)) = \arg(z'(s_0)) + \arg(f'(z_0)) \tag{5.2.3}$$

or arg $dw = $ arg $dz + $ arg $f'(z_0)$, where dw and dz are interpreted as infinitesimal line segments. This concludes the proof that, under the analytic transformation $f(z)$, the directed tangent to any curve through z_0 is rotated by an angle $\arg(f'(z_0))$.

   An immediate consequence of the above geometrical property is that, for points where $f'(z) \neq 0$, analytic transformations preserve angles. Indeed, if two curves intersect at z_0, because the tangent of each curve is rotated by arg $f'(z_0)$, it follows that the angle of intersection (in both magnitude and orientation), being the *difference* of the angles of the tangents, is preserved by such transformations. A transformation with this property is called **conformal**. We state this as a theorem; this theorem is enhanced in Sections 5.3 and 5.5.

Theorem 5.2.1   *Assume that $f(z)$ is analytic and not constant in a domain D of the complex z-plane. For any point $z \in D$ for which $f'(z) \neq 0$, this mapping is* **conformal**, *that is, it preserves the angle between two differentiable arcs.*

Remark A conformal mapping, in addition to preserving angles, has the property that it magnifies distances near z_0 by the factor $|f'(z_0)|$. Indeed, suppose that z is near z_0, and let w_0 be the image of z_0. Then the equation

$$|f'(z_0)| = \lim_{z \to z_0} \frac{|f(z) - f(z_0)|}{|z - z_0|}$$

implies that $|w - w_0|$ is approximately equal to $|f'(z_0)||z - z_0|$.

Example 5.2.2 Let D be the rectangular region in the z-plane bounded by $x = 0$, $y = 0$, $x = 2$ and $y = 1$. The image of D under the transformation $w = (1 + i)z + (1 + 2i)$ is given by the rectangular region D' of the w-plane bounded by $u + v = 3$, $u - v = -1$, $u + v = 7$, $u - v = -3$ (see Figure 5.3).

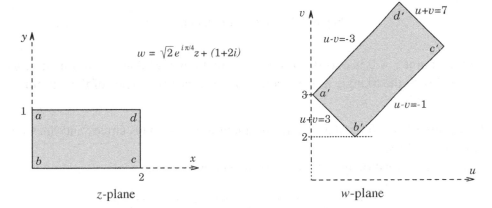

Figure 5.3 The transformation $w = (1 + i)z + (1 + 2i)$

If $w = u + iv$, where $u, v \in \mathbb{R}$, then $u = x - y + 1$, $v = x + y + 2$. Thus, the points a, b, c, and d are mapped to the points $(0, 3)$, $(1, 2)$, $(3, 4)$, and $(2, 5)$ respectively. The line $x = 0$ is mapped to $u = -y + 1$, $v = y + 2$, or $u + v = 3$; similarly for the other sides of the rectangle.

The rectangle D is translated by $(1 + 2i)$, rotated by an angle $\pi/4$ in the counterclockwise direction, and dilated (i.e., contracted) by a factor $\sqrt{2}$. In general, a linear transformation $f(z) = \alpha z + \beta$, translates by β, rotates by $\arg(\alpha)$, and dilates by $|\alpha|$. Because $f'(z) = \alpha \neq 0$, a linear transformation is always conformal. In this example $\alpha = \sqrt{2} \exp(i\pi/4)$, $\beta = 1 + 2i$.

Example 5.2.3 Let D be the triangular region bounded by $x = 1$, $y = 1$, and $x + y = 1$. The image of D under the transformation $w = z^2$, is given by the curvilinear triangle $a'b'c'$ shown in Figure 5.4.

In this example, $u = x^2 - y^2$, $v = 2xy$. The line $x = 1$ is mapped to $u = 1 - y^2$, $v = 2y$, or $u = 1 - \frac{v^2}{4}$; similarly for the other sides of the triangle. Because $f'(z) = 2z$

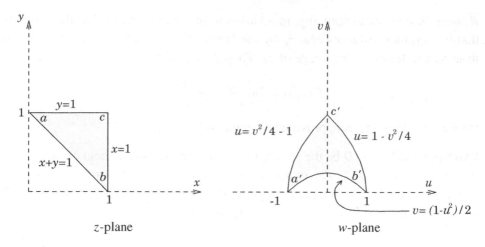

Figure 5.4 The transformation $w = z^2$

and the point $z = 0$ is outside D, it follows that this mapping is conformal; hence the angles of the triangle abc are equal to the respective angles of the curvilinear triangle $a'b'c'$.

Example 5.2.4 Show that the transformation $w = \frac{1}{z}$ maps circles and lines into circles and lines.

The transformation in terms of Cartesian coordinates is

$$u + iv = \frac{1}{x + iy} = \frac{x}{x^2 + y^2} - i\frac{y}{x^2 + y^2} \quad \text{so that:} \quad u = \frac{x}{x^2 + y^2}, v = \frac{-y}{x^2 + y^2}.$$

Hence

$$u^2 + v^2 = \frac{1}{x^2 + y^2}, \qquad \frac{u}{v} = \frac{-x}{y}.$$

Using these equations we find the inverse transformation:

$$x = \frac{u}{u^2 + v^2}, \qquad y = \frac{-v}{u^2 + v^2}.$$

With these formulae we can transform the equation

$$a(x^2 + y^2) + bx + cy + d = 0, \tag{5.2.4}$$

where a, b, c, d are constants, to

$$d(u^2 + v^2) + bu - cv + a = 0. \tag{5.2.5}$$

Equation (5.2.4) is a circle unless $a = 0$ in which case it is a straight line. A similar statement holds for (5.2.5) depending on whether or not $d = 0$. Thus we conclude that the transformation $w = \frac{1}{z}$ maps circles and lines into circles and lines.

5.2.1 Problems for Section 5.2

1. Use the transformation $w = k/(z - z_0)$, $k > 0$ to show that circles and lines map to circles and lines. Hint: See Example 5.2.4.

2. Find the image of the region R_z, bounded by $y = 0$; $x = 2$; $x^2 - y^2 = 1$, for $x \geq 0$ and $y \geq 0$ (see Figure 5.5), under the transformation $w = z^2$. Hint: show that $x = 2$ corresponds to $u = 4 - (\frac{v}{4})^2$, that $y = 0$ corresponds to $v = 0$, that $x \geq 0$, $y \geq 0$ corresponds to $v \geq 0$ and that $x^2 - y^2 = 1$ corresponds to $v = 1$.

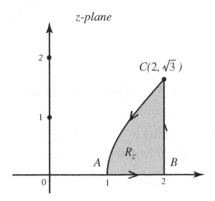

Figure 5.5 Region in Problem 2

3. Find a linear transformation which maps the circle C_1: $|z - 1| = 1$ onto the circle C_2: $|w - 3i/2| = 2$.

4. Show that the function $w = u + iv = e^z$ maps the interior of the rectangle, R_z, $0 < x < 1$, $0 < y < 2\pi$ where $z = x + iy$, onto the interior of the annulus, R_w, $1 < |w| < e$, which has a jump along the positive real axis (see Figure 5.6).

   Hint: First show that $u = e^x \cos y$, $v = e^x \sin y$, hence $|w| = e^x$. Then for ϵ small, show $y = \epsilon$ corresponds to $u = e^x \cos \epsilon \approx e^x$, and $v = e^x \sin \epsilon \approx \epsilon e^x$; and that $y = 2\pi - \epsilon$ corresponds to $u = e^x \cos \epsilon \approx e^x$, and $v = e^x \sin(2\pi - \epsilon) \approx -\epsilon e^x$; note the jump in v.

5. Show that the mapping $w = \sqrt{1 - z^2}$ maps the hyperbola $2x^2 - 2y^2 = 1$ onto itself.

   Hint: show that $u^2 - v^2 = 1 - (x^2 - y^2)$, and $uv = -xy$.

6. (a) Show that transformation $w = 2z + 1/z$ maps the exterior of the unit circle conformally onto the exterior of the ellipse:

$$\left(\frac{u}{3}\right)^2 + v^2 = 1.$$

   Hint: show that $u = 2x + \frac{x}{x^2+y^2}$, and $v = 2y - \frac{y}{x^2+y^2}$ so that when $x^2 + y^2 = 1$ we are on the ellipse and when z is large we are outside it.

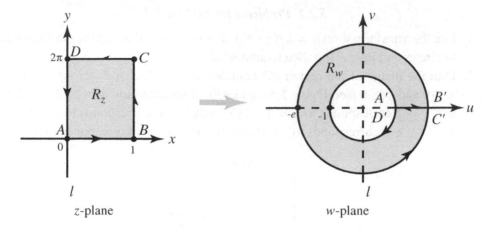

Figure 5.6 Mapping of Problem 4

(b) Show that the transformation $w = \frac{1}{2}(ze^{-\alpha}+e^{\alpha}/z)$, for a real constant α, maps the interior of the unit circle in the z-plane onto the exterior of the ellipse:

$$(u/\cosh\alpha)^2 + (v/\sinh\alpha)^2 = 1$$

in the w-plane.
Hint: show that when $x^2 + y^2 = 1$, then $u = (\cosh\alpha)x$, and $v = (-\sinh\alpha)y$ so that when $x^2 + y^2 = 1$ we are on the ellipse and when $y = 0$, $x \to \infty$, and $v = 0$, $u \to \infty$.

5.3 Critical Points and Inverse Mappings

If $f'(z_0) = 0$, then the analytic transformation $f(z)$ ceases to be conformal. Such a point is called a **critical point** of f. Because critical points are zeroes of the analytic function f', they are isolated. In order to find what happens geometrically at a critical point we use the following heuristic argument. Let $\delta z = z - z_0$, where z is a point near z_0. If the first non-vanishing derivative of $f(z)$ at z_0 is of the nth order, then representing δw by the Taylor series, it follows that

$$\delta w = \frac{1}{n!}f^{(n)}(z_0)(\delta z)^n + \frac{1}{(n+1)!}f^{(n+1)}(z_0)(\delta z)^{n+1} + \cdots, \qquad (5.3.1)$$

where $f^{(n)}(z_0)$ denotes the nth derivative of $f(z)$ at $z = z_0$. Thus, as $\delta z \to 0$,

$$\arg(\delta w) \to n\arg(\delta z) + \arg\left(f^{(n)}(z_0)\right). \qquad (5.3.2)$$

This equation, which is the analogue of Eq. (5.2.3), implies that the angle between any two infinitesimal line elements at the point z_0 is increased by the factor n. This suggests the following result.

Theorem 5.3.1 *Assume that $f(z)$ is analytic and not constant in a domain D of the complex z-plane. Suppose that $f'(z_0) = f''(z_0) = \cdots = f^{(n-1)}(z_0) = 0$, while $f^{(n)}(z_0) \neq 0$, $z_0 \in D$. Then the mapping $z \to f(z)$ magnifies n times the angle between two intersecting differentiable arcs which meet at z_0.*

Proof We now give a proof of this result. Let $z_1(s)$ and $z_2(s)$ be the equations describing the two arcs intersecting at z_0 (see Figure 5.7). If z_1 and z_2 are points on these arcs which have a distance r from z_0, it follows that

$$z_1 - z_0 = re^{i\theta_1}, \quad z_2 - z_0 = re^{i\theta_2}, \quad \text{or} \quad \frac{z_2 - z_0}{z_1 - z_0} = e^{i(\theta_2-\theta_1)}.$$

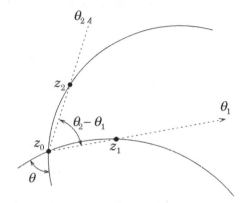

Figure 5.7 The angle between line segments $(\theta_2 - \theta_1)$ tends to the angle between arcs (θ) as $r \to 0$

The angle $\theta_2 - \theta_1$ is the angle formed by the linear segments connecting the points $z_1 - z_0$ and $z_2 - z_0$. As $r \to 0$, this angle tends to the angle formed by the two intersecting arcs in the complex z-plane. Similar considerations apply for the complex w-plane. Hence if θ and φ denote the angles formed by the intersecting arcs in the complex z-plane and w-plane respectively, it follows that

$$\theta = \lim_{r\to 0} \arg\left(\frac{z_2 - z_0}{z_1 - z_0}\right), \qquad \varphi = \lim_{r\to 0} \arg\left(\frac{f(z_2) - f(z_0)}{f(z_1) - f(z_0)}\right). \tag{5.3.3}$$

Hence

$$\varphi = \lim_{r\to 0} \arg\left\{\left(\frac{\frac{f(z_2)-f(z_0)}{(z_2-z_0)^n}}{\frac{f(z_1)-f(z_0)}{(z_1-z_0)^n}}\right)\left(\frac{z_2 - z_0}{z_1 - z_0}\right)^n\right\}. \tag{5.3.4}$$

Using

$$f(z) = f(z_0) + \frac{f^{(n)}(z_0)}{n!}(z - z_0)^n + \frac{f^{(n+1)}(z_0)}{(n + 1)!}(z - z_0)^{n+1} + \cdots, \tag{5.3.5}$$

it follows that

$$\lim_{r \to 0} \frac{f(z_2) - f(z_0)}{(z_2 - z_0)^n} = \lim_{r \to 0} \frac{f(z_1) - f(z_0)}{(z_1 - z_0)^n} = \frac{f^{(n)}(z_0)}{n!}.$$

Hence, Eqs. (5.3.4) and (5.3.3) imply

$$\varphi = \lim_{r \to 0} \arg \left(\frac{z_2 - z_0}{z_1 - z_0} \right)^n = n \lim_{r \to 0} \arg \left(\frac{z_2 - z_0}{z_1 - z_0} \right) = n\theta. \qquad \square$$

Example 5.3.2   Let D be the triangular region bounded by $x = 0$, $y = 0$, and $x + y = 1$. The image of D under the transformation $w = z^2$ is given by the curvilinear triangle $a'b'c'$ shown in Figure 5.8 (note the difference from Example 5.2.3).

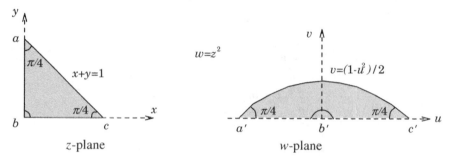

Figure 5.8 The transformation $w = z^2$ for Example 5.3.2

In this example, $u = x^2 - y^2$, $v = 2xy$. The lines $x = 0$; $y = 0$; and $x + y = 1$ are mapped to $v = 0$ with $u \le 0$; $v = 0$ with $u \ge 0$; and $v = \frac{1}{2}(1 - u^2)$, respectively. The transformation $f(z) = z^2$ ceases to be conformal at $z = 0$. Because the second derivative of $f(z)$ at $z = 0$ is the first nonvanishing derivative, it follows that the angle at b (which is $\pi/2$ in the z-plane) should be multiplied by 2. This is indeed the case, as the angle at b' in the w-plane is π.

Critical points are also important in determining whether the function $f(z)$ has an inverse. Finding the inverse of $f(z)$ means solving the equation $w = f(z)$ for z in terms of w.

The following terminology will be useful. An analytic function $f(z)$ is called **univalent** in a domain D if it takes no value more than once in D. It is clear that a univalent function $f(z)$ provides a one-to-one map of D onto $f(D)$; it has a single-valued inverse on $f(D)$.

There are a number of basic properties of conformal maps that are useful and that we now point out to the reader. In this section we only state the relevant theorems; they are proven in the optional Section *5.5.

Theorem 5.3.3   *Let $f(z)$ be analytic and not constant in a domain D of the complex z-plane. The transformation $w = f(z)$ can be interpreted as a mapping of the domain D onto the domain $D^* = f(D)$ of the complex w-plane.*

The proof of this theorem can be found in Section *5.5 but here are some general arguments that will be made rigorous there. Because a domain is an open connected set, this theorem implies that open sets in the domain D of the z-plane map to open sets D^* in the w-plane. A consequence of this fact is that $|f(z)|$ cannot attain a maximum in D^* because any point $w = f(z)$ must be an interior point in the w-plane. This theorem is useful because in practice we first find where the boundaries map. Then, since an open region is mapped to an open region, we need only find how one point is mapped if the boundary is a simple closed curve.

Suppose we try to construct the inverse in the neighborhood of some point z_0. If z_0 is not a critical point then $w - w_0$ is given approximately by $f'(z_0)(z - z_0)$. Hence it is plausible that in this case, for every w there exists a unique z; that is, $f(z)$ is locally invertible. However, if z_0 is a critical point, and the first nonvanishing derivative at z_0 is $f^{(n)}(z_0)$, then $w - w_0$ is given approximately by $f^{(n)}(z_0)(z - z_0)^n/n!$. Hence, now it is natural to expect that for every w there exist n different $z's$; that is, the inverse transformation is not single valued but it has a branch point of order n.

Theorem 5.3.4   (1) *Assume that $f(z)$ is analytic at z_0, and that $f'(z_0) \neq 0$. Then $f(z)$ is univalent in the neighborhood of z_0. More precisely, f has a unique analytic inverse F in the neighborhood of $w_0 \equiv f(z_0)$; that is, if z is sufficiently near z_0, then $z = F(w)$, where $w \equiv f(z)$. Similarly, if w is sufficiently near w_0 and $z \equiv F(w)$, then $w = f(z)$. Furthermore, $f'(z)F'(w) - 1$, which implies that the inverse map is conformal.*
(2) *Assume that $f(z)$ is analytic at z_0 and that it has a zero of order n; that is, the first nonvanishing derivative of $f(z)$ at z_0 is $f^{(n)}(z_0)$. Then to each w sufficiently close to $w_0 = f(z_0)$, there correspond n distinct points z in the neighborhood of z_0, each of which has w as its image under the mapping $w = f(z)$. Actually, this mapping can be decomposed in the form $w - w_0 = \zeta^n$, $\zeta = g(z - z_0)$, $g(0) = 0$, where $g(z)$ is univalent near z_0 and $g(z) = zH(z)$ with $H(0) \neq 0$.*

The proof of this theorem can be found in Section *5.5.

Remark We recall that $w = z^n$ provides a one-to-one mapping of the z-plane onto an n-sheeted Riemann surface in the w-plane (see Section 2.2). If a complex number $w \neq 0$ is given without specification as to the sheet in which it lies, there are n possible values of z that give this w, and so $w = z^n$ has an n-valued inverse. However, when the Riemann surface is introduced, the correspondence becomes one-to-one, and $w = z^n$ has a single-valued inverse.

Theorem 5.3.5 *Let C be a simple closed contour enclosing a domain D, and let $f(z)$ be analytic on C and in D. Suppose $f(z)$ takes no value more than once on C. Then:*
(a) the map $w = f(z)$ takes C enclosing D to a simple closed contour C^* enclosing a domain D^* in the w-plane;
(b) $w = f(z)$ is a one-to-one map from D to D^*;
(c) if z traverses C in the positive direction, then $w = f(z)$ traverses C^* in the positive direction.

The proof of this theorem can be found in Section *5.5.

Remark By examining the mapping of simple closed contours it can be established that conformal maps preserve the connectivity of a domain. For example, the conformal map $w = f(z)$ of a simply connected domain in the z-plane maps into a simply connected domain in the w-plane. Indeed, a simple closed contour in the z-plane can be continuously shrunk to a point, which must also be the case in the w-plane – otherwise, we would violate Theorem 5.3.3.

5.3.1 Problems for Section 5.3

1. Find the families of curves on which Re $z^2 = c_1$ for constant c_1, and Im $z^2 = c_2$, for constant c_2. Show that these two families are orthogonal to each other.
2. Let D be the triangular region of Figure 5.8(left); that is, the region bounded by $x = 0$, $y = 0$, and $x + y = 1$. Find the image of D under the mapping $w = z^3$. (It is sufficient to find a parameterization which describes the mapping of any of the sides.)
3. Express the transformations

$$\text{(a)} \qquad u = 4x^2 - 8y, \qquad v = 8x - 4y^2,$$
$$\text{(b)} \qquad u = x^3 - 3xy^2, \qquad v = 3x^2y - y^3$$

in the form $w = F(z, \bar{z})$, with $z = x + iy$, and $\bar{z} = x - iy$. Which of these transformations can be used to define a conformal mapping?
4. Show that the transformation $w = 2z^{-1/2} - 1$ maps the (infinite) domain exterior of the parabola $y^2 = 4(1 - x)$ conformally onto the domain $|w| < 1$. Explain why this transformation does not map the (infinite) domain interior of the parabola conformally onto the domain $|w| > 1$. Hint: the "intermediate" map $p = -i(1 - w)/(1 + w)$ taking $|w| > 1$ to Im, $p > 0$ is useful. Then, calling $p = R + iS$ show $S = \frac{|w|^2 - 1}{|1+w|^2}$, and $R = \frac{-2v}{|1+w|^2}$ so that $y^2 - 4(1 - x) = 4(R^2 + 1)S(S - 1)$. Hence when $|w| = 1$ we have $y^2 - 4(1 - x) = 0$.

5. Let D denote the domain enclosed by the parabolae $v^2 = 4a(a - u)$ and $v^2 = 4a(a + u)$, with $a > 0$, and $w = u + iv$. Show that the function

$$w = c^2 \left[\int_0^z \frac{dt}{\sqrt{t(1 + t^2)}} \right]^2,$$

where

$$\sqrt{a} = c \int_0^1 \frac{dt}{\sqrt{t(1 + t^2)}}, \qquad c > 0$$

maps the unit circle conformally onto D.

Hint: For the first quadrant of the unit circle, write the integral

$$\int_0^z \frac{dt}{\sqrt{t(t^2 + 1)}} = \int_0^1 \frac{dt}{\sqrt{t(t^2 + 1)}} + \int_1^z \frac{dt}{\sqrt{t(t^2 + 1)}}.$$

Then using the substitution $z = e^{i\theta}$, $t = e^{i\phi}$, $-\pi/2 < \theta < \pi/2$, show that $\int_1^z \frac{dt}{\sqrt{t(t^2+1)}} = iJ$, where $J = \int_0^\theta \frac{d\phi}{\sqrt{2\cos\phi}} > 0$. Then with $c \int_0^1 \frac{dt}{\sqrt{t(t^2+1)}} = \sqrt{a}$, for $c > 0$, show that $u = a - c^2 J^2$, $v = 2c\sqrt{a}J$ which implies $v^2 = 4a(a - u)$. For the other half of the circle use

$$\int_0^z \frac{dt}{\sqrt{t(t^2 + 1)}} = \int_0^{-1} \frac{dt}{\sqrt{t(t^2 + 1)}} + \int_{-1}^z \frac{dt}{\sqrt{t(t^2 + 1)}};$$

following similar steps to the above find $v^2 = 4a(a + u)$.

5.4 Physical Applications

It was shown in Section 2.1 that the real and the imaginary parts of an analytic function satisfy Laplace's equation. This and the fact that the occurrence of Laplace's equation in physics is ubiquitous constitute one of the main reasons for the usefulness of complex analysis in applications. In what follows we first mention a few physical situations that lead to Laplace's equation. Then we discuss how conformal mappings can be effectively used to study the associated physical problems. Some of these ideas were introduced in Chapter 2.

A twice differentiable function $\Phi(x, y)$ satisfying Laplace's equation,

$$\nabla^2 \Phi = \Phi_{xx} + \Phi_{yy} = 0, \tag{5.4.1}$$

in a region R is called harmonic in R. Let $V(z)$, $z = x + iy$, be analytic in R. If $V(z) = u(x, y) + iv(x, y)$, where $u, v \in \mathbb{R}$ and are twice differentiable, then both u and v are harmonic in R. Such functions are called conjugate functions. Given one of them (u or v), the other can be determined uniquely up to an arbitrary additive constant (see Section 2.1).

Let u_1 and u_2 be the components of the vector $\mathbf{u}$ along the positive x- and y-axis respectively. Suppose that the components of the vector $\mathbf{u} = (u_1, u_2)$ satisfy the equation

$$\frac{\partial u_1}{\partial x} + \frac{\partial u_2}{\partial y} = 0. \tag{5.4.2}$$

Suppose further that the vector $\mathbf{u}$ can be derived from a potential; that is, there exists a scalar function Φ such that

$$u_1 = \frac{\partial \Phi}{\partial x}, \qquad u_2 = \frac{\partial \Phi}{\partial y}. \tag{5.4.3}$$

Then Eqs. (5.4.2) and (5.4.3) imply that Φ is harmonic. These equations arise naturally in applications, as shown in the following examples.

Example 5.4.1 (Ideal Fluid Flow))   We consider a two-dimensional, steady, incompressible, irrotational fluid flow (see also the discussion in Section 2.1).

Since the flow is two-dimensional it is the same at all orthogonal planes. So, a flow pattern depicted in this plane can be interpreted as a cross-section of an infinite cylinder perpendicular to this plane. We will take the flow to be studied in the (x, y)-plane. If a flow is steady, it means that the velocity of the fluid at any point depends only on the position (x, y) and not on time. If the flow is incompressible, we take it to mean that the density (i.e., the mass per unit volume) of the fluid is constant. Let ρ and $\mathbf{u}$ denote the density and the velocity of the fluid. The law of conservation of mass implies Eq. (5.4.2). Indeed, consider a rectangle of sides $\Delta x, \Delta y$. See Figure 5.9.

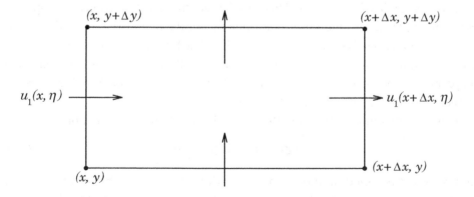

Figure 5.9 Flow through a rectangle of sides $\Delta x, \Delta y$

The rate of accumulation of fluid in this rectangle is given by $\frac{d}{dt}\int_x^{x+\Delta x}\int_y^{y+\Delta y}$ $\rho\,dx\,dy$. The rate of fluid entering along the side located between the points (x, y) and $(x, y + \Delta y)$ is given by $\int_y^{y+\Delta y}(\rho u_1)(x, \eta)\,d\eta$. A similar integral gives the rate of fluid entering the side between (x, y) and $(x + \Delta x, y)$. Letting ρ be a function of x, y and (for the moment) t. Then conservation of mass implies

$$\frac{d}{dt}\int_x^{x+\Delta x}\int_y^{y+\Delta y}\rho\,dx\,dy = \int_y^{y+\Delta y}[(\rho u_1)(x, \eta) - (\rho u_1)(x + \Delta x, \eta)]\,d\eta$$
$$+ \int_x^{x+\Delta x}[(\rho u_2)(\xi, y) - (\rho u_2)(\xi, y + \Delta y)]\,d\xi.$$

Dividing this equation by $\Delta x\Delta y$, taking the limit as Δx and Δy tend to zero, and assuming that ρ, u_1, and u_2 are smooth functions of x, y and t, it follows from calculus that

$$\frac{\partial \rho}{\partial t} + \frac{\partial (\rho u_1)}{\partial x} + \frac{\partial (\rho u_2)}{\partial y} = 0.$$

Because the flow is steady we have $(\partial \rho)/(\partial t) = 0$, and because the flow is incompressible, ρ is constant. Hence this equation yields Eq. (5.4.2).

If the flow is irrotational, it means that the circulation of the fluid along any closed contour C is zero. The circulation around C is given by $\oint_C \mathbf{u} \cdot \mathbf{ds}$, where $\mathbf{ds}$ is the vector element of arc length along C. We could use a derivation similar to the above, or we could use Green's Theorem (see Theorem 2.5.1 with (u, v) replaced by $\mathbf{u} = (u_1, u_2)$ and $\mathbf{ds} = (dx, dy)$), to deduce

$$\frac{\partial u_2}{\partial x} = \frac{\partial u_1}{\partial y}. \tag{5.4.4}$$

This equation is a necessary and sufficient condition for the existence of a potential Φ; that is, Eq. (5.4.3). Therefore Eqs. (5.4.2) and (5.4.4) imply that Φ is harmonic.

Because the function Φ is harmonic, there must exist a conjugate harmonic function, say $\Psi(x, y)$, such that

$$\Omega(z) = \Phi(x, y) + i\Psi(x, y) \tag{5.4.5}$$

is analytic. Differentiating $\Omega(z)$ and using the Cauchy–Riemann Conditions, (2.1.4), it follows that

$$\frac{d\Omega}{dz} = \frac{\partial \Phi}{\partial x} + i\frac{\partial \Psi}{\partial x} = \frac{\partial \Phi}{\partial x} - i\frac{\partial \Phi}{\partial y} = u_1 - iu_2 = \bar{u}, \tag{5.4.6}$$

where $u = u_1 + iu_2$ is the velocity of the fluid. Thus, the "complex velocity" of the fluid is given by

$$u = \overline{\left(\frac{d\Omega}{dz}\right)}. \tag{5.4.7}$$

The function $\Psi(x, y)$ is called the stream function, whereas $\Omega(z)$ is called the complex velocity potential (see also the discussion in Section 2.1). The families of the curves of constant $\Psi(x, y)$ are called streamlines of the flow. These lines represent the actual paths of points in the fluid. Indeed, if the curve C represents a path of such points, then the tangent to C has components $(u_1, u_2) = (\Phi_x, \Phi_y)$. Using the Cauchy–Riemann equations (2.1.4), we have

$$\Phi_x \Psi_x + \Phi_y \Psi_y = 0$$

and it follows that, as vectors,

$$(\Phi_x, \Phi_y) \cdot (\Psi_x, \Psi_y) = 0;$$

that is, the vector perpendicular to C has components (Ψ_x, Ψ_y), which is the gradient of Ψ. Hence we know from vector calculus that the curve C is given by $\Psi = $ a constant.

Example 5.4.2 (Heat Flow)   Discuss two-dimensional, steady heat flow as a potential in the complex plane.

The quantity of heat conducted per unit area per unit time across a surface of a given solid is called heat flux. In many applications the heat flux, denoted by the vector $\mathbf{Q}$, is given by $\mathbf{Q} = -k\nabla T$, where T denotes the temperature of the solid and k is called the thermal conductivity, which is taken to be constant. The conductivity k depends on the material of the solid. Conservation of energy, in steady state, implies that there is no accumulation of heat inside a given simple closed curve C. Hence if we denote $Q_n = \mathbf{Q} \cdot \hat{n}$, where $\hat{n}$ is the unit outward normal, then

$$\oint_C Q_n \, ds = \oint_C (Q_1 \, dy - Q_2 \, dx) = 0.$$

This equation, together with $\mathbf{Q} = -k\nabla T$, i.e.,

$$Q_1 = -k \frac{\partial T}{\partial x}, \qquad Q_2 = -k \frac{\partial T}{\partial y},$$

and Green's Theorem 2.5.1 (see Section 2.5), imply that T satisfies Laplace's equation. If Ψ denotes the harmonic conjugate function of T, then the function

$$\Omega(z) = T(x, y) + i\Psi(x, y) \tag{5.4.8}$$

is analytic. This function is called the complex temperature. The curves of the family $T(x, y) = $ constant are called isothermal lines.

Example 5.4.3 (Electrostatics)   We have seen that the appearance of Laplace's equation in fluid flow is a consequence of the conservation of mass and of the assumption that the circulation of the flow along a closed contour equals zero (irrotationality). Furthermore, conservation of mass is equivalent to the condition that the flux of the fluid across any closed surface is zero. The situation in electrostatics is similar: If **E** denotes the electric field, then the following two laws (consequences of the governing equations of time-independent electromagnetics) are valid:

(a) The flux of **E** through any closed surface enclosing zero charge equals zero. This is a special case of what is known as Gauss' law; that is, $\oint_C E_n \, ds = q/\epsilon_o$, where E_n is the normal component of the electric field, ϵ_o is the dielectric constant of the medium, and q is the net charge enclosed within C.

(b) The electric field is derivable from a potential, or stated differently, the circulation of **E** around a simple closed contour equals zero. If the electric field vector is denoted by $\mathbf{E} = (E_1, E_2)$, then these two conditions imply

$$\frac{\partial E_1}{\partial x} + \frac{\partial E_2}{\partial y} = 0, \qquad \frac{\partial E_2}{\partial x} = \frac{\partial E_1}{\partial y}. \tag{5.4.9}$$

From the second expression in Eq. (5.4.9), we have

$$E_1 = \frac{-\partial \Phi}{\partial x}, \qquad E_2 = \frac{-\partial \Phi}{\partial y} \tag{5.4.10}$$

(the minus signs are standard convention) and thus from Eq. (5.4.9) the function Φ is harmonic; that is, it satisfies Laplace's equation. Let Ψ denote the function that is conjugate to Φ. Then the function

$$\Omega(z) = \Phi(x, y) + i\Psi(x, y) \tag{5.4.11}$$

is analytic in any region not occupied by charge. This function is called the complex electrostatic potential. Differentiating $\Omega(z)$, and using the Cauchy–Riemann conditions, it follows that

$$\frac{d\Omega}{dz} = \frac{\partial \Phi}{\partial x} + i\frac{\partial \Psi}{\partial x} = \frac{\partial \Phi}{\partial x} - i\frac{\partial \Phi}{\partial y} = -\overline{E} \tag{5.4.12}$$

where $E = E_1 + iE_2$ is the complex electric field ($\overline{E} = E_1 - iE_2$). The curves of the families $\Phi(x, y) = $ a constant and $\Psi(x, y) = $ a constant are called equipotential and flux lines, respectively. From Eq. (5.4.12) Gauss' law is equivalent to

$$\text{Im} \oint_C \overline{E} \, dz = \oint_C (E_1 \, dy - E_2 \, dx) = \oint_C E_n \, ds = q/\epsilon_o. \tag{5.4.13}$$

We also note that integrals of the form $\int \overline{E} \, dz$ are invariant under a conformal transformation. More specifically, using Eq. (5.4.12), a conformal transformation $w = f(z)$ transforms the analytic function $\Omega(z)$ to $\Omega(w)$:

$$\int \overline{E} \, dz = -\int \frac{d\Omega}{dz} \, dz = -\int \frac{d\Omega}{dw} \, dw = -\int d\Omega. \tag{5.4.14}$$

In order to find the unique solution $\boldsymbol{\Phi}$ of Laplace's equation (5.4.1), one needs to specify appropriate boundary conditions. Let R be a simply connected region bounded by a simple closed curve C. There are two types of boundary-value problems that arise frequently in applications:

(a) In the **Dirichlet problem**, one specifies $\boldsymbol{\Phi}$ on the boundary C.

(b) In the **Neumann problem** one specifies the normal derivative of $\boldsymbol{\Phi}$ on the boundary C. (There is a third case, the "mixed case" where a combination of $\boldsymbol{\Phi}$ and the normal derivative are given on the boundary. We will not discuss this possibility here.)

If a solution exists for a Dirichlet problem, then it must be unique. Indeed if $\boldsymbol{\Phi}_1$ and $\boldsymbol{\Phi}_2$ are two such solutions then $\boldsymbol{\Phi} = \boldsymbol{\Phi}_1 - \boldsymbol{\Phi}_2$ is harmonic in R and $\boldsymbol{\Phi} = 0$ on C. The well-known vector identity (derivable from Green's Theorem 2.5.1),

$$\oint_C \boldsymbol{\Phi} \left(\frac{\partial \boldsymbol{\Phi}}{\partial x} \, dy - \frac{\partial \boldsymbol{\Phi}}{\partial y} \, dx \right) = \int \int_R \left[\boldsymbol{\Phi} \nabla^2 \boldsymbol{\Phi} + \left(\frac{\partial \boldsymbol{\Phi}}{\partial x} \right)^2 + \left(\frac{\partial \boldsymbol{\Phi}}{\partial y} \right)^2 \right] dx \, dy,$$

$$(5.4.15)$$

implies

$$\int \int_R \left[\left(\frac{\partial \boldsymbol{\Phi}}{\partial x} \right)^2 + \left(\frac{\partial \boldsymbol{\Phi}}{\partial y} \right)^2 \right] dx \, dy = 0. \qquad (5.4.16)$$

Therefore $\boldsymbol{\Phi}$ must be a constant in R, and because $\boldsymbol{\Phi} = 0$ on C, we find that $\boldsymbol{\Phi} = 0$ everywhere. Thus $\boldsymbol{\Phi}_1 = \boldsymbol{\Phi}_2$; that is, the solution is unique. The same analysis implies that if a solution exists for a Neumann problem ($\partial \boldsymbol{\Phi}/\partial n = 0$ on C), then it is unique to within an arbitrary additive constant.

It is possible to obtain the solution of the Dirichlet and Neumann problems using conformal mappings. This involves the following steps:

(a) Use a conformal mapping to transform the region R of the z-plane onto a simple region, such as the unit circle or a half plane of the w-plane.

(b) Solve the corresponding problem in the w-plane.

(c) Use this solution, and the inverse mapping function, to solve the original problem (recall that if $f(z)$ is conformal ($f'(z) \neq 0$), then, according to Theorem 5.3.4, $f(z)$ has a unique inverse).

This procedure is justified because of the following fact. Let $\boldsymbol{\Phi}(x, y)$ be harmonic in the region R of the z-plane. Assume that the region R is mapped onto the region R' of the w-plane by the conformal transformation $w = f(z)$, where $w = u + iv$. Then $\boldsymbol{\Phi}(x, y) = \boldsymbol{\Phi}(x(u, v), y(u, v))$ is harmonic in R'. Indeed, by differentiation and use of the Cauchy–Riemann conditions (2.1.4) we can verify (see also Problem 7 in Section 2.1.3)

$$\frac{\partial^2 \boldsymbol{\Phi}}{\partial x^2} + \frac{\partial^2 \boldsymbol{\Phi}}{\partial y^2} = \left| \frac{df}{dz} \right|^2 \left(\frac{\partial^2 \boldsymbol{\Phi}}{\partial u^2} + \frac{\partial^2 \boldsymbol{\Phi}}{\partial v^2} \right), \qquad (5.4.17)$$

which, because $df/dz \neq 0$, proves the above assertion. We use these ideas in the following example.

Example 5.4.4 Solve Laplace's equation for a function Φ inside the unit circle which on its circumference takes the value Φ_2 for $0 \leq \theta < \pi$, and the value Φ_1 for $\pi \leq \theta < 2\pi$.

This problem can be interpreted as finding the steady state heat distribution inside a disk with a prescribed temperature Φ on the boundary.

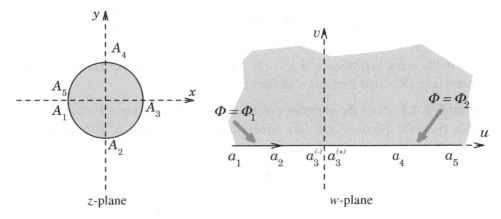

Figure 5.10 The transformation of the unit circle for Example 5.4.4

An important class of conformal transformations are of the form $w = f(z)$ where

$$f(z) = \frac{az + b}{cz + d}, \qquad ad - bc \neq 0. \tag{5.4.18}$$

These transformations are called **bilinear transformations**. They will be studied in detail in Section 5.7. In this problem we can verify that the bilinear transformation (see also the discussion in Section 5.7, especially Eq. (5.7.18))

$$w = i\left(\frac{1-z}{1+z}\right); \tag{5.4.19}$$

that is,

$$u = \frac{2y}{(1+x)^2 + y^2}, \qquad v = \frac{1 - (x^2 + y^2)}{(1+x)^2 + y^2},$$

maps the unit circle onto the upper half of the w-plane. (When z is on the unit circle, $z = e^{i\theta}$, then $w(z) = u = \frac{\sin\theta}{1+\cos\theta}$.) The arcs $A_1A_2A_3$ and $A_3A_4A_5$ are mapped onto the negative and positive real axis, respectively, of the w-plane; see Figure 5.10. Let $w = \rho e^{i\psi}$. The function $a\psi + b$, where a and b are real constants, is the real part of the analytic function $-ai \log w + b$ in the upper half plane and therefore is harmonic. Hence a solution of Laplace's equation in the upper half of the w-plane, satisfying $\Phi = \Phi_1$ for $u < 0$, $v = 0$ (i.e., $\psi = \pi$) and $\Phi = \Phi_2$ for $u > 0$, $v = 0$ (i.e., $\psi = 0$), is given by

$$\Phi = \Phi_2 - (\Phi_2 - \Phi_1)\frac{\psi}{\pi} = \Phi_2 - \frac{\Phi_2 - \Phi_1}{\pi} \tan^{-1}\left(\frac{v}{u}\right).$$

Owing to the uniqueness of solutions to the Dirichlet problem, this is the only solution. Using the expressions for u and v given by Eq. (5.4.19), it follows that in the (x, y)-plane the solution to the problem posed in the unit circle is given by

$$\Phi(x, y) = \Phi_2 - \frac{\Phi_2 - \Phi_1}{\pi} \tan^{-1}\left[\frac{1 - (x^2 + y^2)}{2y}\right].$$

Note, in the above Eq. when $x^2 + y^2 \to 1$, for $y > 0$, $y < 0$, we have $\tan^{-1}[\cdot] \to 0, \pi$ respectively. (See also Problem 9 in Section 2.2.1.)

Example 5.4.5   Find the complex potential and the streamlines of a fluid moving with a constant speed $u_0 \in \mathbb{R}$ in a direction making an angle α with the positive x-axis. (See also Example 2.1.10.)

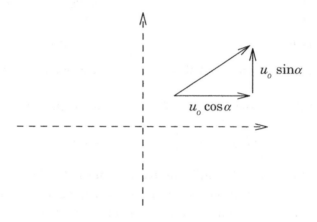

Figure 5.11 Flow velocity in Example 5.4.5

The x- and y-components of the fluid velocity are given by $u_0 \cos \alpha$ and $u_0 \sin \alpha$ respectively. The complex velocity is given by (see Figure 5.11)

$$u = u_0 \cos \alpha + iu_0 \sin \alpha = u_0 e^{i\alpha},$$

and so

$$\frac{d\Omega}{dz} = \bar{u} = u_0 e^{-i\alpha}, \qquad \text{or} \qquad \Omega = u_0 e^{-i\alpha} z,$$

where we have equated the constant of integration to zero. Letting $\Omega = \Phi + i\Psi$, it follows that

$$\Psi(x, y) = u_0(y \cos \alpha - x \sin \alpha) = u_0 r \sin(\theta - \alpha).$$

The streamlines are given by the family of the curves $\boldsymbol{\Psi}$ = constant, which are straight lines making an angle α with the positive x-axis.

Example 5.4.6 Analyze the flow pattern of a fluid emanating at a constant rate from an infinite line source perpendicular to the z-plane at $z = 0$.

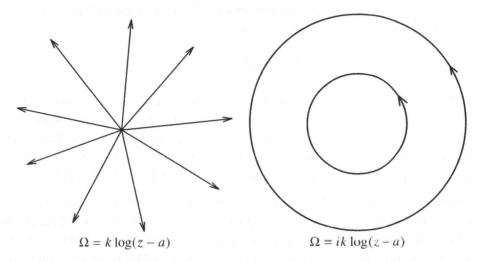

$$\Omega = k \log(z - a) \qquad\qquad \Omega = ik \log(z - a)$$

Figure 5.12 Streamlines associated with potential $\Omega(z)$ for Example 5.4.6

Let ρ and u_r denote the density (constant) and the radial velocity of the fluid, respectively. Let q denote the mass of fluid per unit time emanating from a line source of unit length. Then

$$q = (\text{density})\,(\text{flux}) = \rho(2\pi r u_r).$$

Thus

$$u_r = \frac{q}{2\pi\rho}\frac{1}{r} \equiv \frac{k}{r}, \qquad k > 0,$$

where the constant $k = q/2\pi\rho$ is called the strength of the source. Integrating the equation $u_r = \partial\Phi/\partial r$ and equating the constant of integration to zero ($\boldsymbol{\Phi}$ is cylindrically symmetric), it follows that $\boldsymbol{\Phi} = k \log r$, and hence, with $z = re^{i\theta}$,

$$\Omega(z) = k \log z.$$

The streamlines of this flow are given by $\boldsymbol{\Psi} = \operatorname{Im}\Omega(z) = $ constant; that is, θ is constant; see Figure 5.12 (left). These curves are rays emanating from the origin.

The complex potential $\Omega(z) = k \log(z-a)$ represents a "source" located at $z = a$. Similarly, $\Omega(z) = -k \log(z - a)$ represents a "sink" located at $z = a$ (because of the minus sign the velocity is directed toward $z = 0$).

It is clear that if $\Omega(z) = \Phi + i\Psi$ is associated with a flow pattern of streamlines of constant Ψ, then the function $i\Omega(z)$ is associated with a flow pattern of streamlines of constant Φ. These curves are orthogonal to the curves $\Psi = $ constant; that is, the flows associated with $\Omega(z)$ and $i\Omega(z)$ have orthogonal streamlines.

This discussion implies that in the particular case of the above example the streamlines of $\Omega(z) = ik \log z$ are concentric circles. Because $d\Omega/dz = ikz^{-1}$, it follows that the complex velocity is given by

$$\overline{\left(\frac{d\Omega}{dz}\right)} = \frac{k \sin\theta}{r} - \frac{ik \cos\theta}{r}.$$

This represents the flow of a fluid rotating with a clockwise speed k/r around $z = 0$. This flow is usually referred to as a vortex flow, generated by a vortex of strength k localized at $z = 0$. If the vortex is localized at $z = a$ then the associated complex potential is given by $\Omega = ik \log(z - a)$; the associated streamlines Ψ constant are depicted in Figure 5.12 (right). (See also Problems 7 and 8 of Section 2.2.1.)

Example 5.4.7 (The force due to fluid pressure)   In the physical circumstances we are dealing with, one neglects viscosity; that is, the internal friction of a fluid. It can be shown from the basic fluid equations that in this situation the pressure P of the fluid and the speed $|u|$ of the fluid are related by the so-called Bernoulli equation

$$P + \frac{1}{2}\rho|u|^2 = \alpha, \tag{5.4.20}$$

where α is a constant along each streamline. Let $\Omega(z)$ be the complex potential of some flow and let the simple closed curve C denote the boundary of a cylindrical obstacle of unit length that is perpendicular to the z-plane. We shall show that the force $F = X + iY$ exerted on this obstacle (see Figure 5.13) is given by

$$\overline{F} = \frac{1}{2}i\rho \oint_C \left(\frac{d\Omega}{dz}\right)^2 dz.$$

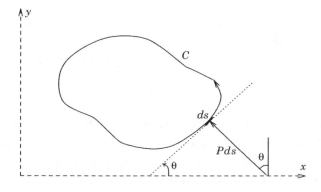

Figure 5.13 Force exerted on a cylindrical object for Example 5.4.7

Let ds denote an infinitesimal element around some point of the curve C, and let θ be the angle of the tangent to C at this point. The infinitesimal force exerted on the part of the cylinder corresponding to ds is perpendicular to ds and has magnitude $P\,ds$. (Recall that force equals pressure times area, and area equals ds, because the cylinder has unit length). Hence

$$dF = dX + i\,dY = -P\,ds\,\sin\theta + iP\,ds\,\cos\theta = iPe^{i\theta}\,ds.$$

Also,

$$dz = dx + i\,dy = ds\,\cos\theta + i\,ds\,\sin\theta = ds\,e^{i\theta}.$$

Without friction, the curve C is a streamline of the flow. The velocity is tangent to this curve, where we denote the complex velocity as $u = |u|e^{i\theta}$; hence

$$\frac{d\Omega}{dz} = |u|\,e^{-i\theta}. \tag{5.4.21}$$

The expression for dF along with Bernoulli's equation (5.4.20) imply

$$F = X + iY = \oint_C i\left(\alpha - \frac{1}{2}\rho|u|^2\right)e^{i\theta}\,ds.$$

The first term in the right-hand side of this equation equals zero because $\oint e^{i\theta}\,ds = \oint dz = 0$. Thus,

$$\overline{F} = \frac{1}{2}i\rho\oint_C |u|^2 e^{-i\theta}\,ds = \frac{1}{2}i\rho\oint_C \left(\frac{d\Omega}{dz}\right)^2 e^{i\theta}\,ds = \frac{1}{2}i\rho\oint_C \left(\frac{d\Omega}{dz}\right)^2\,dz,$$

where we have used Eq. (5.4.21) to replace $|u|$ with $d\Omega/dz$, and dz with $ds\,e^{i\theta}$.

Example 5.4.8 Discuss the flow pattern associated with the complex potential

$$\Omega(z) = u_0\left(z + \frac{a^2}{z}\right) + \frac{i\gamma}{2\pi}\log z.$$

This complex potential represents the superposition of a vortex of circulation of strength γ with a flow generated by the complex potential $u_0(z + a^2 z^{-1})$. (The latter flow was also discussed in Example 2.1.11.) Let $z = re^{i\theta}$; then if $\Omega = \Phi + i\Psi$,

$$\Psi(x, y) = u_0\left(r - \frac{a^2}{r}\right)\sin\theta + \frac{\gamma}{2\pi}\log r, \qquad a > 0,\ u_0, a, \gamma \text{ are real constants.}$$

If $r = a$, then $\Psi(x, y) = \gamma\log a/2\pi = $ constant; therefore $r = a$ is a streamline. Furthermore,

$$\frac{d\Omega}{dz} = u_0\left(1 - \frac{a^2}{z^2}\right) + \frac{i\gamma}{2\pi z},$$

which shows that as $z \to \infty$, the velocity tends to u_0. This discussion shows that the flow associated with $\Omega(z)$ can be considered as a flow with circulation about a circular obstacle. In the special case that $\gamma = 0$, this flow is depicted in Figure 5.14.

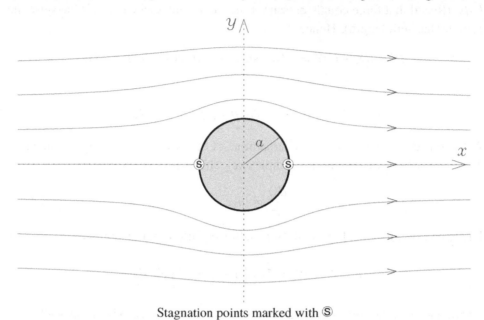

Stagnation points marked with Ⓢ

Figure 5.14 Flow around a circular obstacle ($\gamma = 0$)

Note that when $\gamma = 0$, then $d\Omega/dz = 0$ for $z = \pm a$; that is, there exist two points for which the velocity vanishes. Such points are called stagnation points see Figures 5.14–5.17; the streamline going through these points is given by $\Psi = 0$. In the general case of $\gamma \neq 0$, there also exist two stagnation points given by $d\Omega/dz = 0$, or

$$z = -\frac{i\gamma}{4\pi u_0} \pm \sqrt{a^2 - \frac{\gamma^2}{16\pi^2 u_0^2}}.$$

If $0 \leq \gamma < 4\pi a u_0$, there are two stagnation points on the circle (see Figure 5.15). If $\gamma = 4\pi a u_0$, these two points coincide (see Figure 5.16) at $z = -ia$. If $\gamma > 4\pi a u_0$, then one of the stagnation points lies outside the circle, and one inside (see Figure 5.17).

Using the result of Example 5.4.7 it is possible to compute the force exerted on this obstacle:

$$\overline{F} = \frac{1}{2} i\rho \oint_C \left[u_0 \left(1 - \frac{a^2}{z^2} \right) + \frac{i\gamma}{2\pi z} \right]^2 dz = -i\rho u_0 \gamma;$$

recall that $\oint z^n \, dz = 2\pi i \delta_{n,-1}$, where $\delta_{n,-1}$ is the Kronecker delta function. This

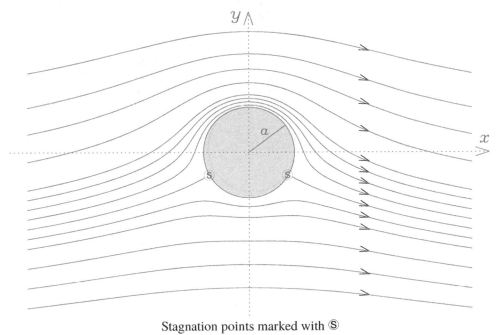

Stagnation points marked with ⑤

Figure 5.15 Separate stagnation points ($\gamma < 4\pi a u_0$)

shows that there exists a net force in the positive y direction of magnitude $\rho u_0 \gamma$. Such a force is known in aerodynamics as **lift**.

Example 5.4.9 Find the complex electrostatic potential due to a line of constant charge q per unit length perpendicular to the z-plane at $z = 0$.

The relevant electric field is radial and has magnitude E_r. If C is the circular basis of a cylinder of unit length located at $z = 0$, it follows from Gauss' law (see Example 5.4.3) that

$$\oint_C E_r \, ds = E_r 2\pi r = 4\pi q, \qquad \text{or} \qquad E_r = \frac{2q}{r},$$

where q is the charge enclosed by the circle C, and here we have normalized by $\epsilon_o = 1/4\pi$. Hence the potential satisfies

$$\frac{\partial \Phi}{\partial r} = -\frac{2q}{r}, \qquad \text{or} \quad \Phi = -2q \log r, \qquad \text{or} \quad \Omega(z) = -2q \log z.$$

This is identical to the complex potential associated with a line source of strength $k = -2q$. From Eq. (5.4.13) we see that

$$\text{Im} \left(\oint \overline{E} \, dz \right) = \text{Im} \left(\oint -\Omega'(z) \, dz \right) = 4\pi q,$$

as expected.

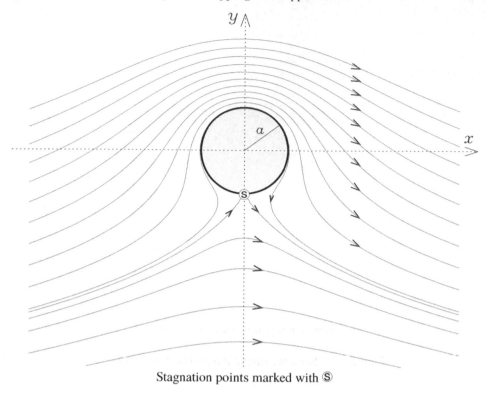

Stagnation points marked with Ⓢ

Figure 5.16 Coinciding stagnation points ($\gamma = 4\pi a u_0$)

Example 5.4.10 Consider two infinite parallel flat plates, separated by a distance d and maintained at zero potential. A line of charge q per unit length is located between the two planes at a distance a from the lower plate (see Figure 5.18). Find the electrostatic potential in the shaded region of the z-plane.

The conformal mapping $w = \exp(\pi z/d)$ maps the shaded strip of the z-plane onto the upper half of the w-plane. So the point $z = ia$ is mapped to the point $w_o = \exp(i\pi a/d)$; the points on the lower plate, $z = x$, and on the upper plate, $z = x + id$, map to the real axis $w = u$ for $u > 0$ and $u < 0$ respectively. Let us consider a line of charge q at w_o and a line of charge $-q$ at $\overline{w_o}$. Consider the associated complex potential (see also the previous Example 5.4.9)

$$\mathbf{\Omega}(w) = -2q \log(w - w_o) + 2q \log(w - \overline{w_o}) = 2q \log\left(\frac{w - \overline{w_o}}{w - w_o}\right).$$

Denoting by C_q a closed contour around the charge q, we see that Gauss' law is satisfied,

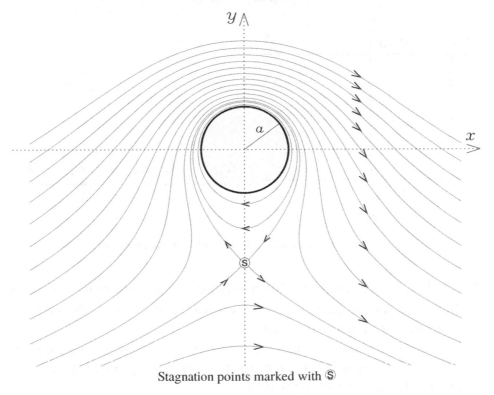

Stagnation points marked with Ⓢ

Figure 5.17 Streamlines and stagnation point ($\gamma > 4\pi a u_0$)

$$\oint_{C_q} E_n \, ds = \text{Im} \oint_{C_q} \bar{E} \, dz = \text{Im} \oint_{\tilde{C}_q} -\Omega'(w) \, dw = 4\pi q,$$

where $\tilde{C}_q$ is the image of C_q in the w-plane. (Again, see Example 5.4.3, with $\epsilon_0 = 1/4\pi$.) Then, writing $\Omega = \Phi + i\Psi$, we see that Φ is zero on the real axis of the w-plane (because $\log A/A^*$ is purely imaginary). Consequently, we have satisfied the boundary condition $\Phi = 0$ on the plates, and hence the electrostatic potential at any point of the shaded region of the z-plane is given by

$$\Phi = 2q \, \text{Re} \log \left[\frac{w - e^{-iv}}{w - e^{iv}} \right] = 2q \, \text{Re} \log \left[\frac{e^{\frac{\pi z}{d}} - e^{-iv}}{e^{\frac{\pi z}{d}} - e^{iv}} \right], \qquad v \equiv \frac{\pi a}{d}.$$

5.4.1 Problems for Section 5.4

1. Consider a source at $z = -a$ and a sink at $z = a$ of equal strengths k.

   (a) Show that the associated complex potential is $\Omega(z) = k \log [(z + a)/(z - a)]$.
   (b) Show that the flow speed is $2ka/\sqrt{a^4 - 2a^2 r^2 \cos 2\theta + r^4}$, where $z = re^{i\theta}$.

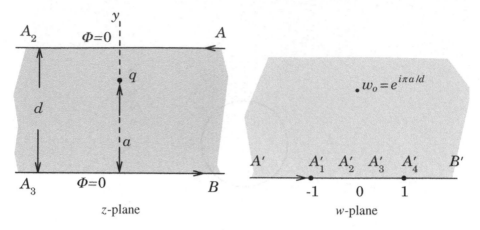

Figure 5.18 Electrostatic potential between parallel plates

2. Use Bernoulli's equation (5.4.20) to determine the pressure at any point of the fluid of the flow studied in Example 5.4.6. Show that $P = P_\infty - \frac{\rho k^2}{2r^2}$ where P_∞ is the pressure far from the source.

3. Consider the flow with the complex potential $\Omega(z) = u_0(z+a^2/z)$, the particular case $\gamma = 0$ of Example 5.4.8. Let p and p_∞ denote the pressure at a point on the cylinder and far from it, respectively.

   (a) Use Eq. (5.4.20) to establish that $p - p_\infty = \frac{1}{2}\rho u_0^2(1 - 4\sin^2\theta)$.
   (b) Show that a vacuum is created at the points $\pm ia$ if the speed of the fluid is equal to or greater than $u_0 = \sqrt{2p_\infty/(3\rho)}$. This phenomenon is usually called **cavitation**.

4. Discuss the fluid flow associated with the complex velocity potential $\Omega(z) = Q_0 z + \frac{\overline{Q_0}a^2}{z} + \frac{i\gamma}{2\pi}\log z$, for $a > 0$, γ real, and $Q_0 = U_0 + iV_0$. Show that the force exerted on the cylindrical obstacle defined by the flow field is given by $\mathbf{F} = i\rho\overline{Q_0}\gamma$. This force is often referred to as the **lift**.

5. Show that the steady-state temperature at any point of the region given in Figure 5.19, where the temperatures are maintained as indicated in the figure, is given by

$$T(r,\theta) = \frac{10}{\pi}\tan^{-1}\left\{\frac{(r^2-1)\sin\theta}{(r^2+1)\cos\theta - 2r}\right\}$$
$$- \frac{10}{\pi}\tan^{-1}\left\{\frac{(r^2-1)\sin\theta}{(r^2+1)\cos\theta + 2r}\right\}.$$

Hint: use the transformation $w = z + 1/z$ to map the above shaded region onto the upper half plane.

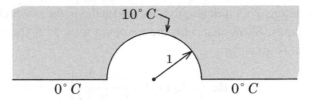

Figure 5.19 Temperature distribution, Problem 5

6. Let $\Omega(z) = z^\alpha$, where α is a real constant and $\alpha > \frac{1}{2}$. If $z = re^{i\theta}$ show that the rays $\theta = 0$ and $\theta = \pi/\alpha$ are streamlines, and hence can be replaced by walls. Show that the speed of the flow is $\alpha r^{\alpha-1}$, where r is the distance from the corner.

7. Two semi-infinite plane conductors meet at an angle $0 < \alpha < \pi/2$, and are charged at constant potentials Φ_1 and Φ_2.

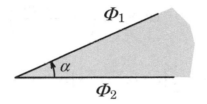

Figure 5.20 Electrostatic potential, Problem 7

Show that the potential Φ and the electric field $\mathbf{E} = (E_r, E_\theta)$ in the region between the conductors are given by

$$\Phi = \Phi_2 + \left(\frac{\Phi_1 - \Phi_2}{\alpha} \right) \theta, \qquad E_\theta = \frac{\Phi_2 - \Phi_1}{\alpha r}, \qquad E_r = 0,$$

where $z = re^{i\theta}$, $0 \leq \theta \leq \alpha$.

8. Two semi-infinite plane conductors intersect at an angle α, $0 < \alpha < \pi$, and are kept at zero potential.

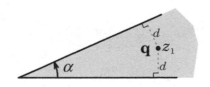

Figure 5.21 Electrostatics, Problem 8

A line of charge q per unit length is located at the point z_1 which is equidistant from both planes. Show that with the mapping $w = z^{\pi/\alpha}$ the potential in the shaded region is given by

$$\text{Re}\left\{-2q\log\left(\frac{z^{\frac{\pi}{\alpha}} - z_1^{\frac{\pi}{\alpha}}}{z^{\frac{\pi}{\alpha}} - \bar{z}_1^{\frac{\pi}{\alpha}}}\right)\right\}.$$

9. Consider the flow past an elliptic cylinder indicated in the following figure.

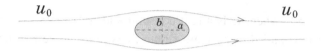

Figure 5.22 Ideal flow, Problem 9

(a) Show that the complex potential is given by

$$\Omega(z) = u_0\left(\zeta + \frac{(a+b)^2}{4\zeta}\right),$$

where

$$\zeta = \frac{1}{2}\left(z + (z^2 - c^2)^{\frac{1}{2}}\right), \qquad c^2 = a^2 - b^2.$$

(b) Show that the fluid speed at the top and bottom of the cylinder is: $u_0(1 + \frac{b}{a})$.

Hint: The flow in the ζ-plane is around a circle radius $\frac{a+b}{2}$. Find $z = \zeta + \frac{c^2}{4\zeta}$ and use it to calculate $\frac{d\Omega}{dz} = \frac{d\Omega}{d\zeta}\frac{d\zeta}{dz}$ on the circle.

10. Two infinitely long cylindrical conductors having cross sections which are confocal ellipses with foci at $(-c, 0)$ and $(c, 0)$ (see Figure 5.23) are kept at constant potentials Φ_1 and Φ_2.

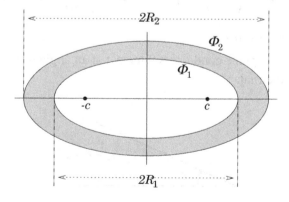

Figure 5.23 Confocal ellipses, Problem 10

(a) Show that the mapping $z = c \sin \zeta = c \sin(\xi + i\eta)$ transforms the confocal ellipses in Figure 5.23 onto two parallel plates such as those depicted in Figure 5.18 where $\Phi = \Phi_j$ on $\eta = \eta_j$, with $\cosh \eta_j = R_j/c$, $j = 1, 2$. Use the transformation $w = \exp(\frac{\pi}{d}(\zeta - i\eta_1))$, where $d = \eta_2 - \eta_1$ (see Example 5.4.10) to show that the complex potential is given by

$$\Omega(w) = \Phi_1 + \frac{\Phi_2 - \Phi_1}{i\pi} \log w = \Phi_1 + \frac{\Phi_2 - \Phi_1}{id} \left[\sin^{-1}\left(\frac{z}{c}\right) - i\eta_1 \right].$$

(b) If the capacitance of two perfect conductors is defined by $C = q/(\Phi_1 - \Phi_2)$, where q is the charge on the inside ellipse, use Gauss' Law to show that the capacitance per unit length is given by

$$C = \frac{1}{2d} = \frac{1}{2\left(\cosh^{-1}\left(\frac{R_2}{c}\right) - \cosh^{-1}\left(\frac{R_1}{c}\right)\right)}.$$

(c) Establish that as $c \to 0$ (two concentric circles),

$$C \to \frac{1}{2\log(R_2/R_1)}.$$

11. A circular cylinder of radius R lies at the bottom of a channel of fluid which, at large distance from the cylinder, has constant velocity u_0.

Figure 5.24 Confocal ellipses, Problem 11

(a) Verify that the complex potential is given by

$$\Omega(z) = \pi R u_0 \coth\left(\frac{\pi R}{z}\right), \qquad z = x + iy.$$

Hint: Show that the bottom $y = 0$ and $z = iR + Re^{i\theta}$ are streamlines. Also show that as $z \to \infty$ uniform flow results.

(b) Show that the magnitude of the difference in pressure between the top and the bottom points of the cylinder is $\rho\pi^4 u_0^2/32$, where ρ is the density of the fluid (see Eqs. (5.4.20)–(5.4.21)).

*5.5 Theoretical Considerations – Mapping Theorems

In Section 5.3, various mapping theorems were stated, but their proofs were postponed to this optional section.

Theorem 5.5.1 (Originally stated as Theorem 5.3.3)
Let $f(z)$ be analytic and not constant in a domain D of the complex z-plane. The transformation $w = f(z)$ can be interpreted as a mapping of the domain D onto the domain $D^* = f(D)$ of the complex w-plane. (Sometimes this theorem is summarized as "open sets map to open sets.")

Proof A point set is a domain if it is open and connected (see Section 1.2). An open set is connected if every two points of this set can be joined by a contour lying in this set. If we can prove that D^* is an open set, its connectivity is an immediate consequence of the fact that, because $f(z)$ is analytic, every continuous arc in D is mapped onto a continuous arc in D^*. The proof that D^* is an open set follows from an application of Rouché's theorem (see Section 4.4) which states: if the functions $g(z)$ and $\tilde{g}(z)$ are analytic in a domain and on the boundary of this domain, and if on the boundary $|g(z)| > |\tilde{g}(z)|$, then in this domain the functions $g(z) - \tilde{g}(z)$ and $g(z)$ have exactly the same number of zeroes. Because $f(z)$ is analytic in D, then $f(z)$ has a Taylor expansion at a point $z_0 \in D$. Assume that $f'(z_0) \neq 0$. Then $g(z) \equiv f(z) - f(z_0)$ vanishes (it has a zero of order 1) at z_0. Because $f(z)$ is analytic, this zero is isolated (see Theorem 3.2.9). That is, there exists a constant $\varepsilon > 0$ such that $g(z) \neq 0$ for $0 < |z - z_0| \leq \varepsilon$. On the circle $|z - z_0| = \varepsilon$, $g(z)$ is continuous; hence there exists a positive constant A such that $A = \min |f(z) - f(z_0)|$ on $|z - z_0| = \varepsilon$. If $\tilde{g}(z) \equiv a$ is a complex constant such that $|a| < A$, then $|g(z)| > |a| = |\tilde{g}(z)|$ on $|z - z_0| = \varepsilon$, and Rouché's theorem implies that $g(z) - \tilde{g}(z)$ vanishes in $|z - z_0| < \varepsilon$.

Hence, for every complex number $a = |a|e^{i\phi}$, $|a| < A$, we find that there is exactly one value z for which $g(z) = w - w_0 = a$ inside $|z - z_0| < \varepsilon$. Therefore, if $z_0 \in D$, $f'(z_0) \neq 0$, and $w_0 = f(z_0)$, then for sufficiently small $\varepsilon > 0$, there exists a $\delta > 0$, such that the image of $|z - z_0| < \varepsilon$ contains the disc $|w - w_0| < \delta$ (here $\delta = A$), and therefore D^* is open. If $f'(z_0) = 0$, a slight modification of the above argument is required. If the first non-vanishing derivative of $f(z)$ at z_0 is of the nth order, then $g(z)$ has a zero of the nth order at $z = z_0$. The rest of the argument goes through as above, but in this case one obtains from Rouché's Theorem the additional information that in $|z - z_0| < \varepsilon$, the values w for which $|w - w_0| < A$ will now be taken n times.   □

Theorem 5.5.2 (originally stated as Theorem 5.3.4)
(1) *Assume that $f(z)$ is analytic at z_0 and that $f'(z_0) \neq 0$. Then $f(z)$ is univalent in the neighborhood of z_0. More precisely, f has a unique analytic inverse F in the neighborhood of $w_0 \equiv f(z_0)$; that is, if z is sufficiently near z_0, then $z = F(w)$,*

where $w \equiv f(z)$. *Similarly, if w is sufficiently near w_0 and $z \equiv F(w)$, then $w = f(z)$. Furthermore, $f'(z)F'(w) = 1$, which implies that the inverse map is conformal.*

(2) *Assume that $f(z)$ is analytic at z_0 and that it has a zero of order n; that is, the first nonvanishing derivative of $f(z)$ at z_0 is $f^{(n)}(z_0)$. Then to each w sufficiently close to $w_0 \equiv f(z_0)$, there correspond n distinct points z in the neighborhood of z_0, each of which has w as its image under the mapping $w = f(z)$. Actually, this mapping can be decomposed in the form $w - w_0 = \zeta^n$, with $\zeta = g(z - z_0)$, $g(0) = 0$, where $g(z)$ is univalent near z_0 and $g(z) = zH(z)$ with $H(0) \neq 0$.*

Proof (1) The first part of the proof follows from Theorem 5.5.1, where it was shown that each w in the disk $|w - w_0| < A$, denoted by P, is the image of a unique point z in the disk $|z - z_0| < \varepsilon$, where $w = f(z)$. This uniqueness implies $z = F(w)$ and $z_0 = F(w_0)$. The equations $w = f(z)$ and $z = F(w)$ imply the usual equation $w = f(F(w))$ satisfied by a function and its inverse. First we show that $F(w)$ is continuous in P, and then show that this implies that $F(w)$ is analytic.

Let $w_1 \in P$ be the image of a unique point z_1 in $|z - z_0| < \varepsilon$. From Theorem 5.5.1, the image of $|z - z_1| < \varepsilon_1$ contains $|w - w_1| < \delta_1$, so for sufficiently small δ_1 we have $|z - z_1| = |F(w) - F(w_1)| < \varepsilon_1$. Now let δ_1 be small enough so that $|w - w_1| < \delta_1$ is in P and $|F(w) - F(w_1)| < \epsilon_1$. Since ϵ_1 is arbitrary and there is a corresponding $\delta_1 > 0$, it follows that $F(w)$ is continuous.

Next assume that w_1 is near w. Then w and w_1 are the images corresponding to $z = F(w)$ and $z_1 = F(w_1)$ respectively. If w is fixed, the continuity of F implies that if $|w_1 - w|$ is small then $|z_1 - z|$ is also small. Thus

$$\frac{F(w_1) - F(w)}{w_1 - w} = \frac{z_1 - z}{w_1 - w} = \frac{z_1 - z}{f(z_1) - f(z)} \longrightarrow \frac{1}{f'(z)}, \qquad (5.5.1)$$

as $|w_1 - w| \to 0$.

Because $f(z) = w$ has only one solution, counting multiplicity, for $|z - z_0| < \varepsilon$, it follows that $f'(z) \neq 0$. Thus Eq. (5.5.1) implies that $F'(w)$ exists and equals $1/f'(z)$. We also see, by the continuity of $f(z)$, that every z near z_0 has as its image a point near w_0. So if $|z - z_0|$ is sufficiently small, $w = f(z)$ is a point in P and $z = F(w)$. Thus $z = F(f(z))$ near z_0, which by the chain rule implies $1 = f'(z)F'(w)$, consistent with Eq. (5.5.1).

(2) Assume for convenience, without loss of generality, that $z_0 = w_0 = 0$. Using the fact that the first $(n - 1)$ derivatives of $f(z)$ vanish at $z = z_0$, we see from its Taylor series that $w = z^n h(z)$, where $h(z)$ is analytic at $z = 0$ and $h(0) \neq 0$. Because $h(0) \neq 0$ there exists an analytic function $H(z)$ such that $h(z) = [H(z)]^n$, with $H(0) \neq 0$. (The function $H(z)$ can be found by taking the logarithm.) Thus, $w = (g(z))^n$, where $g(z) = zH(z)$. The function $g(z)$ satisfies $g(0) = 0$ and $g'(0) \neq 0$, hence it is univalent near 0. The properties of $w = \zeta^n$ together with the fact that $g(z)$ is univalent imply the assertions of part (2) of Theorem 5.3.4. □

Theorem 5.5.3 (Originally stated as Theorem 5.3.5)

Let C be a simple closed contour enclosing a domain D, and let $f(z)$ be analytic on C and in D. Suppose $f(z)$ takes no value more than once on C. Then:

(a) *the map $w = f(z)$ takes C enclosing D to a simple closed contour C^* enclosing a region D^* in the w-plane;*

(b) *$w = f(z)$ is a one-to-one map from D to D^*;*

(c) *if z traverses C in the positive direction, then $w = f(z)$ traverses C^* in the positive direction.*

Proof   (a) The image of C is a simple closed contour C^*, because $f(z)$ is analytic and because $f(z)$ takes on no value more than once for z on C.

(b) Consider the following integral with the transformation $w = f(z)$, where w_0 corresponds to a point $z_0 \in D$ which is not a point on C^*:

$$I = \frac{1}{2\pi i} \oint_C \frac{f'(z)\,dz}{f(z) - w_0} = \frac{1}{2\pi i} \oint_{C^*} \frac{dw}{w - w_0}. \qquad (5.5.2)$$

From the argument principal in Theorem 4.4.1 we find that $I = N - P$, where N and P are the number of zeroes and poles (respectively) of $f(z) - w_0$ enclosed within C. However, because $f(z)$ is analytic, $P = 0$, and $I = N$.

If w_0 lies outside C^*, the right-hand side of Eq. (5.5.2) is 0, and therefore $N = 0$ so that $f(z) \neq w_0$ inside C. If w_0 lies inside C^*, then the right-hand side of Eq. (5.5.2) is 1 (assuming for now, the usual positive convention in $\oint$), and therefore $f(z) = w_0$ once inside C. Finally, w_0 could not lie on C^* because it is an image of some point $z_0 \in D$, and, from Theorem 5.5.1 (open sets map to open sets), some point in the neighborhood of w_0 would need to be mapped to the exterior of C^*, which we have just seen is not possible.

Consequently, each value w_0 inside C^* is attained once and only once, and the transformation $w = f(z)$ is a one-to-one map.

(c) The above proof assumes that both C and C^* are traversed in the positive direction. If C^* is traversed in the negative direction, then the right-hand side of Eq. (5.5.2) would yield -1, which contradicts the fact that N must be positive. Clearly, C and C^* can both be traversed in the negative directions.   □

Finally, we conclude this section with a statement of the Riemann Mapping Theorem.

Theorem 5.5.4 (Riemann Mapping Theorem)   *Let D be a simply connected domain in the z-plane, which is neither the z-plane or the extended z-plane. Then there exists a univalent function $f(z)$, such that $w = f(z)$ maps D onto the disc $|w| < 1$.*

First, we note that the entire finite plane, $|z| < \infty$, is simply connected. However, there exists no conformal map which maps the entire finite plane onto the unit disc. This is a consequence of Liouville's Theorem because an analytic function $w = f(z)$ such that $|f(z)| < 1$ for all finite $z \in \mathbb{C}$ would have to be constant. Similar reasoning shows that there exists no conformal map which maps the extended plane $|z| \leq \infty$ onto the unit disc. By Riemann's Mapping Theorem, these are the only simply connected domains which cannot be mapped onto the unit disc.

The proof of this theorem requires knowledge of the topological concepts of completeness and compactness. It involves considering families of mappings and solving a certain maximum problem for a family of bounded continuous functionals. This proof, which is nonconstructive, can be found in advanced textbooks (see, for example, Nehari (1952)). In the case of a simply connected domain bounded by a smooth Jordan curve, a simpler proof has been given (Garabedian, 1991).

Remarks It should be emphasized that the Riemann Mapping Theorem is a statement about simply connected open sets. It says nothing about the behavior of the mapping function on the boundary. However, for many applications of conformal mappings, such as the solution of boundary value problems, it is essential that one is able to define the mapping function on the boundary. For this reason it is important to identify those bounded regions for which the mapping function can be extended continuously to the boundary. It can be shown (Osgood–Carathéodory Theorem) that if D is bounded by a simple closed contour, then it is possible to extend the function f mapping D conformally onto the open unit disc, in such a way that f extends continuously to the boundary and is also one to one on the boundaries.

A further consequence of all this: if we fix any three points on the boundary of the mapping $w = f(z)$, where the two sets of corresponding points $\{z_1, z_2, z_3\}$, $\{w_1, w_2, w_3\}$ appear in the same order when the two boundaries are described in the positive direction, then this uniquely determines the map. The essential reason for this is that two different maps onto the unit circle can be transformed to one another by a bilinear transformation, which can be shown to be fixed by three points (see Section 5.7). Alternatively, if z_0 is a point in D, fixing $f(z_0) = 0$ with $f'(z_0) > 0$ uniquely determines the map.

We also note that there is a bilinear transformation (e.g. Eq. (5.4.19) and see also Eq. (5.7.18)) that maps the unit circle onto the upper half plane, so in the theorem we could equally well state that $w = f(z)$ maps D onto the upper half w-plane.

5.6 The Schwarz–Christoffel Transformation

As mentioned earlier, one of the most remarkable results in the theory of complex analysis is Riemann's Mapping Theorem, Theorem 5.5.4. This theorem states that any simply connected domain of the complex z-plane, with the exception of the entire z-plane and the extended entire z-plane, can be mapped with a univalent transformation $w = f(z)$ onto the disk $|w| < 1$ or onto the upper half of the complex w-plane. Unfortunately the proof of this celebrated theorem is not constructive; that is, given a specific domain in the z-plane there is no general constructive approach for finding $f(z)$. Nevertheless, as we have already seen, there are many particular domains for which $f(z)$ can be constructed explicitly. One such domain is the interior of a polygon. Let us first consider an example of a very simple polygon.

Example 5.6.1   The interior of an open triangle of angle $\pi\alpha$, with vertex at the origin of the w-plane is mapped to the upper half z-plane, by $w = z^\alpha$, for $0 < \alpha < 2$; see Figure 5.25.

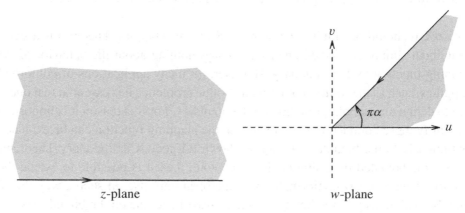

Figure 5.25 The transformation $w = z^\alpha$ for Example 5.6.1

If $z = re^{i\theta}$, $w = \rho e^{i\varphi}$, then the rays $\varphi = 0$ and $\varphi = \pi\alpha$ of the w-plane are mapped to the rays $\theta = 0$ and $\theta = \pi$ of the z-plane. We note that the conformal property – that is, that angles are preserved under the transformation $w = f(z) = z^\alpha$ – doesn't hold at $z = 0$ because $f(z)$ is not analytic there when $\alpha \neq 1$.

The transformation $w = f(z)$ associated with a general polygon is called the Schwarz–Christoffel transformation. In deriving this transformation we will make use of the so-called Schwarz reflection principle. The most basic version of this principle is really based on the following elementary fact. Suppose that $f(z)$ is analytic in a domain D that lies in the upper half of the complex z-plane. Let $\widetilde{D}$ denote the domain obtained from D by reflection with respect to the real axis (obviously $\widetilde{D}$ lies on the lower half of the complex z-plane). Then corresponding to every point $z \in \widetilde{D}$, the function $\tilde{f}(z) = \overline{f(\bar{z})}$ is analytic in $\widetilde{D}$.

   Indeed, if $f(z) = u(x, y) + iv(x, y)$, then $\overline{f(\bar{z})} = u(x, -y) - iv(x, -y)$. This shows that the real and imaginary parts of the function $\tilde{f}(z) \equiv \overline{f(\bar{z})}$ have continuous partial derivatives, and that the Cauchy–Riemann equations (see Section 2.1) for f imply

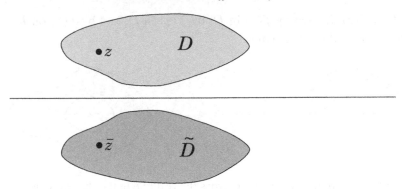

Figure 5.26 Reflection Principle

the Cauchy–Riemann equations for $\tilde{f}$, and hence $\tilde{f}$ is analytic. Call $\tilde{u}(x, y') = u(x, -y)$, $\tilde{v}(x, y') = -v(x, -y)$, $y' = -y$, then the Cauchy–Riemann conditions for $\tilde{u}$ and $\tilde{v}$ in terms of x and y' follow.

Example 5.6.2   The function $f(z) = 1/(z+i)$ is analytic in the upper half z-plane. Use the Schwarz reflection principle to construct a function analytic in the lower half z-plane.

The function $\overline{f(\bar{z})} = 1/(z-i)$ has a pole at $z = i$ as its only singularity; therefore it is analytic for Im $z \leq 0$.

The above idea not only applies to reflection about straight lines, but it also applies to reflections about circular arcs. (This is discussed more fully in Section 5.7.) In a special case, it implies that if $f(z)$ is analytic inside the unit circle, then the function $\overline{f(1/\bar{z})}$ is analytic outside the unit circle. Note that the points in the domains D and $\widetilde{D}$ of Figure 5.26 are distinguished by the property that they are inverse points with respect to the real axis. If one uses a bilinear transformation (see Eqs. (5.4.19) or (5.7.18) for example) to map the real axis onto the unit circle, the corresponding points z and $1/\bar{z}$ will be the "inverse" points with respect to this circle. (Note that the inverse points with respect to a circle are defined in Property (viii) of Section 5.7.)

The Schwarz reflection principle across real line segments is the following. Suppose that the domain D has part of the real axis as part of its boundary. Assume that $f(z)$ is analytic in D and is continuous as z approaches the line segments $L_1, \ldots, L_n$ of the real axis and that $f(z)$ is real on these segments. Then $f(z)$ can be analytically continued across $L_1, \ldots, L_n$ into $\widetilde{D}$ (see also Theorems 3.2.8, 3.2.9, and 3.5.7). Indeed, $f(z)$ is analytic in D, which implies that $\tilde{f}(z) \equiv \overline{f(\bar{z})}$ is analytic in $\widetilde{D}$ because $f(z) = \tilde{f}(z)$ on the line segments (due to the reality condition). These facts, together with the continuity of $f(z)$ as z approaches $L_1, \ldots, L_n$, imply that the function $F(z)$ defined by $f(z)$ in D, by $\tilde{f}(z)$ in $\widetilde{D}$, and by $f(z)$ on $L_1, \ldots, L_n$ is also analytic on these segments.

Figure 5.27 Analytic continuation across the real line in the shaded region when $f(z)$ is real and continuous on the real axis

A particular case of such a situation is shown in Figure 5.27. Although D has two line segments in common with the real axis, let us assume that the conditions of continuity and reality are satisfied only on L_1. Then there exists a function analytic everywhere in the shaded region except on l_1.

The important assumption in deriving the above result was $\text{Im } f = 0$ on $\text{Im } z = 0$. If we think of $f(z)$ as a transformation from the z-plane to the w-plane, this means that a line segment of the boundary of D of the z-plane is mapped into a line segment of the boundary of $f(D)$ in the w-plane which are portions of the real w-axis. By using linear transformations (rotations and translations) in the z and w-plane, one can extend this result to the case that these transformed line segments are not necessarily on the real axis. In other words, the reality condition is modified to the requirement that $f(z)$ maps line segments in the z-plane into line segments in the w-plane. Therefore, if $w = f(z)$ is analytic in D and continuous in the region consisting of D together with the segments $L_1, \ldots, L_n$, and if these segments are mapped into line segments in the w-plane, then $f(z)$ can be analytically continued across $L_1, \ldots, L_n$.

As mentioned above the Schwarz reflection principle can be generalized to the case that line segments are replaced by circular arcs (see also, Nehari (1952)). We discuss this further in Section 5.7.

Theorem 5.6.3 (Schwarz–Christoffel)   *Let Γ be the piecewise linear boundary of a polygon in the w-plane, and let the interior angles at successive vertices be $\alpha_1 \pi, \ldots, \alpha_n \pi$. The transformation defined by the equation*

$$\frac{dw}{dz} = \gamma(z - a_1)^{\alpha_1 - 1}(z - a_2)^{\alpha_2 - 1} \cdots (z - a_n)^{\alpha_n - 1}, \tag{5.6.1}$$

where γ is a complex number and $a_1, \ldots, a_n$ are real numbers, maps Γ onto the real axis of the z-plane and the interior of the polygon to the upper half of the z-plane. The vertices of the polygon, $A_1, A_2, \ldots, A_n$, are mapped to the points $a_1, \ldots, a_n$ on *the real axis. The map is an analytic one-to-one conformal transformation between the upper half z-plane and the interior of the polygon.*

When A_j is finite then $0 < \alpha_j \leq 2$; the case $\alpha_j = 2$ corresponds to the tip of a "slit" (see Example 5.6.6 below). If a vertex A_j is at infinity, then $-2 \leq \alpha_j \leq 0$ (using $z = 1/t$).

In this application we consider both the map and its inverse; that is, mapping the w-plane to the z-plane: $z = F(w)$; or the z-plane to the w-plane: $w = f(z)$.

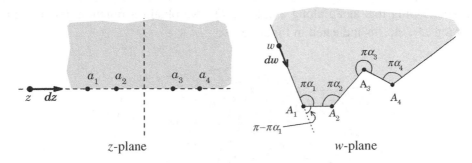

z-plane   w-plane

Figure 5.28 Transformation of a polygon

We first give a heuristic argument of how to derive Eq. (5.6.1). Our goal is to find an analytic function $f(z)$ in the upper half z-plane such that $w = f(z)$ maps the real axis of the z-plane onto the boundary of the polygon. We do this by considering the derivative of the mapping $\frac{dw}{dz} = f'(z)$, or $dw = f'(z)\,dz$. Begin with a point w on the polygon, say to the left of the first vertex A_1 (see Figure 5.28) with its corresponding point z to the left of a_1 in the z-plane. If we think of dw and dz as vectors on these contours, then $\arg(dz) = 0$ (always) and $\arg(dw)$ is constant (always, since this "vector" maintains a fixed direction) until we traverse the first vertex. In fact $\arg(dw)$ only changes when we traverse the vertices. Thus $\arg(f'(z)) = \arg(dw) - \arg(dz) = \arg(dw)$. We see from Figure 5.28 that the *change* in $\arg(f'(z))$ as we traverse (from left to right) the first vertex is $\pi - \pi\alpha_1$, and more generally through any vertex A_ℓ the *change* is $\arg(f'(z)) = \pi(1 - \alpha_\ell)$. This is precisely the behavior of the arguments of the function $(z - a_\ell)^{\alpha_\ell - 1}$: as we traverse a point $z = a_\ell$ we have that $\arg(z - a_\ell) = \pi$ if z is real and on the left of a_ℓ and that $\arg(z - a_\ell) = 0$ if z is real and on the right of a_ℓ. Thus $\arg(z - a_\ell)$ *changes* by $-\pi$ as we traverse the point a_ℓ, and $(\alpha_\ell - 1)\arg(z - a_\ell)$ *changes* by $\pi(1 - \alpha_\ell)$. Because we have a similar situation at each vertex, this suggests that $\frac{dw}{dz} = f'(z)$ is given by the right-hand side of Eq. (5.6.1). Some readers may wish to skip the proof of this theorem, peruse the remarks that follow it, and proceed to the worked examples.

Proof We outline the essential ideas behind the proof. Riemann's mapping theorem, mentioned at the beginning of this section (see also Section *5.5) guarantees that such a univalent map $w = f(z)$ exists. (Actually, one could proceed on the assumption that the mapping function $f(z)$ exists and then verify that the function $f(z)$ defined by (5.6.1) satisfies the conditions of the theorem.) However we prefer to give a constructive proof. We now discuss its construction.

Let us consider the function $w = f(z)$ analytically continued across one of the sides of the polygon D in the w-plane to obtain a function $f_1(z)$ in an adjacent polygon D_1; every point $w \in D$ corresponds to a point in the upper half z-plane, and every point $w \in D_1$ corresponds to a symmetrical point, $\bar{z}$, in the lower half z-plane. Doing this again along a side of D_1, we obtain a function $f_2(z)$ in the polygon D_2, etc., as indicated in Figure 5.29.

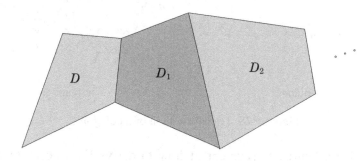

Figure 5.29 Continuation of D

Each reflection of a polygon in the w-plane across, say, the segment $A_k A_{k+1}$, corresponds (by the Schwarz reflection principle) to an analytic continuation of $f(z)$ across the line segment $a_k a_{k+1}$. By repeating this over and over, the Schwarz reflection principle implies that $f(z)$ can be analytically continued to form a single branch of what would be, in general, an infinitely-branched function.

However, because we have reflections about straight sides, geometrical arguments imply that the functions $f(z) \in D$ and $f_2(z) \in D_2$ are linearly related to each other via a rotation and translation; that is, $f_2(z) = A f(z) + B$, $A = e^{i\alpha}$. The same is true for any even number of reflections, $f_4, f_6, \ldots$. But $g(z) = f''(z)/f'(z)$ is invariant under such linear transformations, so any point in the upper half z-plane will correspond to a unique value of $g(z)$; because $g(z) = f_2''(z)/f_2'(z) = f_4''(z)/f_4'(z) = \cdots$, any even number of reflections yields the same value. Similarly, any odd number of reflections returns us a unique value of $g(z)$ corresponding to a point z in the lower half plane. Also from Riemann's Mapping Theorem and the symmetric principle, $f'(z) \neq 0$, and so $f(z)$ is analytic everywhere except possibly at the endpoints $z = a_l$, $l = 1, 2, \ldots$. The only possible locations for singularities correspond to the vertices of the polygon. On the real z-axis, $g(z)$ is real and may be continued by reflection to the lower half plane by $g(z) = \overline{g(\bar{z})}$. In this way, all points z (upper and lower half planes) are determined uniquely and the function $g(z)$ is therefore single valued.

Next, let us consider the points $z = a_\ell$ corresponding to the polygonal vertices A_ℓ. In the neighborhood of a vertex $z = a_\ell$ an argument such as that preceding this theorem shows that the mapping has the form

$$w - w_0 = f(z) - f(z_0) = (z - a_\ell)^{\alpha_\ell} \left[c_\ell^{(0)} + c_\ell^{(1)}(z - a_\ell) + c_\ell^{(2)}(z - a_\ell)^2 + \cdots \right].$$

Consequently $g(z) = f''(z)/f'(z)$ is analytic in the extended z-plane except for poles at the points $a_1, \ldots, a_n$ with residues $(\alpha_1 - 1), \ldots, (\alpha_n - 1)$. It follows from Liouville's Theorem that

$$\frac{f''(z)}{f'(z)} - \sum_{l=1}^{n} \frac{(\alpha_l - 1)}{z - a_l} = c, \tag{5.6.2}$$

where c is some complex constant. But $f(z)$ is analytic at $z = \infty$ (assuming no vertex, $z = a_\ell$ is at infinity), so $f(z) = f(\infty) + b_1/z + b_2/z^2 + \cdots$; hence $f''(z)/f'(z) \to 0$ as as $z \to \infty$, which implies that $c = 0$. Integration of Eq. (5.6.2) yields Eq. (5.6.1). □

Remarks (1) For a closed polygon, $\sum_{l=1}^{n}(1 - \alpha_l) = 2$, and hence $\sum_{l=1}^{n} \alpha_l = (n - 2)$ where n is the number of sides. This is a consequence of the well-known geometrical property that the sum of the exterior angles of any closed polygon is 2π.

(2) It is shown in Section 5.7 that for bilinear transformations, the correspondence of three (and only three) points on the boundaries of two domains can be prescribed arbitrarily. Actually it can also be shown that this is true for any univalent transformation between the boundary of two simply-connected domains. In particular, any of the three vertices of the polygon, say A_1, A_2 and A_3, can be associated with any three points on the real axis, a_1, a_2, a_3, (of course preserving order and orientation). More than three of the vertices a_ℓ cannot be prescribed arbitrarily, and the actual determination of $a_4, a_5, \ldots, a_n$ (sometimes called accessory parameters) might be difficult. In application, symmetry or other considerations usually are helpful, though numerical computation is usually the only means to evaluate the constants $a_4, a_5, \ldots, a_n$. Sometimes it is useful to fix more independent real conditions instead of fixing three points, (e.g. map a point $w_0 \in D$ to a fixed point z_0 in the z-plane, and fix a direction of $f'(z_0)$; that is, fix arg $f'(z_0)$).

(3) The integration of Eq. (5.6.1) usually leads to multivalued functions. A single branch is chosen by the requirement that $0 < \arg(z - a_l) < \pi, l = 1, \ldots, n$. The function $f(z)$ is analytic in the semiplane Im $z > 0$; it has branch points at $z = a_\ell$.

(4) Formula (5.6.1) holds when none of the points coincide with the point at infinity. However, using the transformation $z = a_n - 1/\zeta$ which transforms the point

$z = a_n$ to $\zeta = \infty$ but transforms all other points a_ℓ to finite points $\zeta_\ell = 1/(a_n - a_\ell)$, we see that Eq. (5.6.1) yields

$$\frac{df}{d\zeta}\left(\zeta^2\right) = \gamma \left(a_n - a_1 - \frac{1}{\zeta}\right)^{\alpha_1 - 1} \cdots \left(a_n - a_{n-1} - \frac{1}{\zeta}\right)^{\alpha_{n-1} - 1}\left(-\frac{1}{\zeta}\right)^{\alpha_n - 1}.$$

Using Remark (1) we have

$$\frac{df}{d\zeta} = \hat{\gamma}\left(\zeta - \zeta_1\right)^{\alpha_1 - 1} \cdots \left(\zeta - \zeta_{n-1}\right)^{\alpha_{n-1} - 1}, \tag{5.6.3a}$$

where $\zeta_\ell = 1/(a_n - a_\ell)$ and $\hat{\gamma}$ is a new constant. Thus formula (5.6.1) holds with the point at ∞ removed. If the point $z = a_n$ is mapped to $\zeta = \infty$, then, by virtue of Remark (2), only two other vertices can be arbitrarily prescribed. Using Remark (1), we find that as $\zeta \to \infty$,

$$\frac{df}{d\zeta} = \hat{\gamma}\zeta^{-\alpha_n - 1}\left[1 + \frac{c_1}{\zeta} + \cdots\right].$$

(5) The formula (5.6.1) also holds for the mapping of a unit circle to a polygon. Using the bilinear transformation

$$z = i\left(\frac{1 + \zeta}{1 - \zeta}\right), \qquad \zeta = \frac{z - i}{z + i}, \qquad \frac{dz}{d\zeta} = \frac{2i}{(1 - \zeta)^2},$$

which transforms the upper half z-plane onto the unit circle $|\zeta| < 1$, it follows that Eq. (5.6.1) holds with z replaced by ζ with suitable constants, γ replaced by $\hat{\gamma}$, and a_ℓ replaced by $\zeta_\ell, \ell = 1, 2, \ldots, n$), being on the unit circle.

(6) These ideas can be used to map the complete exterior of a closed polygon (with n vertices) in the w-plane to the upper half z-plane; also see Carrier et al. (1966). We note that at first glance one might not expect this to be possible because an annular region (not simply connected) cannot be mapped onto a half plane. In fact, the exterior of a polygon, which contains the point at infinity, is simply connected. A simple closed curve surrounding the closed polygon can be continuously deformed to the point at infinity. In order to obtain the formula in this case, we note that all of the interior angles $\pi\alpha_\ell, \ell = 1, 2, \ldots, n$, must be transformed to exterior angles $2\pi - \pi\alpha_\ell$ because we traverse the polygon in the opposite direction, keeping the exterior of the polygon to our left. Thus the change in arg $f'(z)$ at a vertex A_ℓ is $-(\pi - \pi\alpha_\ell)$ and therefore in Eq. (5.6.1) $(\alpha_\ell - 1) \to (1 - \alpha_\ell)$. We write the transformation in the form

$$\frac{dw}{dz} = g(z)(z - a_1)^{1-\alpha_1}(z - a_2)^{1-\alpha_2} \cdots (z - a_n)^{1-\alpha_n}.$$

The function $g(z)$ is determined by properly mapping the point $w = \infty$, which is now an *interior* point of the domain to be mapped. Let us map $w = \infty$ to a point in the upper half plane, say, $z = ia_0$, $a_0 > 0$. We require $w(z)$ to be a conformal transformation at infinity, so near $z = ia_0$, $g(z)$ must be single valued, and $w(z)$ should transform like

$$w(\zeta) = \gamma_1 \zeta + \gamma_0 + \cdots, \qquad \zeta = \frac{1}{z - ia_0} \quad \text{as} \quad \zeta \to \infty, \quad \text{or} \quad z \to ia_0.$$

Similar arguments pertain to the mapping of the lower half plane by using the symmetry principle. Using the fact that the polygon is closed, $\sum_{\ell=1}^{n}(1 - \alpha_\ell) = 2$, and conformal at $z = \infty$, we deduce that

$$g(z) = \frac{\gamma}{(z - ia_0)^2(z + ia_0)^2};$$

note that $g(z)$ is real for real z. Thus, the Schwarz–Christoffel formula mapping the exterior of a closed polygon to the upper half z-plane is given by

$$\frac{dw}{dz} = \frac{\gamma}{(z - ia_0)^2(z + ia_0)^2}(z - a_1)^{1-\alpha_1}(z - a_2)^{1-\alpha_2} \cdots (z - a_n)^{1-\alpha_n}. \quad (5.6.3b)$$

We also note that using the bilinear transformation of Remark (5) with $a_0 = 1$, we find that the Schwarz–Christoffel transformation from the *exterior* of a polygon to the *interior* of a unit circle $|\zeta| < 1$ is given by

$$\frac{dw}{d\zeta} = \frac{\hat{\gamma}}{\zeta^2}(\zeta - \zeta_1)^{1-\alpha_1}(\zeta - \zeta_2)^{1-\alpha_2} \cdots (\zeta - \zeta_n)^{1-\alpha_n}, \qquad (5.6.3c)$$

where the points ζ_i, $i = 1, 2, \ldots, n$ lie on the unit circle.

Example 5.6.4   Determine the function that maps the half strip indicated in Figure 5.30 onto the upper half of the z-plane.

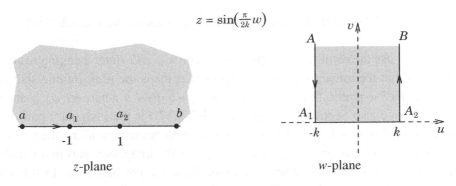

$$z = \sin\left(\tfrac{\pi}{2k} w\right)$$

Figure 5.30 Transformation of a polygon for Example 5.6.4

We associate $A(\infty)$ with $a(\infty)$, $A_1(-k)$ with $a_1(-1)$, and $A_2(k)$ with $a_2(1)$. Then, from symmetry, we have that $B(\infty)$ is associated with $b(\infty)$. Equation (5.6.1) with $\alpha_1 = \alpha_2 = \frac{1}{2}$, $a_1 = -1$, $a_2 = 1$ yields

$$\frac{dw}{dz} = \gamma(z+1)^{-\frac{1}{2}}(z-1)^{-\frac{1}{2}} = \frac{\tilde{\gamma}}{\sqrt{1-z^2}}.$$

Integration implies

$$w = \tilde{\gamma}\sin^{-1}z + c,$$

when $z = 1$, $w = k$, and when $z = -1$, $w = -k$. Thus

$$k = \tilde{\gamma}\sin^{-1}(1) + c, \qquad -k = \tilde{\gamma}\sin^{-1}(-1) + c.$$

Using $\sin^{-1}(1) = \pi/2$ and $\sin^{-1}(-1) = -\pi/2$, these equations yield $c = 0$, $\tilde{\gamma} = 2k/\pi$. Thus $w = (2k/\pi)\sin^{-1}z$, and $z = \sin(\pi w/2k)$.

Example 5.6.5 A semi-infinite slab has its vertical boundaries maintained at temperature T_o and $2T_o$, and its horizontal boundary at a temperature 0 (see Figure 5.31). Find the steady state temperature distribution inside the slab.

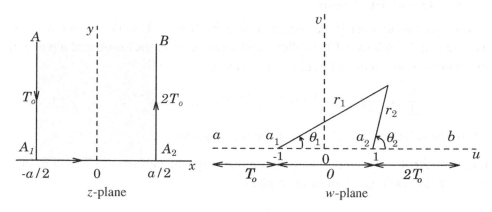

Figure 5.31 Constant temperature boundary conditions for Example 5.6.5

We shall use the result of Example 5.6.4, with $k = a/2$ (interchanging w and z). It follows that the transformation $w = \sin(\pi z/a)$ maps the semi-infinite slab onto the upper half w-plane. The function $T = \alpha_1\theta_1 + \alpha_2\theta_2 + \alpha$ where $\alpha_1, \alpha_2, \alpha$ are real constants (see also Section 2.3, Eqs. (2.3.13–2.3.17)) is the imaginary part of the function: $\alpha_1\log(w+1) + \alpha_2\log(w-1) + i\alpha$ with branch cut along the real axis. It is therefore analytic in the upper half strip. As the imaginary part of an analytic function it also satisfies Laplace's equation in the upper half strip. In this strip, $w + 1 = r_1 e^{i\theta_1}$ and $w - 1 = r_2 e^{i\theta_2}$, where $0 \le \theta_1, \theta_2 \le \pi$. To determine α_1, α_2 and α, we use the boundary conditions. If $\theta_1 = \theta_2 = 0$, then $T = 2T_o$; if $\theta_1 = 0$ and $\theta_2 = \pi$, then $T = 0$; if $\theta_1 = \theta_2 = \pi$, then $T = T_o$. Hence

$$T = \frac{T_o}{\pi}\theta_1 - \frac{2T_o}{\pi}\theta_2 + 2T_o = \frac{T_o}{\pi}\tan^{-1}\frac{v}{u+1} - \frac{2T_o}{\pi}\tan^{-1}\frac{v}{u-1} + 2T_o.$$

Using

$$w = u + iv = \sin\left(\frac{\pi z}{a}\right),$$

in other words

$$u = \sin\left(\frac{\pi x}{a}\right)\cosh\left(\frac{\pi y}{a}\right) \quad \text{and} \quad v = \cos\left(\frac{\pi x}{a}\right)\sinh\left(\frac{\pi y}{a}\right),$$

we find

$$T = \frac{T_o}{\pi}\tan^{-1}\left[\frac{\cos(\frac{\pi x}{a})\sinh(\frac{\pi y}{a})}{\sin(\frac{\pi x}{a})\cosh(\frac{\pi y}{a}) + 1}\right] - \frac{2T_o}{\pi}\tan^{-1}\left[\frac{\cos(\frac{\pi x}{a})\sinh(\frac{\pi y}{a})}{\sin(\frac{\pi x}{a})\cosh(\frac{\pi y}{a}) - 1}\right] + 2T_o.$$

Example 5.6.6 Determine the function that maps the "slit" of height s depicted in Figure 5.32 onto the upper half of the z-plane.

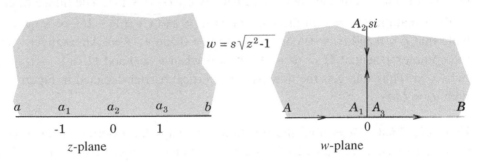

Figure 5.32 Transformation of cut half plane onto a slit for Example 5.6.6

We associate $A_1(0-)$ with $a_1(-1)$, $A_2(si)$ with $a_2(0)$, and $A_3(0+)$ with $a_3(1)$. Then, Eq. (5.6.1) with $\alpha_1 = \frac{1}{2}$, $a_1 = -1$, $\alpha_2 = 2$, $a_2 = 0$, $\alpha_3 = \frac{1}{2}$, and $a_3 = 1$ yields

$$\frac{dw}{dz} = \gamma(z+1)^{-\frac{1}{2}}z(z-1)^{-\frac{1}{2}} = \frac{\tilde{\gamma}z}{\sqrt{1-z^2}}.$$

Thus

$$w = \delta\sqrt{z^2 - 1} + c.$$

When $z = 0$, $w = si$, and when $z = 1$, $w = 0$. Hence,

$$w = s\sqrt{z^2 - 1}. \tag{5.6.4}$$

Example 5.6.7 Find the flow past a vertical slit of height s, which far away from this slit is moving with a constant velocity u_o in the horizontal direction.

It was shown in Example 5.6.6 that the transformation $w = s\sqrt{z^2 - 1}$ maps the vertical slit of height s in the w-plane onto the real axis of the z-plane. The flow field over a slit in the w-plane is therefore transformed into a uniform flow in the z-plane with complex velocity $\Omega(z) = U_o z = U_o(x+iy)$, and with constant velocity

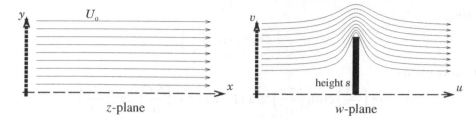

Figure 5.33 Flow over vertical slit of height s for Example 5.6.7

U_o. The streamlines of the uniform flow in the z-plane correspond to $y = c$ (recall that $\Omega = \Phi + i\Psi$ where Φ and Ψ are the velocity potential and stream function, respectively), where c is a positive constant and the flow field in the w-plane is obtained from the complex potential $\Omega(w) = U_o((\frac{w}{s})^2 + 1)^{\frac{1}{2}}$. The image of each of the streamlines $y = c$ in the w-plane is $w = s\sqrt{(x + ic)^2 - 1}$, $-\infty < x < \infty$. Note that $c = 0$ and $c \to \infty$ correspond to $v = 0$ and $v \to \infty$. Alternatively, from the complex potential $\Omega = \Phi + i\Psi$, $\Psi = 0$ when $v = 0$ and $\Omega'(w) \to U_o/s$ as $|w| \to \infty$. Thus, one gets the flow past the vertical barrier depicted in Figure 5.33 with $u_0 = U_0/s$.

Example 5.6.8   Determine the function that maps the exterior of an isoceles triangle located in the upper half of the w-plane onto the upper half of the z-plane.

We note that this is not a mapping of a complete exterior of a closed polygon; in fact, this problem is really a modification of Example 5.6.6. We will show that a limit of this example as $k \to 0$ (see Figure 5.34) reduces to the previous one. Note $\tan(\pi\alpha) = s/k$.

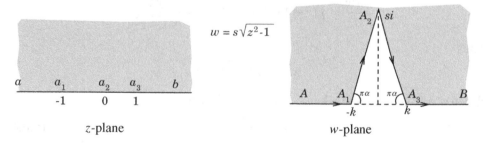

Figure 5.34 Transformation of the exterior of an isoceles triangle for Example 5.6.8

We associate $A_1(-k)$ with $a_1(-1)$, $A_2(si)$ with $a_2(0)$ and $A_3(k)$ with $a_3(1)$. The angles at A_1, A_2, and A_3 are given by $\pi - \pi\alpha$, $2\pi - (\pi - 2\pi\alpha)$, and $\pi - \pi\alpha$, respectively. Thus, $\alpha_1 = 1 - \alpha$, $\alpha_2 = 1 + 2\alpha$, $\alpha_3 = 1 - \alpha$, and Eq. (5.6.1) yields

$$\frac{dw}{dz} = \gamma(z+1)^{-\alpha}z^{2\alpha}(z-1)^{-\alpha} = \tilde{\gamma}\frac{z^{2\alpha}}{(1-z^2)^{\alpha}}.$$

Hence,

$$w = \tilde{\gamma}\int_0^z \frac{\zeta^{2\alpha}}{(1-\zeta^2)^{\alpha}}d\zeta + c. \tag{5.6.5}$$

When $z = 0$, we have $w = si$, and when $z = 1$, $w = k$. Thus, $c = si$, and

$$k = \tilde{\gamma}\int_0^1 \frac{\zeta^{2\alpha}}{(1-\zeta^2)^{\alpha}}d\zeta + si.$$

The integral $\int_0^1$ can be expressed in terms of gamma functions by writing $t = \zeta^2$ and using a well-known result for integrals:

$$B(p,q) = \int_0^1 t^{p-1}(1-t)^{q-1}\,dt = \frac{\Gamma(p)\Gamma(q)}{\Gamma(p+q)};$$

here $B(p,q)$ denotes the beta function (see Eq. (4.5.30) for the definition of the gamma function, $\Gamma(z)$.) Using this equation with $p = \alpha + 1/2$, $q = 1 - \alpha$, and $\Gamma(\frac{3}{2}) = \frac{1}{2}\Gamma(\frac{1}{2})$, where $\Gamma(\frac{1}{2}) = \sqrt{\pi}$, it follows that $\tilde{\gamma}\Gamma(\alpha+1/2)\Gamma(1-\alpha) = (k-si)\sqrt{\pi}$. Thus,

$$w = \frac{(k-si)\sqrt{\pi}}{\Gamma(\alpha+1/2)\Gamma(1-\alpha)}\int_0^z \frac{\zeta^{2\alpha}}{(1-\zeta^2)^{\alpha}}\,d\zeta + si. \tag{5.6.6}$$

We note that Example 5.6.4 corresponds to the limit $k \to 0$, $\alpha \to \frac{1}{2}$ in Eq. (5.6.6); that is, under this limit, Eq. (5.6.6) reduces to Eq. (5.6.4):

$$w = si - si\int_0^z \frac{\zeta}{\sqrt{1-\zeta^2}}d\zeta = si\sqrt{1-z^2} = s\sqrt{z^2-1}.$$

An interesting application of the Schwarz–Christoffel construction is the mapping of a rectangle. Despite the fact that it is a simple closed polygon, the function defined by the Schwarz–Christoffel transformation is not elementary. (Neither is the function elementary in the case of triangles.) In the case of a rectangle, we find that the mapping functions involve elliptic integrals and elliptic functions.

Example 5.6.9 Find the function which maps the interior of a rectangle onto the upper half of the z-plane. See Figure 5.35.

We associate $A_1(-1+si)$ with $a_1(-1/k)$, $A_2(-1)$ with $a_2(-1)$, and $z = 0$ with $w = 0$. Then by symmetry A_3, A_4 are associated with a_3, a_4 respectively. In this example we regard k as given and assume that $0 < k < 1$. Our goal is to determine both the transformation $w = f(z)$ and the constant s as a functions of k. In this case $\alpha_1 = \alpha_2 = \alpha_3 = \alpha_4 = \frac{1}{2}$, $a_1 = -\frac{1}{k}$, $a_2 = -1$, $a_3 = 1$, and $a_4 = 1/k$. Furthermore,

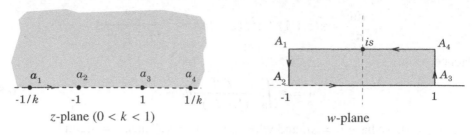

Figure 5.35 Transformation of rectangle for Example 5.6.8

because $f(0) = 0$ (symmetry), the constant of integration is zero; thus Eq. (5.6.1) yields

$$\frac{dw}{dz} = \gamma(z-1)^{-1/2}(z+1)^{-1/2}(z-1/k)^{-1/2}(z+1/k)^{-1/2}.$$

Then, by integration manipulation and by redefining γ (to $\tilde{\gamma}$):

$$w = \tilde{\gamma} \int_0^z \frac{d\zeta}{\sqrt{(1-\zeta^2)(1-k^2\zeta^2)}} = \tilde{\gamma}F(z,k). \tag{5.6.7}$$

The integral appearing in Eq. (5.6.7), with the choice of the branch defined in Remark (3) (we fix $\sqrt{1} = 1$ and note w is real for real z, $|z| > 1/k$), is the so-called elliptic integral of the first kind; it is usually denoted by $F(z, k)$. (Note, from the integral in Eq. (5.6.7), that $F(z, k)$ is an odd function, i.e., $F(-z, k) = -F(z, k)$.) When $z = 1$, this becomes $F(1, k)$ which is referred to as the complete elliptic integral, usually denoted by $K(k) \equiv F(1, k) = \int_0^1 d\zeta / \sqrt{(1-\zeta^2)(1-k^2\zeta^2)}$. The association of $z = 1$ with $w = 1$ implies that $\tilde{\gamma} = 1/K(k)$. The association of $z = 1/k$ with $w = 1 + is$ yields

$$1 + is = \frac{1}{K} \int_0^{\frac{1}{k}} \frac{d\zeta}{\sqrt{(1-\zeta^2)(1-k^2\zeta^2)}}$$

$$= \frac{1}{K}\left(K + \int_1^{\frac{1}{k}} \frac{d\zeta}{\sqrt{(1-\zeta^2)(1-k^2\zeta^2)}}\right),$$

or $Ks = K'$, where K' denotes the associated elliptic integral (not the derivative), which is defined by

$$K'(k) = \int_1^{\frac{1}{k}} \frac{d\xi}{\sqrt{(\xi^2-1)(1-k^2\xi^2)}}.$$

This expression takes an alternative, standard form if one uses the substitution $\xi = (1 - k'^2\xi'^2)^{-1/2}$, where $k' = \sqrt{1-k^2}$ (see Eq. (5.6.9)).

In summary, the transformation $f(z)$ and the constant s are given by

$$w = \frac{F(z,k)}{K(k)}, \qquad s = \frac{K'(k)}{K(k)}, \tag{5.6.8}$$

where (the symbol $\equiv$ denotes "by definition")

$$F(z,k) \equiv \int_0^z \frac{d\zeta}{\sqrt{(1-\zeta^2)(1-k^2\zeta^2)}}, \qquad K(k) \equiv F(1,k),$$

$$K'(k) \equiv \int_0^1 \frac{d\xi}{\sqrt{(1-\xi^2)[1-(1-k^2)\xi^2]}}. \tag{5.6.9}$$

The parameter k is called the modulus of the elliptic integral. The inverse of Eq. (5.6.8) gives z as a function of w via one of the so-called Jacobian elliptic functions (see e.g., Nehari (1952))

$$w = \frac{F(z,k)}{K(k)} \Rightarrow z = \mathrm{sn}(wK, k). \tag{5.6.10}$$

We note that Example 5.6.4 (with $k = 1$ in that example) corresponds to the following limit in this example: $s \to \infty$, which implies that $k \to 0$, because $\lim_{k\to 0} K'(k) = \infty$, and $\lim_{k\to 0} K(k) = \pi/2$. Then the rectangle becomes an infinite strip, and the left-hand equality of (5.6.8) reduces to the equation

$$w = \frac{2}{\pi} \int_0^z \frac{d\zeta}{\sqrt{1-\zeta^2}} = \frac{2}{\pi} \sin^{-1} z.$$

Remark The fundamental properties of elliptic functions are their "double periodicity" and single valuedness. We illustrate this for one of the elliptic functions, the Jacobian "sn" function, which we have already seen in Eq. (5.6.10). A standard "normalized" definition is (replacing wK by w in Eq. (5.6.10)):

$$w = F(z,k) \Rightarrow z = F^{-1}(w,k) = \mathrm{sn}(w,k), \tag{5.6.11}$$

where again

$$F(z,k) \equiv \int_0^z \frac{d\zeta}{\sqrt{(1-\zeta^2)(1-k^2\zeta^2)}}$$

so that $\mathrm{sn}(0,k) = 0$ and $\mathrm{sn}'(0,k) = 1$ (in the latter we used $\frac{dz}{dw} = 1/\frac{dw}{dz}$).

We shall show the double periodicity

$$\mathrm{sn}(w + n\omega_1 + im\omega_2, k) = \mathrm{sn}(w,k) \tag{5.6.12}$$

where m and n are integers, $\omega_1 = 4K(k)$, and $\omega_2 = 2K'(k)$. Given the normalization of Eq. (5.6.11) we have that the "fundamental" rectangle in the w-plane corresponding to the upper half z-plane is $A_1 = -K + iK'$, $a_1 = -1/k$, $A_2 = -K$, and $a_2 = -1$, with A_3, a_3, A_4, and a_4 as the points symmetric to these (in Figure 5.35 all points on the w-plane are multiplied by K; this is now the rectangle R in Figure 5.36). The function $z = \text{sn}(w, k)$ can be analytically continued by the symmetry principle.

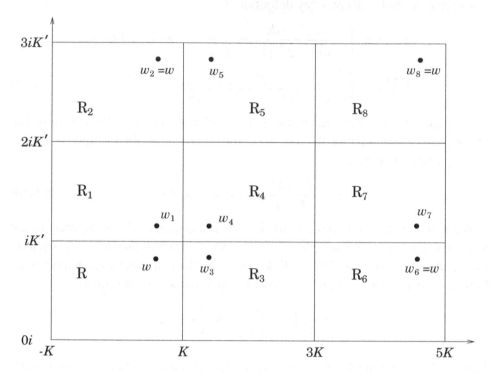

Figure 5.36 Reflecting in the w-plane

Beginning with any point w in the fundamental rectangle R we must obtain the same point w by symmetrically reflecting twice about a horizontal side of the rectangle, or twice about a vertical side of the rectangle, etc., which corresponds to returning to the same point in the upper half z-plane each time. This yields the double periodicity relationship (5.6.12).

These symmetry relationships also imply that the function $z = \text{sn}(w, k)$ is single valued. Any point z in the upper half plane is uniquely determined and corresponds to an even number of reflections, and similarly, any symmetric point $\bar{z}$ in the lower half plane is found uniquely by an odd number of reflections. Analytic continuation of $z = \text{sn}(w, k)$ across any boundary therefore uniquely determines a value in the z-plane. The only singularities of the map $w = f(z)$ are at the vertices of the rectangle, and near the vertices of the rectangle we have

$$w - A_i = C_i(z - a_i)^{\frac{1}{2}} \left[1 + c_i^{(1)}(z - a_i) + \cdots \right],$$

so we see that $z = \text{sn}(w, k)$ is single valued there as well. The "period rectangle" consists of any four rectangles meeting at a corner, such as R, R_1, R_3, R_4 in Figure 5.36. All other such period rectangles are periodic extensions of the fundamental rectangle. Two of the rectangles map to the upper half z-plane and two map to the lower half z-plane. Thus, a period rectangle covers the z-plane twice; that is, for $z = \text{sn}(w, k)$ there are two values of w that correspond to a fixed value of z.

For example, the zeroes of $\text{sn}(w, k)$ are located at $w = 2nK + 2miK'$ for integers m and n. From the definition of $F(z, k)$ we see that $F(0, k) = 0$. If we reflect the rectangle R to R_1, this zero is transformed to the location $w = 2iK'$, while reflecting to R_3 transforms the zero to $w = 2K$, etc. Hence, two zeroes are located in each period rectangle. From the definition we also find the pairs $w = -K$, $z = -1$; $w = -K + iK'$, $z = -\frac{1}{k}$; and $w = iK'$, $z = \infty$. The latter is a simple pole.

Schwarz–Christoffel transformations with more than four vertices usually require numerical computation (see Trefethan, 1986; Driscoll and Trefethen, 2002).

5.6.1 Problems for Section 5.6

1. Use the Schwarz–Christoffel transformation to obtain a function which maps each of the indicated regions in Figures 5.37(a), and 5.37(b) in the w-plane onto the upper half of the z-plane.

   Hint: In the case of Figure 5.37(a) take $w = 0$ to $z = 0$, $w = 1$ to $z = 1$. For Figure 5.37(b), treat $z = \infty$ as having zero interior angle and take $w = 0$ to $z = 1$, $w = i\pi$ to $z = -1 = e^{i\pi}$.

2. Find a function which maps the indicated region of the w-plane in Figure 5.38 onto the upper half of the z-plane, such that $(P, Q, R) \mapsto (-\infty, 0, \infty)$.

3. Show that the function $w = \int_0^z dt / \left(1 - t^6\right)^{\frac{1}{3}}$ maps the interior of the unit circle in z to a regular hexagon in w.

   Hint: Note: $(1 - t^6) = (1 - t^3)(1 + t^3) = \Pi_{j=1}^6 (-1)(\omega_j - t)^j$, where ω_j, $j = 1, \ldots, 6$ are the zeroes of $t^3 = \pm 1$.

4. Find the Schwarz–Christoffel transformation that maps the upper half plane onto the triangle with vertices $(0, 0)$, $(0, 1)$, $(1, 0)$ (See Figure 5.39).

   Hint: map $w = 0$ to $z = 0$, $w = 1$ to $z = 1$ and $w = i$ to $z = \infty$. It is also useful to use the formula $B(p, q)$ below Eq. (5.6.5).

5. A fluid flows with initial velocity u_0 through a semi-infinite channel of width d and emerges through the opening AB of the channel (see Figure 5.40). Find the speed of the fluid.

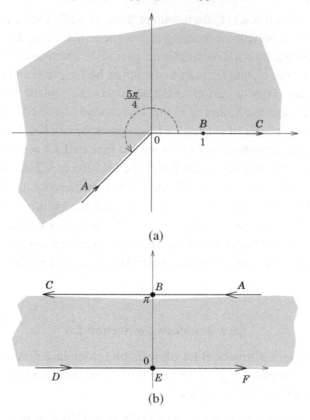

(a)

(b)

Figure 5.37 Schwarz–Christoffel transformations for Problem 1

Hint: first show that the conformal mapping $w = z + e^{2\pi z/d}$ maps the channel $|y| < d/2$ onto the w-plane excluding slits, as indicated in Figure 5.41. To calculate the speed, it is useful to use

$$\left|\frac{d\Omega}{dw}\right|^2 = \left|\frac{d\Omega/dz}{dw/dz}\right|^2$$

and note that the channel flow is uniform.

6. The shaded region of Figure 5.42 represents a semi-infinite conductor with a vertical slit of height h in which the boundaries AD, DE, and DB are maintained at temperatures T_1, T_2, and T_3, respectively. Find the temperature everywhere. Hint: use the conformal mapping studied in Example 5.6.7.

7. Utilize the Schwarz–Christoffel transformation in order to find the complex potential $F(w)$ governing the flow of a fluid over a step with velocity at infinity equal to q, where q is real. The step $A_1(-\infty)A_2(ih)A_3(0)$ is shown in Figure 5.43. The step is taken to be a streamline.

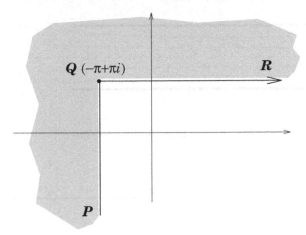

Figure 5.38 Schwarz–Christoffel transformations for Problem 2

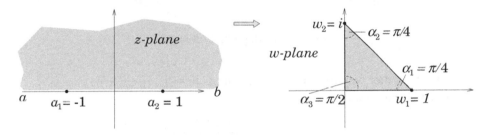

Figure 5.39 Schwarz–Christoffel transformations for Problem 4

8. Find the domain onto which the function

$$w = \int_0^z \frac{(1+t^3)^\alpha}{(1-t^3)^{\frac{2}{3}+\alpha}}, \qquad -1 < \alpha < \frac{1}{3}$$

maps the unit disk.

9. Find the mapping of a triangle with angles $\alpha_1\pi, \alpha_2\pi, \alpha_3\pi$, where the side opposite the angle $\alpha_3\pi$ has length l.

10. Show that the mapping $w = \int_0^z dt/(\sqrt{t}(t^2-1)^{1/2})$ maps the upper half plane conformally onto the interior of a square. Hint: show that the vertices of the square are $w(0) = 0$, $w(1) = -iA$, $w(-1) = A$, and $w(\infty) = A - iA$, where A is given by the real integral: $A = \int_0^1 dt/(\sqrt{t}(1-t^2)^{1/2})$.

11. Use the Schwarz–Christoffel transformation to show that

$$w = \frac{c}{\pi} \log\left(2\sqrt{z^2+z} + 2z + 1\right)$$

maps the upper half plane conformally onto the interior of a semi-infinite strip.

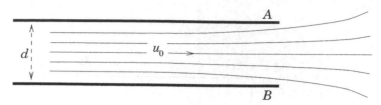

Figure 5.40 Fluid flow – Problem 5

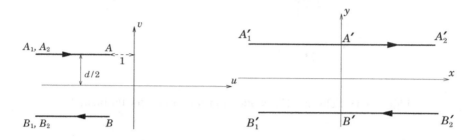

Figure 5.41 Mapping Problem 5

Hint: another way to write the mapping is $w = \int_0^z \frac{dt}{t^{1/2}(t+1)^{1/2}}$.

12. Show that the function

$$w = \int_1^z \frac{(1-\zeta^4)^{1/2}}{\zeta^2} \, d\zeta$$

maps the interior of the unit circle in the z-plane to the exterior of a square in the w-plane.

Hint: Use Equation (5.6.3c).

13. Show that a necessary (but not sufficient) condition for which $z = z(w)$ from the Schwarz–Christoffel formula is single valued (i.e., for which polygons the inverse mapping is defined and single-valued) is that $\alpha_\ell = 1/n_\ell$, where n_ℓ is an integer for all α_ℓ, $\ell = 1, 2, \ldots, n$.

5.7 Bilinear Transformations

An important class of conformal mappings is given by the particular choice of $f(z)$,

$$w = f(z) = \frac{az + b}{cz + d}, \qquad ad - bc \neq 0, \tag{5.7.1}$$

where a, b, c, and d are complex numbers. This transformation is called **bilinear**, or Möbius, or sometimes   linear fractional.

Bilinear transformations have a number of remarkable properties. Furthermore, these properties are global. In particular, they are valid for any z including $z = \infty$;

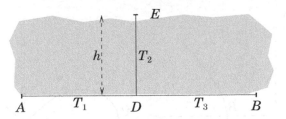

Figure 5.42 Temperature distribution for Problem 6

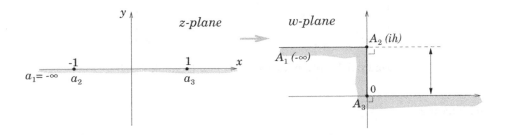

Figure 5.43 Fluid flow for Problem 7

that is, valid in the entire extended complex z-plane. Throughout this section, we derive the most important properties of bilinear transformations, labeled (i)–(viii).

(i) Conformality *Bilinear transformations are conformal.* Indeed, if $c = 0$, $f'(z) = \frac{a}{d} \neq 0$. If $c \neq 0$,

$$f'(z) = \frac{ad - bc}{(cz + d)^2}$$

which shows that $f'(z)$ is well defined for $z \neq -d/c$, z finite. In order to analyze the point $z = \infty$ we let $\tilde{z} = 1/z$, then $w = f(z)$ becomes $w = (b\tilde{z} + a)/(d\tilde{z} + c)$ which is well behaved at $\tilde{z} = 0$. The image of the point $z = -d/c$ is $w = \infty$, which motivates the transformation $\tilde{w} = 1/w$. Then $\tilde{w} = (cz + d)/(az + b)$, and because the derivative of the right-hand side is not zero at $z = -d/c$, it follows that $w = f(z)$ is conformal at $\tilde{w} = 0$.

Example 5.7.1 Find the image of $x^2 - y^2 = 1$ under inversion; that is, under the transformation $w = 1/z$.

Using $z = x + iy$ and $\bar{z} = x - iy$ it follows that the hyperbola $x^2 - y^2 = 1$ can be written as

$$(z + \bar{z})^2 + (z - \bar{z})^2 = 4, \quad \text{or} \quad z^2 + \bar{z}^2 = 2.$$

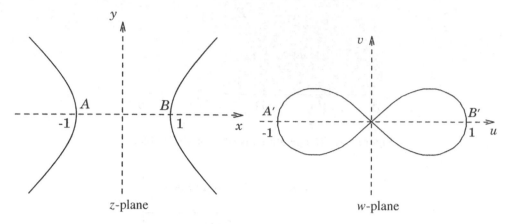

Figure 5.44 Inversion ($w = 1/z$) for Example 5.7.1

This becomes

$$\frac{1}{w^2} + \frac{1}{\bar{w}^2} = 2, \quad \text{or} \quad w^2 + \bar{w}^2 = 2|w|^4, \quad \text{or} \quad u^2 - v^2 = (u^2 + v^2)^2 \qquad (5.7.2)$$

under $z = 1/w$, $w = u + iv$. For small u and v, Eq. (5.7.2) behaves like $u \approx \pm v$. This, together with conformality at the points A and B, suggests the lemniscate graph depicted in Figure 5.44. (In polar coordinates $w = \rho e^{i\phi}$, the lemniscate above is $\rho^2 = \cos 2\phi$.) Note that the left (right) branch of the hyperbola transforms to the left (right) lobe of the lemniscate.

(ii) Decomposition *Any bilinear transformation that is not linear can be decomposed into two linear transformations and an inversion.* Indeed, if $c = 0$ the transformation is linear. If $c \neq 0$, then it can be written in the form

$$w = \frac{a}{c} + \frac{bc - ad}{c(cz + d)}.$$

This shows that a general bilinear transformation can be decomposed into the following three successive transformations:

$$z_1 = cz + d, \qquad z_2 = \frac{1}{z_1}, \qquad w = \frac{a}{c} + \frac{bc - ad}{c} z_2. \qquad (5.7.3)$$

(iii) Group Property *Bilinear transformations form a group.* This means that bilinear transformations contain the identity transformation, and that the inverse as well as the "product" of bilinear transformations are also bilinear transformations (i.e., closure under the operation product).

Indeed, the choice $b = c = 0$, $a = d = 1$, reduces $w = f(z)$ to $w = z$; that is, to the identity transformation. The inverse of the transformation $f(z)$ is obtained by solving the equation $w = f(z)$ for z in terms of w. Hence the inverse of $f(z)$ is given by

$$f^{-1}(z) = \frac{dz - b}{-cz + a}.$$
(5.7.4)

This corresponds to Eq. (5.7.1) with $a \to d, b \to -b, c \to -c$, and $d \to a$; therefore $ad - bc \to ad - bc$, which shows that the inverse of $f(z)$ is also bilinear. In order to derive the product of two transformations we let $z_2 = f_2(z)$ and $w = f_1(z_2)$, where f_1 and f_2 are defined by Eq. (5.7.1) with all the constants replaced by constants with subscripts 1 and 2, respectively. Computing $w = f_1(f_2(z)) = f_3(z)$, one finds

$$f_3(z) = \frac{a_3 z + b_3}{c_3 z + d_3}; \quad \text{where} \quad \begin{aligned} a_3 &= a_1 a_2 + b_1 c_2, & b_3 &= a_1 b_2 + b_1 d_2, \\ c_3 &= c_1 a_2 + d_1 c_2, & d_3 &= c_1 b_2 + d_1 d_2. \end{aligned}$$
(5.7.5)

It can be verified that $(a_3 d_3 - b_3 c_3) = (a_1 d_1 - b_1 c_1)(a_2 d_2 - b_2 c_2) \neq 0$, which together with Eq. (5.7.5) establishes that $f_3(z)$ is a bilinear transformation. Actually the operation product in this case is composition.

There exists an alternative, somewhat more elegant formulation of bilinear transformations. Associate with $f(z)$ the matrix

$$T = \begin{pmatrix} a & b \\ c & d \end{pmatrix}.$$
(5.7.6)

The condition $ad - bc \neq 0$ implies that $\det T \neq 0$; that is, the matrix T is nonsingular. The inverse of T is denoted by

$$T^{-1} = \begin{pmatrix} d & -b \\ -c & a \end{pmatrix}.$$

Equation (5.7.4) implies that the inverse of f is associated with the inverse of T. (If $ad - bc = 1$, then T^{-1} is exactly the matrix inverse.) Furthermore, Eq. (5.7.5) can be used to show that the product (i.e., composition) of two bilinear transformations $f_3(z) \equiv f_1(f_2(z))$ is associated with $T_1 T_2$. Using this notation and Property (ii), it follows that any bilinear transformation is either linear or can be decomposed in the form $T_1 T_2 T_3$ where T_1 and T_3 are associated with linear transformations and T_2 is associated with inversion; that is,

$$T_1 = \begin{pmatrix} a_1 & b_1 \\ 0 & 1 \end{pmatrix}, \quad T_2 = \begin{pmatrix} 0 & 1 \\ 1 & 0 \end{pmatrix}, \quad T_3 = \begin{pmatrix} a_3 & b_3 \\ 0 & 1 \end{pmatrix}.$$

(iv) Invariance *The cross ratio of four points is an invariant.* This means that

$$\frac{(w_1 - w_4)(w_3 - w_2)}{(w_1 - w_2)(w_3 - w_4)} = \frac{(z_1 - z_4)(z_3 - z_2)}{(z_1 - z_2)(z_3 - z_4)},$$
(5.7.7)

where w_i is associated with z_i in Eq. (5.7.1).

This property can be established by manipulating expressions of the form

$$w_1 - w_2 = \frac{(ad - bc)(z_1 - z_2)}{(cz_1 + d)(cz_2 + d)}. \tag{5.7.8}$$

An alternative proof is as follows. Let $X(z_1, z_2, z_3, z_4)$ denote the right-hand side of Eq. (5.7.7). Because every bilinear transformation can be decomposed into linear transformations and an inversion it suffices to show that these transformations leave X invariant. Indeed, if z_l is replaced by $az_l + b$, then both the numerator and the denominator of X are multiplied by a^2 and hence X is unchanged. Similarly, if z_l is replaced by $1/z_l$, X is again unchanged.

If one of the points w_l, say w_1, is ∞, then the left-hand side of Eq. (5.7.7) becomes $(w_3 - w_2)/(w_3 - w_4)$. Therefore this ratio should be regarded as the ratio of the points ∞, w_2, w_3, w_4.

Equation (5.7.7) has the following important consequence: letting $w_4 = w$, $z_4 = z$, Eq. (5.7.7) becomes

$$\frac{(w_1 - w)(w_3 - w_2)}{(w_1 - w_2)(w_3 - w)} = \frac{(z_1 - z)(z_3 - z_2)}{(z_1 - z_2)(z_3 - z)}. \tag{5.7.9}$$

This equation can be written in the form $w = f(z)$, where $f(z)$ is some bilinear transformation uniquely defined in terms of the points z_l and w_l. This allows the interpretation that bilinear transformations take any three distinct points z_l, into any three distinct point w_l. Furthermore, a bilinear transformation is uniquely determined by these three associations.

The fact that a bilinear transformation is completely determined by how it transforms three points, is consistent with the fact that $f(z)$ depends at most on three complex parameters (because one of c, d is different from zero and hence can be divided out).

Example 5.7.2 Find the bilinear transformation that takes the three points $z_1 = 0$, $z_2 = 1$, and $z_3 = -i$, into $w_1 = 1$, $w_2 = 0$, and $w_3 = i$.

Equation (5.7.9) implies

$$\frac{(1 - w)i}{i - w} = \frac{(-z)(-i - 1)}{(-1)(-i - z)} \quad \text{or} \quad w = \frac{1 - z}{1 + z}.$$

(v) Fixed-Point Property *Bilinear transformations have one or two fixed points.* By fixed points we mean those points of the complex z-plane which do not change their position if $z \rightarrow f(z)$. In other words now we interpret the transformation $f(z)$ not as a transformation of one plane onto another, but as a transformation of the plane onto itself. The fixed points are the invariant points of this transformation.

We can find these fixed points by solving the equation $z = (az + b)/(cz + d)$. It follows that we have a quadratic equation and, except for the trivial cases ($c = 0$

and $a = d$, or $b = c = 0$), there exist two fixed points. If $(d - a)^2 + 4bc = 0$, these two points coincide. We exclude these cases and denote the two fixed points by α and β. Equation (5.7.9) with $w_1 = z_1 = \alpha$, $w_3 = z_3 = \beta$ implies

$$\frac{w - \alpha}{w - \beta} = \lambda\left(\frac{z - \alpha}{z - \beta}\right), \tag{5.7.10}$$

where λ is a constant depending on w_2, z_2, α, and β. Thus, the general form of a bilinear transformation with two given fixed points α and β depends only on one extra constant λ.

Example 5.7.3 Find all bilinear transformations which map 0 and 1 to 0 and 1, respectively.

Equation (5.7.10) implies

$$\frac{w - 1}{w} = \lambda\left(\frac{z - 1}{z}\right) \quad \text{or} \quad w = \frac{z}{(1 - \lambda)z + \lambda}.$$

(vi) Mapping Property *Bilinear transformations map circles and lines into circles or lines.* In particular, we will show that under inversion a line through the origin goes into a line through the origin, a line not through the origin goes into a circle, a circle through the origin goes into a line, and a circle not through the origin goes into a circle.

The derivation of these results follows. The mathematical expression of a line is

$$ax + by + c = 0, \quad a, b, c \in \mathbb{R}.$$

Using $z = x + iy$, $\bar{z} = x - iy$, this becomes

$$Bz + \bar{B}\bar{z} + c = 0, \quad B = \frac{a}{2} - \frac{ib}{2}. \tag{5.7.11}$$

The mathematical expression of a circle with center z_0 and radius ρ is $|z - z_0| = \rho$ or $(z - z_0)(\bar{z} - \bar{z}_0) = \rho^2$, or

$$z\bar{z} + \bar{B}z + B\bar{z} + c = 0; \quad B = -z_0, \quad c = |B|^2 - \rho^2. \tag{5.7.12}$$

Under the inversion transformation $w = 1/z$, the line given by Eq. (5.7.11), replacing $z = 1/w$ and $\bar{z} = 1/\bar{w}$, becomes

$$cw\bar{w} + \bar{B}w + B\bar{w} = 0.$$

If $c = 0$ (which corresponds to the line (5.7.11) going through the origin), this equation defines a line going through the origin. If $c \neq 0$, using Eq. (5.7.12), the above equation is a circle of radius $|B|/|c|$ and with center at $-B/c$.

Under the inversion transformation, the circle defined by Eq. (5.7.12) becomes

$$cw\bar{w} + Bw + \bar{B}\bar{w} + 1 = 0.$$

If $c = 0$ (which corresponds to the circle (5.7.12) going through the origin), this equation defines a line. If $c \neq 0$, the above equation is the equation of a circle.

Remark It is natural to group circles and lines together, because a line in the extended complex plane can be thought of as a circle through the point $z = \infty$. Indeed, consider the line (5.7.11), and let $z = 1/\zeta$. Then, Eq. (5.7.11) reduces to the equation $c\zeta\bar{\zeta} + B\bar{\zeta} + \bar{B}\zeta = 0$, which is the equation of a circle going through the point $\zeta = 0$, that is, $z = \infty$. Using the terminology of this remark, we say that the inversion transformation maps circles into circles.

The linear transformation $az + b$ translates by b, rotates by $\arg a$, and dilates (or contracts) by $|a|$ (see Example 5.2.2). Therefore, a linear transformation also maps circles into circles. These properties of the linear and inversion transformations, together with Property (ii), imply that a general bilinear transformation maps circles into circles.

The interior and the exterior of a circle are called the complementary domains of the circle. Similarly, the complementary domains of a line are the two half planes, one on each side of the line. Let K, K_c denote the complementary domains of a circle in the z-plane and K^*, K_c^* denote the complementary domains of the corresponding circle in the w-plane obtained under the bilinear transformation. With this terminology, for a circle we include the degenerate case of a line.

(vii) Complementarity *If K, K_c denote the complementary domains of a circle, then under a bilinear transformation either $K^* = K$ and $K_c^* = K_c$ or $K^* = K_c$ and $K_c^* = K$.* The derivation of this property can be achieved in a manner similar to the derivation of Property (vi), where one uses inequalities instead of the equality (5.7.12). These inequalities follow from the fact that the interior and exterior of the circle are defined by $|z - z_0| < \rho$ and $|z - z_0| > \rho$, respectively. We will not go through the details here.

Example 5.7.4 The inversion $w = 1/z$ maps the interior (exterior) of the unit circle in the z-plane to the exterior (interior) of the unit circle in the w-plane.

Indeed, if $|z| < 1$, then $|w| = 1/|z| > 1$.

(viii) Inverse Property *Bilinear transformations map inverse points (with respect to a circle) to inverse points* The points p and q are called inverse with respect to the circle of radius ρ and center z_0, if z_0, p, q lie, in that order, on the same line, and the distances $|z_0 - p|$ and $|z_0 - q|$ satisfy $|z_0 - p||z_0 - q| = \rho^2$ (see Figure 5.45).

If the points z_0, p, and q lie on the same line it follows that $p = z_0 + r_1 \exp(i\varphi)$, $q = z_0 + r_2 \exp(i\varphi)$. If these points are inverse, then $r_1 r_2 = \rho^2$, or $(p - z_0)(\bar{q} - \bar{z}_0) =$

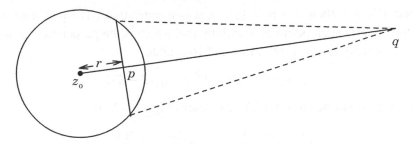

Figure 5.45 Inverse points p, q

ρ^2. Thus the mathematical description of two inverse points i8

$$p = z_0 + re^{i\varphi}, \quad q = z_0 + \frac{\rho^2}{r}e^{i\varphi}, \quad r \neq 0. \tag{5.7.13}$$

As $r \to 0$, $p = z_0$ and $q = \infty$. This is consistent with the geometrical description of the inverse points that shows that as q recedes to ∞, p tends to the center. When a circle degenerates into a line, then the inverse points with respect to the line may be viewed as the points which are perpendicular to the line and are at equal distances from it.

Using Eq. (5.7.13) and $z = z_0 + \rho e^{i\theta}$ (the equation for points on a circle), we have

$$\frac{z-p}{z-q} = \frac{\rho e^{i\theta} - re^{i\varphi}}{\rho e^{i\theta} - \frac{\rho^2}{r}e^{i\varphi}} = \frac{r}{\rho} \cdot \frac{re^{-i\theta} - \rho e^{-i\varphi}}{re^{i\theta} - \rho e^{i\varphi}} \cdot \left(-e^{i\varphi}e^{i\theta}\right), \tag{5.7.14}$$

whereupon

$$\left|\frac{z-p}{z-q}\right| = \frac{r}{\rho}. \tag{5.7.15}$$

We also have the following. Let p and q be distinct complex numbers, and consider the equation

$$\left|\frac{z-p}{z-q}\right| = k, \quad 0 < k \leq 1. \tag{5.7.16}$$

It will be shown below that if $k = 1$, this equation represents a line and the points p and q are inverse points with respect to this line. If $k \neq 1$, this equation represents a circle with center at z_0 and radius ρ, given by

$$z_0 = \frac{p - k^2 q}{1 - k^2}, \quad \rho = \frac{k|p - q|}{1 - k^2}, \tag{5.7.17}$$

the points p and q are inverse points with respect to this circle, and the point p is inside the circle.

Indeed, if $k = 1$, then $|z - p| = |z - q|$ which states that z is equidistant from p and q; the locus of such points is a straight line, which is the perpendicular bisector of the segment pq. If $k \neq 1$, then Eq. (5.7.16) yields

$$(z - p)(\bar{z} - \bar{p}) = k^2(z - q)(\bar{z} - \bar{q}).$$

This equation simplifies to Eq. (5.7.12), describing a circle with

$$B = \frac{k^2 q - p}{1 - k^2}, \qquad c = \frac{|p|^2 - k^2|q|^2}{1 - k^2}.$$

These equations together with $z_0 = -B$, $\rho^2 = |B|^2 - c$ imply Eqs. (5.7.17). Equations (5.7.17) can be written as $|p - z_0| = k\rho$ and $|q - z_0| = \rho/k$, which shows that p and q are inverse points. Furthermore, because $k < 1$ the points p and q are inside and outside the circle, respectively.

Using Eq. (5.7.16), we demonstrate that bilinear transformations map inverse points to inverse points. Recall that bilinear transformations are compositions of linear transformations and inversion. If $w = az + b$, we must show that the points $\tilde{p} = ap + b$, $\tilde{q} = aq + b$ are inverse point with respect to the circle in the complex w-plane. But

$$\left| \frac{w - \tilde{p}}{w - \tilde{q}} \right| = \frac{|z - p|}{|z - q|} = k,$$

which shows that indeed $\tilde{p}$ and $\tilde{q}$ are inverse points in the w-plane. Similarly, if $w = 1/z$, $\tilde{p} = 1/p$, and $\tilde{q} = 1/q$, we have

$$\left| \frac{w - \tilde{p}}{w - \tilde{q}} \right| = \left| \frac{\frac{p-z}{pz}}{\frac{q-z}{qz}} \right| = \frac{|q|}{|p|} k,$$

which shows that again $\tilde{p}$ and $\tilde{q}$ are inverse points, though $\tilde{p}$ might now lie inside or outside the circle. (Here we consider the generic case $p \neq 0$, $q \neq \infty$; the particular cases of $p = 0$ or $q = \infty$ are handled in a similar way.)

Example 5.7.5   A necessary and sufficient condition for a bilinear transformation to map the upper half plane Im $z > 0$ onto the unit disk $|w| < 1$, is that it is of the form

$$w = \beta \frac{z - \alpha}{z - \bar{\alpha}}, \qquad |\beta| = 1, \text{ Im } \alpha > 0. \tag{5.7.18}$$

Sufficiency: We first show that this transformation maps the upper half z-plane onto $|w| < 1$. If z is on the real axis, then $|x - \alpha| = |x - \bar{\alpha}|$. Thus, the real axis is mapped to $|w| = 1$; hence $y > 0$ is mapped onto one of the complementary domains of $|w| = 1$. Because $z = \alpha$ is mapped into $w = 0$, this domain is $|w| < 1$.

Necessity: We now show that the most general bilinear transformation mapping $y > 0$ onto $|w| < 1$ is given by Eq. (5.7.18). Because $y > 0$ is mapped onto one of the complementary domains of either a circle or of a line, $y = 0$ is to be mapped onto $|w| = 1$. Let α be a point in the upper half z-plane that is mapped to the center of the unit circle in the w-plane (i.e., to $w = 0$). Then $\bar{\alpha}$, which is the inverse point of α with respect to the real axis, must be mapped to $w = \infty$ (which is the inverse point of $w = 0$ with respect to the unit circle). Hence

$$w = \frac{a}{c}\frac{z - \alpha}{z - \bar{\alpha}}.$$

Because the image of the real axis is $|w| = 1$, it follows that $|\frac{a}{c}| = 1$, and the above equation reduces to Eq. (5.7.18), where $\beta = \frac{a}{c}$.

Example 5.7.6 A necessary and sufficient condition for a bilinear transformation to map the disk $|z| < 1$ onto $|w| < 1$ is that it is of the form

$$w = \beta\frac{z - \alpha}{\bar{\alpha}z - 1}, \qquad |\beta| = 1, \ |\alpha| < 1. \tag{5.7.19}$$

Sufficiency: We first show that this transformation maps $|z| < 1$ onto $|w| < 1$. If z is on the unit circle $z = e^{i\theta}$, then

$$|w| = |\beta|\left|\frac{e^{i\theta} - \alpha}{\bar{\alpha}e^{i\theta} - 1}\right| = \frac{|\alpha - e^{i\theta}|}{|\bar{\alpha} - e^{-i\theta}|} = 1.$$

Hence $|z| < 1$ is mapped onto one of the complementary domains of $|w| = 1$. Because $z = 0$ is mapped into $\beta\alpha$, and $|\beta\alpha| < 1$, this domain is $|w| < 1$.

Necessity: We now show that the most general bilinear transformation mapping $|z| < 1$ onto $|w| < 1$ is given by Eq. (5.7.19). Because $|z| < 1$ is mapped onto $|w| < 1$, then $|z| = 1$ is to be mapped onto $|w| = 1$. Let α be the point in the unit circle which is mapped to $w = 0$. Then, from Eq. (5.7.13), $1/\bar{\alpha}$ (which is the inverse point of α with respect to $|z| = 1$) must be mapped to $w = \infty$. Hence, if $\alpha \neq 0$,

$$w = \frac{a}{c}\frac{z - \alpha}{z - \frac{1}{\bar{\alpha}}} = \frac{a\bar{\alpha}}{c}\frac{z - \alpha}{\bar{\alpha}z - 1}.$$

Because the image of $|z| = 1$ is $|w| = 1$, it follows that $|\frac{a\bar{\alpha}}{c}| = 1$, and the above equation reduces to Eq. (5.7.19), with $\beta = a\bar{\alpha}/c$. If $\alpha = 0$ and $\beta \neq 0$, then the points $0, \infty$ map into the points $0, \infty$, respectively, and $w = \beta z$, $|\beta| = 1$. Thus, Eq. (5.7.19) is still valid.

It is worth noting that the process of successive inversions about an even number of circles is expressible as a bilinear transformation, as the following example illustrates.

Example 5.7.7 Consider a point z inside a circle C_1 of radius r and centered at the origin and another circle C_2 of radius R centered at z_0 (see Figure 5.46), containing

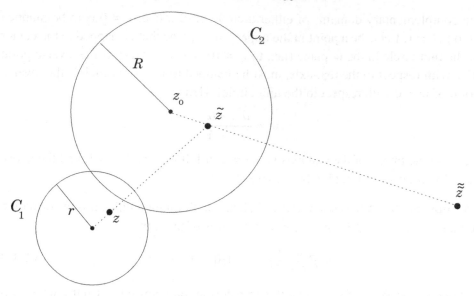

Figure 5.46 Two successive inversions of point z

the inverse point to z with respect to C_1. Show that two successive inversions of the point z about C_1 and C_2, respectively, can be expressed as a bilinear transformation.

The point $\tilde{z}$ is the inverse of z about the circle C_1, and is given by

$$z\bar{\tilde{z}} = r^2 \qquad \text{or} \qquad \tilde{z} = \frac{r^2}{\bar{z}}.$$

The second inversion satisfies, for the point $\tilde{\tilde{z}}$

$$\left(\bar{\tilde{\tilde{z}}} - \overline{z_0}\right)\left(\tilde{\tilde{z}} - z_0\right) = R^2$$

or

$$\tilde{\tilde{z}} = z_0 + \frac{R^2}{\left(\bar{\tilde{z}} - \overline{z_0}\right)} = z_0 + \frac{R^2}{\frac{r^2}{z} - \overline{z_0}} = \frac{r^2 z_0 + \left(R^2 - |z_0|^2\right)z}{r^2 - \overline{z_0}z},$$

which is a bilinear transformation.

In addition to yielding a conformal map of the entire extended z-plane, the bilinear transformation is also distinguished by the interesting fact that it is the only univalent function in the entire extended z-plane.

Theorem 5.7.8   *The bilinear transformation (5.7.1) is the only univalent function which maps $|z| \leq \infty$ onto $|w| \leq \infty$.*

Proof Equation (5.7.8) shows that if $z_1 \neq z_2$ then $w_1 \neq w_2$; that is, a bilinear transformation is univalent. We shall now prove that a univalent function that maps $|z| \leq \infty$ onto $|w| \leq \infty$ must necessarily be bilinear.

To achieve this, we shall first prove that a univalent function that maps the finite complex z-plane onto the finite complex w-plane must be necessarily linear. We first note that if $f(z)$ is univalent in some domain D, then $f'(z) \neq 0$ in D. This is a direct consequence of Theorem 5.3.4, because if $f'(z_0) = 0$, $z_0 \in D$, then $f(z) - f(z_0)$ has a zero of order $n \geq 2$, and hence equation $f(z) = w$ has at least two distinct roots near z_0 for w near $f(z_0)$. It was shown in Theorem 5.3.3 (i.e., Theorem 5.5.1) that the image of $|z| < 1$ contains some disk $|w - w_0| < A$. This implies that ∞ is not an essential singularity of $f(z)$. Because if ∞ were an essential singularity, then as $z \to \infty$, $f(z)$ would come arbitrarily close to w_0 (see Theorem 3.5.6); hence some values of f corresponding to $|z| > 1$ would also lie in the disk $|w - w_0| < A$, which would contradict the fact that f is univalent. It cannot have a branch point; therefore, $z = \infty$ is at worst a pole of f; that is, f is polynomial. But because $f'(z) \neq 0$ for $z \in D$, this polynomial must be linear. Having established the relevant result in the finite plane, we can now include infinities. Indeed, if $z = \infty$ is mapped into $w = \infty$, $f(z)$ being linear is satisfactory, and the theorem is proved. If $z_0 \neq \infty$ is mapped to $w = \infty$, then the transformation $\zeta = 1/(z - z_0)$ reduces this case to the case of the finite plane discussed above, in which case $w(\zeta)$ being linear (i.e., $w(\zeta) = a\zeta + b$) corresponds to $w(z)$ being bilinear. $\qquad\square$

Example 5.7.9 Consider the region bounded by two cylinders perpendicular to the z-plane; the bases of these cylinders are the discs bounded by the two circles $|z| = R$ and $|z - a| = r$, $0 < a < R - r$ ($R, r, a \in \mathbb{R}$). The inner cylinder is maintained at a potential V, while the outer cylinder is maintained at a potential zero. Find the electrostatic potential in the region between these two cylinders (see Figure 5.47).

Recall from Eq. (5.7.16) that the equation $|z - \alpha| = k|z - \beta|$, $k > 0$ is the equation of a circle with respect to which the points α and β are inverse to one another. If α and β are fixed, while k is allowed to vary, the above equation describes a family of nonintersecting circles. The two circles in the z-plane can be thought of as members of this family, provided that α and β are chosen so that they are inverse points with respect to both of these circles (by symmetry considerations we take them to be real), i.e., $\alpha\beta = R^2$ and $(\alpha - a)(\beta - a) = r^2$. Solving for α and β we find

$$\beta = \frac{R^2}{\alpha} \quad \text{and} \quad \alpha = \frac{1}{2a}(R^2 + a^2 - r^2 - A),$$
$$A^2 \equiv [(R^2 + a^2 - r^2)^2 - 4a^2 R^2],$$

where the choice of the sign of A is fixed by taking α inside, and β outside both circles. The bilinear transformation $w = \kappa(z - \alpha)/(\alpha z - R^2)$ maps the above

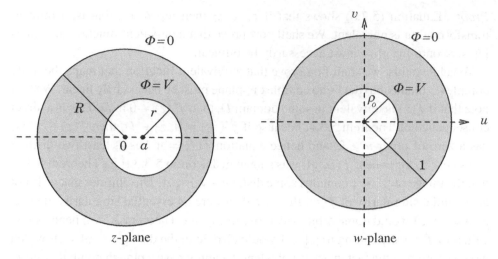

z-plane   w-plane

Figure 5.47 Region between two cylinders for Example 5.7.9

family of nonintersecting circles into a family of concentric circles. We choose the constant $\kappa = -R$ so that $|z| = R$ is mapped onto $|w| = 1$. (Here $z = Re^{i\theta}$ maps onto $w = (Be^{i\theta})/(\bar{B})$, where $B = R - \alpha e^{-i\theta}$, so that $|w| = 1$.) By using Eq. (5.7.17), for the circle $|z| = R$, $k_1^2 = \alpha/\beta$, and for the circle $|z - a| = r$, $k_2^2 = \frac{\alpha - a}{\beta - a}$. Thus, from Eqs. (5.7.16)–(5.7.17) we see that the transformation

$$w = R\frac{z - \alpha}{R^2 - \alpha z} = \left(\frac{-R}{\alpha}\right)\left(\frac{z - \alpha}{z - \beta}\right)$$

maps $|z| = R$ onto $|w| = 1$ and maps $|z - a| = r$ onto $|w| = \rho_0$, where ρ_0 is given by

$$\rho_0 = k_2\left|\frac{R}{\alpha}\right| = \left|\frac{R}{\alpha}\right|\sqrt{\frac{\alpha - a}{\beta - a}}.$$

From this information, we can now find the solution of the Laplace equation that satisfies the boundary conditions. Writing $w = \rho e^{i\phi}$, this solution is $\Phi = V \log \rho/\log \rho_0$. Thus,

$$\Phi = \frac{V}{\log \rho_0} \log |w| = \frac{V}{\log \rho_0} \log \left|R\frac{z - \alpha}{R^2 - \alpha z}\right|.$$

From the mapping, we conclude that when $|z| = R$, $|w| = 1$, hence $\Phi = 0$; and when $|z - a| = r$, $|w| = \rho_0$, and therefore $\Phi = V$. Hence the real part of the analytic function $\Omega(w) = \frac{V}{\log \rho} \log w$ leads to a solution Φ ($\Omega = \Phi + i\Psi$) of Laplace's equation with the required boundary conditions.

In conclusion, we mention without proof, the Schwarz symmetry principle, also referred to as the reflection principle, pertaining to analytic continuation across arcs of circles. This is a generalization of the symmetry or reflection principle mentioned in conjunction with the Schwarz–Christoffel transformation in Section 5.6, which required the analytic continuation of a function across straight line segments, for example, the real axis.

Theorem 5.7.10 (Schwarz Symmetry Principle – Circles) *Let $z \in D$ and $w \in D_w$ be points in the domains D and D_w, which contain circular arcs γ and γ_w respectively; these arcs could degenerate into straight lines. Let $f(z)$ be analytic in D and continuous in $D \cup \gamma$. If $w = f(z)$ maps D onto D_w so that the arc γ is mapped to γ_w, then $f(z)$ can be analytically continued across γ into the domain $\widetilde{D}$ obtained from D by inversion with respect to the circle C of which γ is a part.*

Let γ, γ_w be part of the circles $C: |z - z_0| = r$, $C_w: |w - w_0| = R$, then the analytic continuation is given by

$$\tilde{z} - z_0 = \frac{r^2}{\overline{z} - \overline{z_0}}, \qquad \tilde{w} - w_0 = f(\tilde{z}) - w_0 = \frac{R^2}{\overline{f(z)} - \overline{w_0}}.$$

Consequently, if z and $\tilde{z}$ are inverse points with respect to C, where $z \in D$ and $\tilde{z} \in \widetilde{D}$, then the analytic continuation is given by $f(\tilde{z}) = \tilde{f}(z)$, where $\tilde{f}(z) = \tilde{w}$ is the inverse point to w with respect to circle C_w (see Figure 5.48).

Note: when $\tilde{z}$ is on γ we have $\tilde{w} = w$ and $w - w_0 = R^2/(\overline{w} - \overline{w_0})$.

In fact, the proof of the symmetry principle can be reduced to that of symmetry across the real axis by transforming the circles C and C_w to the real axis, by bilinear transformations. We will not go into further details here.

Thus, for example, let γ be the unit circle centered at the origin in the z-plane and let $f(z)$ be analytic within γ and continuous on γ. Then if $|f(z)| = R$ (i.e., γ_w is a circle of radius R in the w-plane centered at the origin) on γ, then $f(z)$ can be analytically continued across γ by means of the formula $f(z) = R^2/\overline{f(1/\overline{z})}$ because $R^2/\overline{f}$ is the inverse point of f with respect to the circle of radius R centered at the origin and $1/\overline{z}$ is the inverse point to the point z inside the unit circle γ. On the other hand, suppose $f(z)$ maps to a real function on γ. By transforming z to $z - a$, and f to $f - b$ we find $z - a = r^2/(\overline{z} - \overline{a})$, $f(\overline{z}) - b = R^2/(\overline{f}(z) - \overline{b})$ as the inverse points of circles, radii r, R centered at $z = a$, $w = b$ respectively Then, the formula for analytic continuation is given by $f(z) = \overline{f}(\frac{1}{\overline{z}})$, because $\overline{f}$ is the inverse point of f with respect to the real axis.

We note the "symmetry" in this continuation formula; that is, $\tilde{w} = \tilde{f}(z)$ is the inverse point to $w = f(z)$ with respect to the circle C_w of which γ_w is a part, and $\tilde{z}$ is the inverse point to z with respect to the circle C of which γ is a part. As indicated in Section 5.6, in the case where γ and γ_w degenerate into the real axis, this formula yields the continuation of a function $f(z)$ where $f(z)$ is real for real z from the upper half plane to the lower half plane: $f(\overline{z}) = \overline{f}(z)$. Similar specializations apply

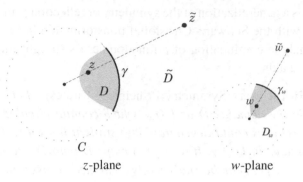

z-plane   *w*-plane

Figure 5.48 Schwarz Symmetry Principle

when the circles reduce to arbitrary straight lines. We also note that in Section 5.8 the symmetry principle across circular arcs is used in a crucial way.

5.7.1 Problems for Section 5.7

1. Show that the "cross ratios" associated with the points $(z, 0, 1, -1)$ and $(w, i, 2, 4)$ are $(z + 1)/2z$ and $(w - 4)(2 - i)/2(i - w)$, respectively. Use these to find the bilinear transformation which maps $0, 1, -1$ to $i, 2, 4$, respectively.

2. Show that the transformation $w_1 = ((z + 2)/(z - 2))^{1/2}$ maps the z-plane with a cut $-2 \le \operatorname{Re} z \le 2$ to the right half plane. Show that the latter is mapped onto the interior of the unit circle by the transformation $w = (w_1 - 1)/(w_1 + 1)$. Thus, deduce the overall transformation which maps the simply connected region containing all points of the plane (including ∞), except the real points z in $-2 \le z \le 2$, onto the interior of the unit circle.

   Hint: for all angles use $-\pi/2 \le \theta \le \pi/2$. Writing $w_1 = u_1 + iv_1$, $w = u + iv$ show that

   $$u = \frac{u_1^2 + v_1^2 - 1}{u_1^2 + v_1^2 + 1}, \qquad v = \frac{2v_1}{u_1^2 + v_1^2 + 1}.$$

   Hence when $u_1 = 0$ show that $u^2 + v^2 = 1$.

3. Show that the transformation $w = (z - a)/(z + a)$, $a = \sqrt{c^2 - \rho^2}$ (where c and ρ are real, $0 < \rho < c$), maps the domain bounded by the circle $|z - c| = \rho$ and the imaginary axis onto the annular domain bounded by $|w| = 1$ and an inner concentric circle (see Figure 5.49). Find the radius, δ, of the inner circle.

   Hint: show that $x = 0$ maps to $u^2 + v^2 = 1$ and $|z - c| = \rho$ maps to $u^2 + v^2 = \frac{c - a}{c + a} = \frac{\rho^2}{(c + a)^2} = \delta^2$.

4. Show that the transformation $w_1 = [(1 + z)/(1 - z)]^2$ maps the upper half unit circle to the upper half plane, and $w_2 = (w_1 - i)/(w_1 + i)$ maps the latter to

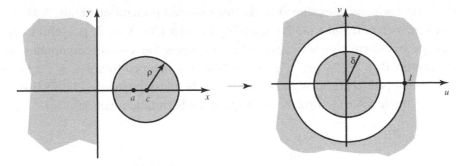

Figure 5.49 Mapping of Problem 3

the interior of the unit circle. Use these results to find an elementary conformal mapping that maps a semicircular disk onto a full disk.

5. Let C_1 be the circle with center $i/2$ passing through 0, and let C_2 be the circle with center $i/4$ passing through 0 (see Figure 5.50). Let D be the region enclosed by C_1 and C_2. Show that the inversion $w_1 = 1/z$ maps D onto the strip $-2 < \text{Im } w_1 < -1$, and the transformation $w_2 = e^{\pi w_1}$ maps this strip to the upper half plane. Use these results to find a conformal mapping which maps D onto the unit disk.

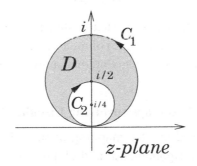

Figure 5.50 Mapping of Problem 5

6. Find a conformal map f which maps the region between two circles $|z| = 1$ and $|z - \frac{1}{4}| = \frac{1}{4}$ onto an annulus $\rho_0 < |z| < 1$, and find ρ_0.

Hint: Show that the bilinear transformation that maps $\{z, 0, 1/2, 1\}$ to $\{w, -\rho_0, \rho_0, 1\}$ leads to

$$w = \frac{z\rho_0(3 - \rho_0) + \rho_0(\rho_0 - 1)}{z(3\rho_0 - 1) + 1 - \rho_0}.$$

Then mapping the point $w = -1$ to $z = -1$ determines ρ_0.

7. Find the harmonic function ϕ in the lens-shaped domain of Figure 5.51, that takes the values 0 and 1 on the bounding circular arcs. Hint: It is useful to note that the transformation $w = z/(z - (1 + i))$ maps the lens-shaped domain into the region $R_w : \frac{3\pi}{4} \leq \arg w \leq \frac{5\pi}{4}$ with $\phi = 1$ mapped to $\arg w = 3\pi/4$ and $\phi = 0$ mapped to $\arg w = 5\pi/4$. Then use the ideas introduced in Section 5.4 (see Example 5.4.4) to find the corresponding harmonic function $\phi(w)$.

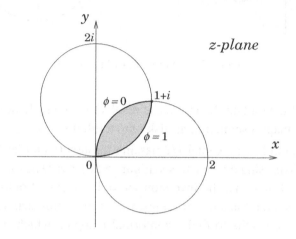

Figure 5.51 Mapping of Problem 7

Note: ϕ can be interpreted as the steady-state temperature inside an infinitely long strip (perpendicular to the plane) of material having this lens-shaped region as its cross section, with its sides maintained at the given temperatures.

*5.8 Mappings Involving Circular Arcs

In Section 5.6 we showed that the mapping of special polygonal regions to the upper half plane involved trigonometric and elliptic functions. In this section we investigate the mapping of a region whose boundary consists of a **curvilinear polygon**; that is, a polygon whose sides are made up of circular arcs. We outline the main ideas, and in certain important special cases we will be led to an interesting class of functions called automorphic functions, which can be considered generalizations of elliptic functions. We will study a class of automorphic functions known as Schwarzian triangle functions, of which the best known (with zero angles) is the so-called elliptic modular function.

Consider a domain of the w-plane bounded by circular arcs. Our aim is to find the transformation $w = f(z)$ which maps this domain onto the upper half of the z-plane (see Figure 5.52).

The relevant construction is conceptually similar to the one used for linear polygons (i.e., the Schwarz–Christoffel transformation). We remind the reader that the

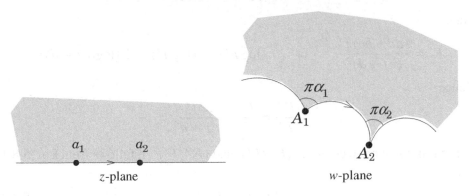

Figure 5.52 Mapping of a region whose boundary contains circular arcs

crucial step in that construction is the introduction of the ratio f''/f'. The Riemann Mapping Theorem ensures that there is a conformal $(f'(z) \neq 0)$ map onto the upper half plane. The Schwarz reflection principle implies that this ratio is analytic and one-to-one in the entire z-plane except at the points corresponding to the vertices of the polygon; near these vertices in the z-plane; that is, near $z = a_\ell$

$$f(z) = (z - a_\ell)^{\alpha_\ell} \left[c_\ell^{(0)} + c_\ell^{(1)}(z - a_\ell) + \cdots \right];$$

therefore f''/f' has simple poles. These two facts and Liouville's Theorem imply the Schwarz–Christoffel transformation. The distinguished property of f''/f' is that it is invariant under a linear transformation; that is, if we transform $f = A\hat{f} + B$, where A and B are constants, then $f''/f' = \hat{f}''/\hat{f}'$. The fact that the mapping is constructed from a given polygon through an even number of Schwarz reflections implies that the most general form of the mapping is given by $f(z) = A\hat{f}(z) + B$ where A and B are constants.

The generalization of the above construction to the case of circular arcs is as follows. In Section 5.7 the Schwarz symmetry principle across circular arcs was discussed. We also mentioned there that an even succession of inversions across circles can be expressed as a bilinear transformation. It is then natural to expect that the role which was played by $f''(z)/f'(z)$ in the Schwarz–Christoffel transformation will now be generalized to an operator which is invariant under bilinear transformations. This quantity is the so-called **Schwarzian derivative**, defined by

$$\{f, z\} \equiv \left(\frac{f''(z)}{f'(z)} \right)' - \frac{1}{2} \left(\frac{f''(z)}{f'(z)} \right)^2. \tag{5.8.1}$$

Indeed, let

$$F = \frac{af + b}{cf + d}, \qquad ad - bc \neq 0. \tag{5.8.2}$$

Then,

$$F' = \frac{(ad - bc)f'}{(cf + d)^2} \qquad \text{or} \qquad (\log F')' = (\log f')' - 2\,(\log(cf + d))'.$$

Hence,

$$\frac{F''}{F'} = \frac{f''}{f'} - \frac{2cf'}{cf + d}.$$

Using this equation to compute $(F''/F')'$ and $(F''/F')^2$, it follows from Eq. (5.8.1) that

$$\{f, z\} = \{F, z\}. \tag{5.8.3}$$

The single-valuedness of $\{f, z\}$ follows in much the same way as the derivation of the single-valuedness of $f''(z)/f'(z)$ in the Schwarz–Christoffel derivation. Riemann's Mapping Theorem establishes the existence of a conformal map to the upper half plane. From the Schwarz reflection principle, $f(z)$ is analytic and one-to-one everywhere except possibly the endpoints $z = a_l$, for $l = 1, 2 \ldots$. In the present case any even number of inversions across circles is a bilinear transformation (see Example 5.7.7). Because the Schwarzian derivative is invariant under a bilinear transformation, it follows that the function $\{f, z\}$ corresponding to any point in the upper half z-plane is uniquely obtained. Similar arguments hold for an odd number of inversions and points in the lower half plane. Moreover, the function $\{f, z\}$ takes on real values for real values of z. Hence we can analytically continue $\{f, z\}$ from the upper half to lower half z-plane by Schwarz reflection. Consequently, there can be no branches whatsoever and the function $\{f, z\}$ is single valued. Thus the Schwarzian derivative is analytic in the entire z-plane except possibly at the points a_ℓ, $\ell = 1, \ldots, n$. The behavior of $f(z)$ at a_ℓ can be found by noting that (after a bilinear transformation) $f(z)$ maps a piece of the real z-axis containing $z = a_\ell$ onto two linear segments forming an angle $\pi \alpha_\ell$. Therefore, in the neighborhood of $z = a_\ell$,

$$f(z) = (z - a_\ell)^{\alpha_\ell} g(z), \tag{5.8.4}$$

where $g(z)$ is analytic at $z = a_\ell$, $g(a_\ell) \neq 0$, and $g(z)$ is real when z is real. This implies that the behavior of $\{f, z\}$ near a_ℓ is given by

$$\{f, z\} = \frac{1}{2} \frac{1 - \alpha_\ell^2}{(z - a_\ell)^2} + \frac{\beta_\ell}{z - a_\ell} + h(z), \qquad \beta_\ell \equiv \frac{1 - \alpha_\ell^2}{\alpha_\ell} \frac{g'(a_\ell)}{g(a_\ell)}, \tag{5.8.5}$$

where $h(z)$ is analytic at $z = a_\ell$, and where we have left to the reader the verification of the intermediate step

$$\frac{f''}{f'}(z) = \frac{\alpha_\ell - 1}{z - a_\ell} + \frac{1 + \alpha_\ell}{\alpha_\ell} \frac{g'(a_\ell)}{g(a_\ell)} + \cdots.$$

Using these properties of $\{f, z\}$ and Liouville's Theorem, it follows that

$$\{f, z\} = \frac{1}{2} \sum_{\ell=1}^{n} \frac{(1 - \alpha_\ell^2)}{(z - a_\ell)^2} + \sum_{\ell=1}^{n} \frac{\beta_\ell}{z - a_\ell} + c, \tag{5.8.6}$$

where $\alpha_1, \ldots, \alpha_n, \beta_1, \ldots, \beta_n, a_1, \ldots, a_n$, are real numbers and c is a constant. We recall that in the case of the Schwarz–Christoffel transformation the analogous constant c was determined by analyzing $z = \infty$. We now use the same idea. If we assume that none of the points $a_1, \ldots, a_n$ coincide with ∞, then $f(z)$ is analytic at $z = \infty$; that is, $f(z) = f(\infty) + c_1/z + c_2/z^2 + \cdots$ near $z = \infty$. Using this expansion in equation Eq. (5.8.1) it follows that $\{f, z\} = k_4/z^4 + k_5/z^5 + \cdots$ near $z = \infty$. This implies that by expanding the right-hand side of (5.8.6) in a power series in $1/z$, and equating to zero the coefficients of z^0, $1/z$, $1/z^2$, and $1/z^3$, we find that c (the coefficient of z^0) is zero, and for the coefficients of $1/z$, $1/z^2$ and $1/z^3$,

$$\sum_{\ell=1}^{n} \beta_\ell = 0, \qquad \sum_{\ell=1}^{n} (2a_\ell \beta_\ell + 1 - \alpha_\ell^2) = 0, \qquad \sum_{\ell=1}^{n} [\beta_l a_\ell^2 + a_\ell (1 - \alpha_\ell^2)] = 0. \tag{5.8.7}$$

In summary, let $f(z)$ be a solution of the third-order differential equation (5.8.6) with $c = 0$, where $\{f, z\}$ is defined by Eq. (5.8.1) and where the real numbers appearing in the right-hand side of Eq. (5.8.6) satisfy the relations given by Eq. (5.8.7). Then the transformation $w = f(z)$ maps the domain of the w-plane bounded by circular arcs forming vertices with angles $\pi\alpha_1, \ldots, \pi\alpha_n$, where $0 \le \alpha_\ell \le 2$ and $\ell = 1, \ldots, n$, onto the upper half of the z-plane. The vertices are mapped to the points $a_1, \ldots, a_n$ of the real z-axis.

It is significant that the third-order nonlinear differential equation (5.8.6) can be reduced to a second-order linear differential equation. Indeed, if $y_1(z)$ and $y_2(z)$ are two linearly independent solutions of the equation

$$y''(z) + \frac{1}{2} P(z) y(z) = 0, \tag{5.8.8}$$

then

$$f(z) \equiv \frac{y_1(z)}{y_2(z)} \tag{5.8.9a}$$

solves

$$\{f, z\} = P(z). \tag{5.8.9b}$$

The proof of this fact is straightforward. Substituting $y_1 = y_2 f$ into Eq. (5.8.8), demanding that both y_1 and y_2 solve Eq. (5.8.8), and noting that the Wronskian $W = y_2 y_1' - y_1 y_2'$ is a constant for Eq. (5.8.8), it follows that

$$\frac{f''}{f'} = -2\frac{y_2'}{y_2}$$

which implies Eq. (5.8.9b). This concludes the derivation of the main results of this section, which we express as a theorem.

Theorem 5.8.1 (Mapping of Circular Arcs)   *If $w = f(z)$ maps the upper half of the z-plane onto a domain of the w-plane bounded by n circular arcs, and if the points $z = a_\ell$, $\ell = 1, \ldots, n$, on the real z-axis are mapped to the vertices of angle $\pi\alpha_\ell$, $\ell = 1, \ldots, n$, then*

$$w = f(z) = \frac{y_1(z)}{y_2(z)}, \qquad (5.8.10)$$

where $y_1(z)$ and $y_2(z)$ are two linearly independent solutions of the linear differential equation

$$y''(z) + \left[\sum_{\ell=1}^{n} \frac{(1 - \alpha_\ell^2)}{4(z - a_\ell)^2} + \frac{1}{2}\sum_{\ell=1}^{n} \frac{\beta_\ell}{z - a_\ell}\right] y(z) = 0 \qquad (5.8.11)$$

and the real constants β_ℓ, $\ell = 1, \ldots, n$ satisfy the relations (5.8.7).

Remarks (1) The three identities (5.8.7) are the only general relations that exist between the constants entering Eq. (5.8.11). Indeed, the relevant domain is specified by n circular arcs, that is, $3n$ real parameters (each circle is prescribed by the radius and the two coordinates of the center). However, as mentioned in Section 5.7, three arbitrary points on the real z-axis can be mapped to any three vertices (i.e., six real parameters) in the w-plane. This reduces the number of parameters describing the w domain to $3n - 6$. On the other hand, the transformation $f(z)$ involves $3n - 3$ independent parameters: $3n$ real quantities $\{\alpha_\ell, \beta_\ell, a_\ell\}_{\ell=1}^{n}$, minus the three constraints (5.8.7). Because three of the values a_ℓ can be arbitrarily prescribed, we see that the $f(z)$ also depends on $3n - 6$ parameters.

(2) The procedure of actually constructing a mapping function $f(z)$ in terms of a given curvilinear polygon is further complicated by the determination of the constants in Eq. (5.8.11) in terms of the given geometrical configuration. In Eq. (5.8.11) we know the angles $\{\alpha_\ell\}_{\ell=1}^{n}$. We require that the points a_ℓ on the real z-axis correspond to the vertices A_ℓ of the polygon. Characterizing the remaining $n - 3$ constants – that is, the n values β_ℓ (the so-called accessory parameters) minus three constraints – by geometrical conditions is in general unknown. The cases of $n = 2$ (a crescent; see also Problem 5 in this section) and $n = 3$ (a curvilinear triangle; see Figure 5.53) are the only cases in which the mapping is free of the determination of accessory parameters. Mapping with more than three vertices generally requires numerical computation (cf. Trefethen, 1986, Driscoll and Trefethen, 2002).

(3) The Schwarz–Christoffel transformation discussed in Section 5.6 (see (5.6.1)–(5.6.2)) can be deduced from Equations (5.8.6)–(5.8.7) with suitable choices for β_l.

(4) If one of the points a_l, say a_1, is taken to be ∞ then the sum in Equations (5.8.6) and (5.8.11) is taken from 2 to n. The conditions (5.8.7) must then be altered since $f(z)$ is not analytic at ∞ (see Example 5.8.2 following).

Example 5.8.2 Consider a domain of the w-plane bounded by three circular arcs with interior angles $\pi\alpha$, $\pi\beta$, $\pi\gamma$. Find the transformation which maps this domain to the upper half of the z-plane. Specifically, map the vertices with angles $\pi\alpha$, $\pi\beta$, and $\pi\gamma$ to the points ∞, 0, and 1.

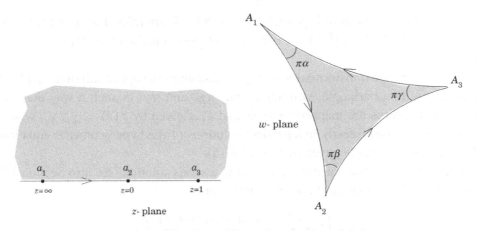

Figure 5.53 Mapping from three circular arcs for Example 5.8.2

We associate with the vertices A_1, A_2, and A_3 the points $a_1(\infty)$, $a_2(0)$, and $a_3(1)$. Writing $\alpha_2 = \beta$, $\alpha_3 = \gamma$, $a_2 = 0$, and $a_3 = 1$, Eq. (5.8.6) with $c = 0$ becomes

$$\{f, z\} = \frac{1 - \beta^2}{2z^2} + \frac{1 - \gamma^2}{2(z - 1)^2} + \frac{\beta_2}{z} + \frac{\beta_3}{z - 1}. \tag{5.8.12}$$

When one point, in this case $w = A_1$, is mapped to $z = \infty$, then the terms involving a_1 drop out of the right-hand side of Eq. (5.8.6), and from Eq. (5.8.4), recalling the transformation $z - a_1 \to 1/z$, we find that $f(z) = \gamma z^{-\alpha}[1 + c_1/z + \cdots]$ for z near ∞. Similarly, owing to the identification $a_1 = \infty$, one must reconsider the derivation of the relations (5.8.7). These equations were derived under the assumption that $f(z)$ is analytic at ∞. However, in this example the above behavior of $f(z)$ implies that $\{f, z\} = ((1 - \alpha^2)/2z^2)[1 + D_1/z + \cdots]$ as $z \to \infty$. Expanding the right-hand side of Eq. (5.8.12) in powers of $1/z$ and equating the coefficients of $1/z$ and $1/z^2$ to 0 and $(1 - \alpha^2)/2$, respectively, we find $\beta_2 + \beta_3 = 0$ and $\beta_3 \equiv (\beta^2 + \gamma^2 - \alpha^2 - 1)/2$. Using these values for β_2 and β_3 in Eq. (5.8.12), we deduce that $w = f(z) = y_1/y_2$, where y_1 and y_2 are two linearly independent solutions of Eq. (5.8.11):

$$y''(z) + \frac{1}{4}\left[\frac{1 - \beta^2}{z^2} + \frac{1 - \gamma^2}{(z - 1)^2} + \frac{\beta^2 + \gamma^2 - \alpha^2 - 1}{z(z - 1)}\right] y(z) = 0. \tag{5.8.13}$$

Equation (5.8.13) is related to an important differential equation known as the hypergeometric equation, which is defined as in Eq. (3.7.35c):

$$z(1 - z)\chi''(z) + [c - (a + b + 1)z]\chi'(z) - ab\chi(z) = 0, \tag{5.8.14}$$

where a, b, and c are, in general, complex constants. It is easy to verify that if

$$a = \frac{1}{2}(1 + \alpha - \beta - \gamma), \quad b = \frac{1}{2}(1 - \alpha - \beta - \gamma), \quad c = 1 - \beta \tag{5.8.15}$$

(all real), then solutions of Eqs. (5.8.13) and (5.8.14) are related by $\chi = u(z)y(z)$, where $u(z) = z^A/(1 - z)^B$, $A = -c/2$, $B = \frac{a+b-c+1}{2}$, and therefore $f(z) = y_1/y_2 = \chi_1/\chi_2$.

In summary, the transformation $w = f(z)$ that maps the upper half of the z-plane onto a curvilinear triangle with angles $\pi\alpha$, $\pi\beta$, and $\pi\gamma$, in such a way that the associated vertices are mapped to ∞, 0, and 1, is given by $f(z) = \chi_1/\chi_2$, where χ_1 and χ_2 are two linearly independent solutions of the hypergeometric equation (5.8.14) with a, b, and c given by Eqs. (5.8.15).

The hypergeometric equation (5.8.14) has a series solution (see also Section 3.7 and Nehari, 1952) that can be written in the form

$$\chi_1(z; a, b, c) = k\left(1 + \frac{ab}{c}z + \frac{a(a + 1)b(b + 1)}{c(c + 1)2!}z^2 + \cdots\right), \tag{5.8.16a}$$

where k is constant as can be directly verified. This function can also be expressed as an integral:

$$\chi_1(z; a, b, c) = \int_0^1 t^{b-1}(1 - t)^{c-b-1}(1 - zt)^{-a}\, dt, \tag{5.8.16b}$$

where the conditions $b > 0$ and $c > b$ (a, b, c assumed real) are necessary for the existence of the integral.

We shall assume that $\alpha + \beta + \gamma < 1$, $\alpha, \beta, \gamma > 0$; then we see that Eq. (5.8.15) ensures that the conditions $b > 0$, $c > 0$ hold. Moreover, expanding $(1 - tz)^{-a}$ in a power series in z leads to Eq. (5.8.16a), apart from a multiplicative constant. (To verify this one can use $\int_0^1 t^{b-1}(1 - t)^{c-b-1}dt = \Gamma(b)\Gamma(c - b)/\Gamma(c) = k$.) To obtain $w = f(z)$, we need a second linearly independent solution of Eq. (5.8.14). We note that the transformation $z' = 1 - z$ transforms Eq. (5.8.14) to

$$z'(1 - z')\chi'' + [a + b - c + 1 - (a + b + 1)z']\chi' - ab\chi = 0$$

and we see that the parameters of this hypergeometric equation are $a' = a$, $b' = b$, $c' = a + b - c + 1$, whereupon a second linearly independent solution can be written in the integral form

$$\chi_2(z; a, b, c) = \chi_1(1 - z, a, b, a + b - c + 1)$$

$$= \int_0^1 t^{b-1}(1 - t)^{a-c}(1 - (1 - z)t)^{-a} \, dt. \tag{5.8.16c}$$

Once again, the condition $\alpha + \beta + \gamma < 1$ ensures the existence of the integral (5.8.16c) because we have that $b > 0$ and $a > c - 1$.

Consequently, the mapping

$$w = f(z) = \frac{\chi_1(z; a, b, c)}{\chi_2(z; a, b, c)}, \tag{5.8.16d}$$

taking the upper half z-plane to the w-plane, is now fixed with χ_1 and χ_2 specified as above. The real z-axis maps to the circular triangle as depicted in Figure 5.53. So, for example, the straight line on the real axis from $z = 0$ to $z = 1$ maps to a circular arc between A_2 and A_3 in the w-plane.

We note that the case of $\alpha + \beta + \gamma = 1$ can be transformed into a triangle with straight sides (note that the sum of the angles is π) and therefore can be considered by the methods of Section 5.6. In the case of $\alpha + \beta + \gamma > 1$, one needs to employ different integral representations of the hypergeometric function (see Whittaker and Watson, 1927).

In the next example we discuss the properties of $f(z)$ and the analytic continuation of the inverse of $w = f(z)$, or alternatively, the properties of the map and its inverse as we continue from the upper half z-plane to the lower half z-plane and repeat this process over and over again. This is analogous to the discussion of the elliptic function in Example 5.6.9.

Example 5.8.3 (The Schwarzian Triangle Functions)   In Example 5.8.2 we derived the function $w = f(z)$ which maps the upper half of the z-plane onto a curvilinear triangle in the w-plane. Such functions are known as Schwarzian s-functions, $w = s(z)$, or as Schwarzian triangle functions. Now we shall further study this function and the inverse of this function, which is important in applications such as the solution to certain differential equations (e.g. Chazy's Equation (3.7.52) the Darboux–Halphen system (3.7.53)) which arise in relativity and integrable systems. These inverse functions $z = S(w)$ are also frequently called Schwarzian S-functions (capital S), or Schwarzian triangle functions.

We recall from Example 5.6.9 that although the function $w = f(z) = F(z, k)$, which maps the upper half of the z-plane onto a rectangle in the w-plane, is multivalued; nevertheless its inverse $z = \text{sn}(w, k)$ is single valued. Similarly, the Schwarzian function $f(z) = s(z)$, which maps the upper half of the z-plane onto a curvilinear triangle in the w-plane, is also multivalued. While the inverse of this function is not in general single valued, we shall show that in the particular case that the angles of the curvilinear triangle satisfy $\alpha + \beta + \gamma < 1$ and

$$\alpha = \frac{1}{l}, \quad \beta = \frac{1}{m}, \quad \gamma = \frac{1}{n}; \qquad l, m, n \in \mathbf{Z}^+; \ \alpha, \beta, \gamma \neq 0, \tag{5.8.17}$$

where $\mathbf{Z}^+$ denotes the set of positive integers, the inverse function is single valued.

For convenience we shall assume that two of the sides of the triangle are formed by straight line segments (a special case of a circle is a straight line) meeting at the origin, and that one of these segments coincides with part of the positive real axis.

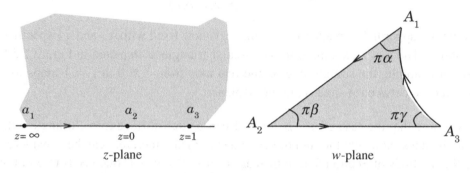

Figure 5.54 Two straight segments, one circular arc for Example 5.8.3

This is without loss of generality. Indeed, let C_1 and C_2 be two circles that meet at $z = A_2$ at an angle $\pi\beta$. Because $\beta \neq 0$, these circles intersect also at another point, say A. The transformation $\tilde{w} = (w - A_2)/(w - A)$ maps all the circles through A into straight lines. (Recall from Section 5.7 that bilinear transformations map circles into either circles or lines, but because $w = A$ maps to $\tilde{w} = \infty$, it must be the latter.) In particular, the transformation maps C_1 and C_2 into two straight lines through A_2. By an additional rotation it is possible to make one of these lines to coincide with the real axis (see Figure 5.54, w-plane).

It turns out that if

$$\alpha + \beta + \gamma < 1, \tag{5.8.18}$$

then there exists a circle that intersects at right angles the three circles that make up a curvilinear triangle. Indeed, as discussed above we may without loss of generality consider a triangle with two straight sides (see Figure 5.54). Any circle centered at the origin, which we will call C_o, is obviously orthogonal to the two straight sides of the triangle. Let C denote the circle whose part forms the third side of the triangle (see Figure 5.55, and note specifically the arc between A_1 and A_3 which extends to the circle C). Equation (5.8.18) implies that the origin of C_o lies exterior to C because the arc (A_1, A_3) is convex. Hence, it is possible to draw tangents from the origin of C_o to C. If P and Q denote the points of contact of the tangents with C, then the circle C_o is orthogonal to C. This circle, C_o, is called the orthogonal circle of the triangle. Given the angles $\alpha\pi$, $\beta\pi$, and $\gamma\pi$ and the points A_1 and A_3 (point A_2 is the origin) that are determined by the properties of the equation (i.e., the hypergeometric equation; we discuss this issue later in this example), the circle C and the orthogonal circle C_o are then fixed.

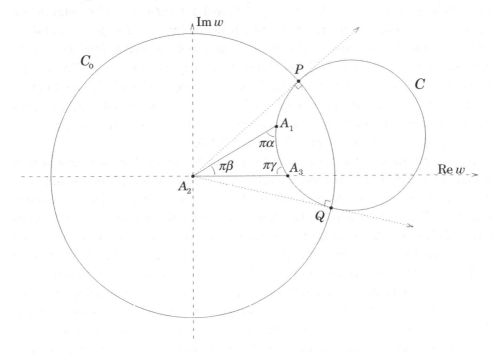

Figure 5.55 Orthogonal circle C_o

If either of the angles α or γ are zero then the lines A_2A_1 or A_2A_3, respectively, correspond to a tangent to the circle C, in which case the vertex A_1 or A_3, respectively lies on the orthogonal circle C_o.

Let $w = s(z; \alpha, \beta, \gamma)$ denote the transformation that maps the upper half of the z-plane onto the triangle depicted in Figure 5.54. If the angles of this triangle satisfy the conditions (5.8.17) and (5.8.18), then it turns out that the inverse of this transformation, denoted by S,

$$w = s(z; \alpha, \beta, \gamma), \qquad z = S(w; \alpha, \beta, \gamma) \tag{5.8.19}$$

is single valued in the interior of the orthogonal circle associated with this triangle. We next outline the main ideas needed to establish this result.

The derivation is based on two facts. First, the function S has no singularities in the entire domain of its existence except poles of order α^{-1}. Second, the domain of existence of S is simply connected. Using these facts (see the Monodromy Theorem,

Section 3.5), it follows that S is single valued. The singularity structure of S follows from the fact that for the original triangle, $s(z) = z^{-\alpha} g_1(z)$, $s(z) = z^{\beta} g_2(z)$, and $s(z) = (z - 1)^{\gamma} g_3(z)$ near $z = \infty$, $z = 0$, and $z = 1$, respectively, where the $g_i(z)$, $i = 1, 2, 3$, are analytic and nonzero. This follows from the properties of the mapping of two line segments meeting at an angle to the real z-axis (see Section 5.6). Hence, $S(w)$ behaves like $c_1 (w - A_1)^{-1/\alpha}$, $c_2 w^{1/\beta}$, and $1 + c_3 (w - A_3)^{1/\gamma}$, with c_1, c_2, c_3 respective constants, which shows that if the reciprocal of α, β, γ are positive integers then the only singularity of $S(w)$ is a pole of order $1/\alpha$ at $w = A_1$ corresponding to $z = \infty$ (recall that vertex A_1 corresponds to $z = \infty$). Because the transformation is conformal, there can be no other singularities inside the triangle. All possible analytic continuations of $z = S(w; \alpha, \beta, \gamma)$ to points outside the original triangle can be obtained by reflections about any of the sides of the triangles, using the Schwarz reflection Principle (Theorem 5.7.10).

By the properties of bilinear transformations we know that an inversion with respect to a circular arc transforms circles into circles, preserves angles, and maps the orthogonal circle onto itself. It follows that any number of inversions of a circular triangle will again lead to a circular triangle situated in the interior of the orthogonal circle. This shows that S cannot be continued to points outside the orthogonal circle and that the vertices are the locations of the only possible singularities. Any point within the orthogonal circle can be reached by a sufficient number of inversions. Indeed, at the boundary of the domain covered by these triangles, there can be no circular arcs of positive radius, because otherwise it would be possible to extend this domain by another inversion. Hence, the boundary of this domain, which is the orthogonal circle, is made up of limit points of circular arcs whose radii tend to zero. This discussion implies that the domain of existence of $S(w)$ is the interior of the orthogonal circle, which is a simply connected domain. The function $S(w)$ cannot be continued beyond the circumference of this circle, so the circumference of the orthogonal circle is a **natural boundary** of $S(w; \alpha, \beta, \gamma)$. The boundary is a dense set of singularities, in this case a dense set of poles of order $1/\alpha$.

Next let us find the analytic expression of the function $w = s(z; \alpha, \beta, \gamma)$. Recall that because $f(z) = s(z)$ satisfies Eq. (5.8.12) then $s(z) = \hat{\chi}_1 / \hat{\chi}_2$, where $\hat{\chi}_1$ and $\hat{\chi}_2$ are any two linearly independent solutions of the hypergeometric equation (5.8.14), and the numbers a, b, c are related to α, β, γ by Eq. (5.8.15). A solution $\hat{\chi}_2$ of the hypergeometric equation which we will now use is given by Eq. (5.8.16b): $\hat{\chi}_2 = \chi_1(z; a, b, c)$. A second solution can be obtained by the observation that if χ is a solution of the hypergeometric equation (5.8.4), then $z^{1-c} \chi(z; a', b', c')$ is also a solution of the same equation, where $a' = a - c + 1$, $b' = b - c + 1$, and $c' = 2 - c$. Because the value of $\chi_1(z; a, b, c)$ at $z = 0$ is a nonzero constant, while the value of $z^{1-c} \chi_1(z; a', b', c')$ at $z = 0$ is zero (for $0 < c < 1$) it follows that the Wronskian is nonzero and that these two solutions are linearly independent. Hence

$$w = s(z; \alpha, \beta, \gamma) = \frac{z^{1-c} \chi_1(z; a - c + 1, b - c + 1, 2 - c)}{\chi_1(z; a, b, c)}, \qquad (5.8.20)$$

where a, b and c are given by Eq. (5.8.15). (This is a different representation than that discussed in Example 5.8.2. It is more convenient for this case, two sides being straight lines.) The vertices of the triangle with angles β, γ, α correspond to the points $z = 0, 1, \infty$, respectively. Because $c < 1$, $s(0; \alpha, \beta, \gamma) = 0$; that is, the origin of the z-plane corresponds to the origin (vertex A_2 in Figure 5.54) in the w-plane. We choose the branch of z^{1-c} in such a way that z^{1-c} is real for positive z. Thus, $s(z)$ is real if z varies along the real axis from $z = 0$ to $z = 1$; hence one side of the curvilinear triangle is part of the positive real axis of the w-plane. For negative z, we find (using $c = 1 - \beta$)

$$z^{1-c} = (e^{i\pi}|z|)^{1-c} = |z|^\beta e^{i\pi\beta},$$

which shows that $-\infty < z < 0$ is mapped by the transformation (5.8.20) onto another side of the triangle, the linear segment which makes the angle $\pi\beta$ with the real axis at the origin.

The remaining portions of the circular triangle in the w-plane are fixed by knowledge of the hypergeometric equation and formula (5.8.20) (details of which we do not go into here; the interested reader can consult one of the many references to properties of the hypergeometric equation such as Whittaker and Watson (1927) or Nehari (1952)). So, for example, the vertices $S(1; \alpha, \beta, \gamma)$ corresponding to point A_3 and $S(\infty; \alpha, \beta, \gamma)$ corresponding to point A_1, may be calculated from Eq. (5.8.20) and using the properties of the hypergeometric functions. This yields

$$S(1; \alpha, \beta, \gamma) = \frac{\Gamma(2 - c)\Gamma(c - a)\Gamma(c - b)}{\Gamma(c)\Gamma(1 - a)\Gamma(1 - b)},$$

which is real, and

$$S(\infty; \alpha, \beta, \gamma) = \frac{e^{i\pi(1-c)}\Gamma(b)\Gamma(c - a)\Gamma(2 - c)}{\Gamma(c)\Gamma(b - c + 1)\Gamma(1 - a)},$$

where, given any positive α, β, γ satisfying $\alpha + \beta + \gamma < 1$, we determine a, b, c from Eq. (5.8.15). The fundamental triangle obtained by the map $w = S(z; \alpha, \beta, \gamma)$ is the one depicted in Figures 5.54 and 5.55 (also see Figure 5.56).

The single-valuedness of the inverse function $z = S(w; \alpha, \beta, \gamma)$, $\alpha = 1/\ell$, $\beta = 1/m$, $\gamma = 1/n$, in the case when ℓ, m, n are integers, makes their study particularly important. Successive continuations of the fundamental triangle in the w-plane across their sides correspond to reflections from the upper to lower half z-plane, and this corresponds to analytically continuing the solution in terms of the hypergeometric functions given by Eq. (5.8.20). Inverting an infinite number of times allows us to eventually "tile" the orthogonal circle C_o. A typical situation is illustrated in Figure 5.56, with $m = 3$ and $\ell = n = 4$; that is, $\beta = \pi/3$ and $\alpha = \gamma = \pi/4$. The shaded and white triangles correspond to the upper and lower half planes, respectively.

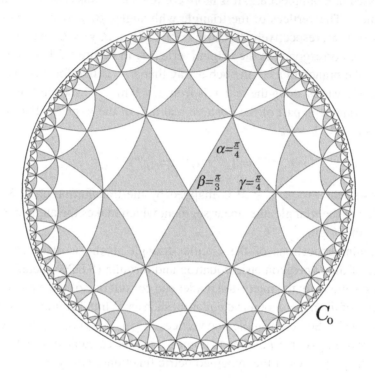

Figure 5.56 Tiling the orthogonal circle with circular triangles

Finally, we note that when $\beta \to 0$ (A_2 remaining at the origin) and thus $c \to 1$, the triangle degenerates to a line segment. We note that when $c \to 1$, the representation (5.8.20) breaks down because $\hat{\chi}_1$ and $\hat{\chi}_2$ are not linearly independent. The important special case $\alpha = \beta = \gamma = 0$, corresponding to three points lying on the orthogonal circle, is discussed below in Example 5.8.4.

Remark An even number of reflections with respect to circular arcs is a bilinear transformation (see Example 5.7.7). An inversion (reflection) with respect to a circular arc maps circles into circles, preserves the magnitude of an angle, and inverts its orientation. Hence an even number of inversions preserves the magnitude of an angle and its orientation and is a conformal transformation mapping circles into circles; that is, it is equivalent to a bilinear transformation (see Example 5.7.7). Because these inversions are symmetric with respect to the real axis in the z-plane, it follows that an even number of inversions in the w-plane will return us to the original position in the z-plane. Hence $z = S(w)$ satisfies the functional equation

$$S\left(\frac{aw + b}{cw + d}\right) = S(w). \tag{5.8.21}$$

Functions that satisfy this equation are usually referred to as **automorphic functions**. Such functions can be viewed as generalizations of periodic functions, for example, elliptic functions. Equation (5.8.21) can also be ascertained by studying the Schwarzian equation (5.8.12) for $w = s(z)$ and its inverse $z = S(w)$.

In order to determine the precise form of the bilinear transformation associated with the curvilinear triangles, we first note that these transformations must leave the orthogonal circle invariant. Suppose we normalize this circle to have radius 1. We recall that the most general bilinear transformation taking the unit circle onto itself is (see Example 5.7.6)

$$w = B\left(\frac{z - A}{\bar{A}z - 1}\right), \qquad |B| = 1, \; |A| < 1. \tag{5.8.22}$$

This shows that under the conditions (5.8.17) and (5.8.18), $S(w; \alpha, \beta, \gamma)$ is a single-valued automorphic function in $|w| < 1$ satisfying the functional equation

$$S\left(B\left(\frac{w - A}{\bar{A}w - 1}\right); \alpha, \beta, \gamma\right) = S(w; \alpha, \beta, \gamma). \tag{5.8.23}$$

The bilinear transformations associated with an automorphic function form a group. Indeed, if T and $\widetilde{T}$ denote two such bilinear transformations – that is, if S is an automorphic function satisfying Eq. (5.8.21), $S(Tw) = S(w)$ and $S(\widetilde{T}w) = S(w)$ – then

$$S(T\widetilde{T}w) = S[T(\widetilde{T}w)] = S[\widetilde{T}w] = S(w).$$

Furthermore, if T^{-1} denotes the inverse of T, then

$$S(w) = S(TT^{-1}w) = S(T^{-1}Tw).$$

Example 5.8.4 Find the transformation which maps a curvilinear triangle with zero angles onto the upper half of the z-plane.

In this case $\alpha = \beta = \gamma = 0$, so that $a = b = \frac{1}{2}$, $c = 1$, and the hypergeometric equation (5.8.14) reduces to

$$z(1 - z)\chi''(z) + [(1 - z) - z]\chi'(z) - \frac{1}{4}\chi(z) = 0.$$

From this equation we see that if $\chi(z)$ is a solution, then $\chi(1 - z)$ is also a solution (see also Example 5.8.2). We use $w = \frac{\hat{\chi}_2}{\hat{\chi}_1}$, where $\hat{\chi}_1$ and $\hat{\chi}_2$ are two linearly independent solutions of the above hypergeometric equation, specifically,

$$\hat{\chi}_2 = \chi(1 - z), \qquad \hat{\chi}_1 = \chi(z),$$

where χ is given by Eq. (5.8.16b) with $\alpha = \beta = \gamma = 0$, $a = b = \frac{1}{2}$, $c = 1$; that is,

$$\chi(z) = \int_0^1 \frac{dt}{\sqrt{t(1-t)(1-zt)}}.$$

Using the change of variables $t = \tau^2$, as well as $z = k^2$, we find

$$w = f(z) = \frac{\chi(1-z)}{\chi(z)} = \frac{K'(k)}{K(k)}, \qquad z = k^2,$$

where the functions K' and K are defined as in Example 5.6.9. The variable k is usually referred to as the modulus of the elliptic function $z = F^{-1}(w, k) = \mathrm{sn}(w, k)$. Because of this connection, the *inverse* of the function $w = K'(k)/K(k)$ is often referred to as the **elliptic modular function**. This function is usually denoted by J. It is actually customary to give this name to the inverse of iK'/K (note the extra factor of i, which is a standard normalization):

$$w = s(z; 0, 0, 0) = \frac{iK'(k)}{K(k)} \Rightarrow z = k^2 = J(w) = S(w; 0, 0, 0). \qquad (5.8.24)$$

It is useful to determine the location of the "zero angle triangle" in the w-plane consistent with Eq. (5.8.24), which we will see has degenerated into a strip. Note that when $z = 0$, $K(0) = \pi/2$ and $K'(0) = \infty$, and when $z = 1$, $K(1) = \infty$ and $K'(1) = 0$. As z increases from $z = 0$ to $z = 1$, w decreases along the imaginary axis from $i\infty$ to 0. We know that the angles of the "triangle" are zero so at the vertex corresponding to $z = 0$ we have a half circle, cutting out a piece of the upper half w-plane, which begins at the origin and intersects the real w-axis somewhere to the right of the origin. The fundamental triangle must lie in the first quadrant to be consistent with an orientation that has the upper half z-plane to the left as we proceed from $z = 0$ to $z = 1$ to $z = \infty$. Finally, the last critical point has value $w(z = \infty) = s(\infty; 0, 0, 0) = 1$ (see Figure 5.57). We could determine this from the properties of the hypergeometric equation, or from the following. From Eq. (5.8.24) we have the relationship

$$s(z; 0, 0, 0)\, s(1 - z; 0, 0, 0) = -1. \qquad (5.8.25)$$

Writing $s(z; 0, 0, 0) = s(z)$, for simplicity of notation, and letting $z = \frac{1}{2} + iy$, the Schwarz reflection principle about $y = 0$ implies $s(\frac{1}{2} + iy) = -\bar{s}(\frac{1}{2} - iy)$ because $s(z)$ is pure imaginary for $0 < z < 1$; hence Eq. (5.8.25) yields

$$\left| s\left(\frac{1}{2} + iy \right) \right|^2 = 1$$

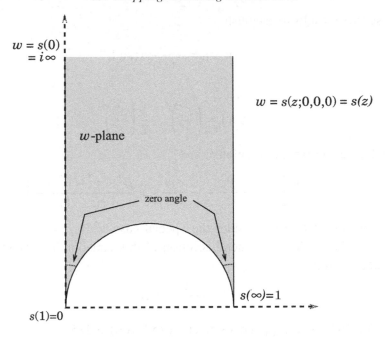

$w = s(0) = i\infty$

$w = s(z;0,0,0) = s(z)$

w-plane

zero angle

$s(\infty)=1$

$s(1)=0$

Figure 5.57 Fundamental domain of a "zero angle triangle" for Example 5.8.4

and for $y \to \infty$, $|s(\infty)| = 1$. But $s(\infty)$ must lie on the positive real axis, whereupon $s(\infty) = 1$.

Successive inversions (i.e., reflections) about the z-axis correspond to inversions of the fundamental strip in the w-plane. It turns out that after an infinite number of inversions we fill the entire upper half w-plane (see Nehari (1952) for a more detailed discussion). In this formulation the real w-axis is a natural boundary; that is, the orthogonal circle described in the previous example is now the real w-axis.

5.8.1 Problems for Section 5.8

1. In this problem we study the equation

$$\{w, z\} = \frac{1 - \beta^2}{2z^2} + \frac{1 - \gamma^2}{2(z - 1)^2} + \frac{\beta^2 + \gamma^2 - \alpha^2 - 1}{2z(z - 1)}$$

(cf. Equation (5.8.12) where $w = f(z)$). Frequently it is useful to consider $z = z(w)$ instead of $w = w(z)$.

(a) Show that

$$\frac{d}{dz} = \frac{1}{z'} \frac{d}{dw}, \qquad \frac{d^2w}{dz^2} = -\frac{z''}{z'^3},$$

where $z' \equiv \frac{dz}{dw}$.

(b) Use these results to establish

$$\{w, z\} = -\frac{1}{(z')^2}\{z, w\},$$

where

$$\{z, w\} = \left(\frac{z''}{z'}\right)' - \frac{1}{2}\left(\frac{z''}{z'}\right)^2,$$

and hence derive the equation

$$\{z, w\} + \frac{(z')^2}{2}\left(\frac{1 - \beta^2}{z^2} + \frac{1 - \gamma^2}{(z - 1)^2} + \frac{\beta^2 + \gamma^2 - \alpha^2 - 1}{z(z - 1)}\right) = 0.$$

2. In the previous problem consider the special case $\alpha = \beta = \gamma = 1$ so that $\{z, w\} = 0$. Show that the general solution of this equation, and hence of the mapping, is given by

$$z = \frac{A}{w - w_0} + B.$$

3. Using (5.8.16d), where χ_1 and χ_2 given by (5.8.16b) and (5.8.16c) respectively, show that the vertices with the $\pi\beta$ and $\pi\gamma$ (corresponding to $z = 0$ and $z = 1$) are mapped to

$$w_0 = \frac{\sin \pi\beta}{\cos\left[\frac{\pi}{2}(\alpha - \beta + \gamma)\right]} \quad \text{and} \quad w_1 = \frac{\cos\left[\frac{\pi}{2}(\alpha + \beta - \gamma)\right]}{\sin \pi\gamma}.$$

Hint: Use the identities

$$\int_0^1 t^{r-1}(1 - t)^{s-1}dt = \frac{\Gamma(r)\Gamma(s)}{\Gamma(r + s)}, \qquad r > 0, \ s > 0,$$

and

$$\Gamma(r)\Gamma(1 - r) = \frac{\pi}{\sin \pi r}.$$

4. If $\beta = \gamma$, show that the function $f(z) = \chi_1/\chi_2$ where χ_1 and χ_2 are given by Equations (5.8.16b) and (5.8.16c) respectively, satisfy the functional equation

$$f(z)f(1 - z) = 1.$$

5. Consider the crescent-shaped region shown in Figure 5.58.

(a) Show that in this case Equation (5.8.6) reduces to

$$\{f, z\} = \frac{(1 - \alpha^2)(a - b)^2}{2(z - a)^2(z - b)^2},$$

where a and b are the points on the real axis associated with the vertices.

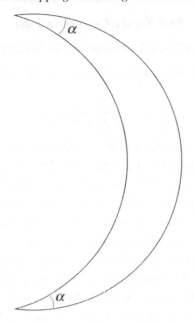

Figure 5.58 Crescent region, Problem 5

(b) Show that the associated linear differential equation (see (5.8.13)) is

$$y'' + \frac{(1 - \alpha^2)(a - b)^2}{4(z - a)^2(z - b)^2} y = 0.$$

(c) Show that the above equation is equivalent to the differential equation

$$g'' + (1 - \alpha)\left[\frac{1}{z - a} + \frac{1}{z - b}\right]g' - \frac{\alpha(1 - \alpha)}{(z - a)(z - b)}g = 0,$$

which admits $(z - a)^\alpha$ and $(z - b)^\alpha$ as particular solutions.

Hint: show that $y(z) = [(z - a)(z - b)]^{(1-\alpha)/2}g(z)$.

(d) Deduce that

$$f(z) = \frac{c_1(z - a)^\alpha + c_2(z - b)^\alpha}{c_3(z - a)^\alpha + c_4(z - b)^\alpha},$$

where $c_1, \ldots, c_4$ are constants, for which $c_1 c_4 \neq c_2 c_3$.

*5.9 Other Considerations

*5.9.1 Rational Functions of the Second Degree

The most general rational function of the second degree is of the form

$$f(z) = \frac{az^2 + bz + c}{a'z^2 + b'z + c'}, \tag{5.9.1}$$

where a, b, c, a', b', c' are complex numbers. This function remains invariant if both the numerator and the denominator are multiplied by a nonzero constant; therefore $f(z)$ depends only on five arbitrary constants. The equation $f(z) - w_o = 0$ is of second degree in z, which shows that under the transformation $w = f(z)$, every value w_o is taken twice. This means that this transformation maps the complex z-plane onto the doubly covered w-plane, or equivalently that it maps the z-plane onto a two-sheeted Riemann surface whose two sheets cover the entire w-plane. The branch points of this Riemann surface are those points w which are common to both sheets. These points correspond to points z such that either $f'(z) = 0$ or $f(z)$ has a double pole. From Eq. (5.9.1) we can see that there exist precisely two such branch points. We distinguish two cases:

(a) $f(z)$ has a double pole; that is, $w = \infty$ is one of the two branch points.

(b) $f(z)$ has two finite branch points.

It will turn out that in the case (a), $f(z)$ can be decomposed into two successive transformations, a bilinear one, and one of the type z^2 + constant; in the case (b), $f(z)$ can be decomposed into three successive transformations, a linear one, a bilinear one, and one of the type $z + 1/z$.

We first consider the case (a). Let $w = \infty$ and $w = \lambda$ be the two branch points of $w = f(z)$, and let $z = z_1$ and $z = z_2$ be the corresponding points in the z-plane. The expansions of $f(z)$ near these points are of the form

$$f(z) = \frac{\alpha_{-2}}{(z - z_1)^2} + \frac{\alpha_{-1}}{(z - z_1)} + \alpha_0 + \alpha_1(z - z_1) + \cdots, \qquad \alpha_{-2} \neq 0,$$

and

$$f(z) - \lambda = \beta_2(z - z_2)^2 + \beta_3(z - z_2)^3 + \cdots, \qquad \beta_2 \neq 0,$$

respectively. The function $(f(z) - \lambda)^{1/2}$, takes no value more than once (because $f(z)$ takes no value more than twice), and its only singularity in the *entire* z-plane is a simple pole at $z = z_1$. Hence from Liouville's Theorem this function must be of the bilinear form (5.7.1). Therefore

$$f(z) = \lambda + \left(\frac{Az + B}{Cz + D}\right)^2; \tag{5.9.2}$$

that is,

$$w = \lambda + z_1^2, \qquad z_1 \equiv \frac{Az + B}{Cz + D}. \tag{5.9.3}$$

We now consider the case (b). Write $w = \lambda$ and $w = \mu$ for the two finite branch points. Using a change of variables from $f(z)$ to $g(z)$ these points can be normalized to be at $g(z) = \pm 1$, hence

$$f(z) = \frac{\lambda - \mu}{2} g(z) + \frac{\lambda + \mu}{2}. \tag{5.9.4}$$

Let $z = z_1$ and $z = z_2$ be the points in the z-plane corresponding to the branch points λ and μ, respectively. Series expansions of $g(z)$ near these points are of the form

$$g(z) - 1 = \alpha_2(z - z_1)^2 + \alpha_3(z - z_1)^3 + \cdots, \qquad \alpha_2 \neq 0,$$

and

$$g(z) + 1 = \beta_2(z - z_2)^2 + \beta_3(z - z_2)^3 + \cdots, \qquad \beta_2 \neq 0,$$

respectively. The function $f(z)$ has two simple poles: therefore the function $g(z)$ also has two simple poles, which we shall denote by $z = \zeta_1$ and $z = \zeta_2$. Using a change of variables from $g(z)$ to $h(z)$, it is possible to construct a function that has only one simple pole

$$g(z) = \frac{1}{2}\left(h(z) + \frac{1}{h(z)}\right). \tag{5.9.5}$$

Indeed, the two poles of $g(z)$ correspond to

$$h(z) = \gamma(z - \zeta_1)[1 + c_1(z - \zeta_1) + \cdots]$$

and to

$$h(z) = \delta(z - \zeta_2)^{-1}[1 + d_1(z - \zeta_2) + \cdots];$$

that is, they correspond to one zero and one pole of $h(z)$. Furthermore, the expansions of $g(z)$ near ± 1, together with Eq. (5.9.5) imply that $h(z)$ is regular at the points $z = z_1$ and $z = z_2$. The only singularity of $h(z)$ in the entire z-plane is a pole, hence $h(z)$ must be of the bilinear form (5.7.1). Renaming functions and constants, Eqs. (5.9.4) and (5.9.5) imply

$$w = A'\zeta_2 + B', \qquad \zeta_2 = \frac{1}{2}\left(\zeta_1 + \frac{1}{\zeta_1}\right), \qquad \zeta_1 = \frac{Az + B}{Cz + D}. \tag{5.9.6}$$

The important consequence of the above discussion is that the study of the transformation (5.9.1) reduces to the study of the bilinear transformation (which was discussed in Section 5.7) of the transformation $w = z^2$ and of the transformation $w = (z + z^{-1})/2$.

Let us consider the transformation

$$w = z^2; \quad w = u + iv, \ z = x + iy; \quad u = x^2 - y^2, \quad v = 2xy. \quad (5.9.7)$$

Example 5.9.1 Find the curves in the z-plane that, under the transformation $w = z^2$, give rise to horizontal lines in the w-plane.

Because horizontal lines in the w-plane are $v =$ constant, it follows that the relevant curves in the z-plane are the hyperbolae $xy =$ constant. We note that because the lines $u =$ constant are orthogonal to the lines $v =$ constant, it follows that the family of the curves $x^2 - y^2 =$ constant is orthogonal to the family of the curves $xy =$ constant. (Indeed, the vectors obtained by taking the gradient of the functions $F_1(x, y) = (x^2 - y^2)/2$ and $F_2(x, y) = xy$, $(x, -y)$, and (y, x), are perpendicular to these curves, and clearly these two vectors are orthogonal).

Example 5.9.2 Find the curves in the z-plane that, under the transformation $w = z^2$, give rise to circles in the w-plane.

Let $c \neq 0$ be the center and R be the radius of the circle. Then $|w - c| = R$, or if we call $c = C^2$, then $w = z^2$ implies

$$|z - C||z + C| = R. \quad (5.9.8)$$

Hence, the images of circles are the loci of points whose distances from two fixed points have a constant product. These curves are called **Cassinians**. The cases of $R > |C|^2$, $R = |C|^2$, and $R < |C|^2$ correspond to one closed curve, to the lemniscate, and to two separate closed curves, respectively. These three cases are depicted in Figure 5.59, when C is real. Otherwise we obtain a rotation of angle θ when $C = |C|e^{i\theta}$.

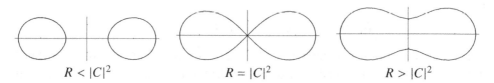

$$R < |C|^2 \qquad\qquad R = |C|^2 \qquad\qquad R > |C|^2$$

Figure 5.59 Cassinians from equation (5.9.8) for Example 5.9.2

We now consider the transformation

$$w = \frac{1}{2}\left(z + \frac{1}{z}\right); \quad u = \frac{1}{2}\left(r + \frac{1}{r}\right)\cos\theta, \quad v = \frac{1}{2}\left(r - \frac{1}{r}\right)\sin\theta, \quad (5.9.9)$$

where $z = r\exp(i\theta)$.

Example 5.9.3 Find the image of a circle centered at the origin in the z-plane under the transformation (5.9.9).

Let $r = \rho$ be a circle in the z-plane. Equation (5.9.9) implies

$$\frac{u^2}{\left[\frac{1}{2}(\rho + \rho^{-1})\right]^2} + \frac{v^2}{\left[\frac{1}{2}(\rho - \rho^{-1})\right]^2} = 1, \qquad \rho \text{ a constant.}$$

This shows that the transformation (5.9.9) maps the circle $r = \rho$ onto the ellipse of semiaxes $(\rho + \rho^{-1})/2$ and $(\rho - \rho^{-1})/2$ as depicted in Figure 5.60. Because

$$\frac{1}{4}(\rho + \rho^{-1})^2 - \frac{1}{4}(\rho - \rho^{-1})^2 = 1,$$

all such ellipses have the same foci located on the u-axis at ± 1.

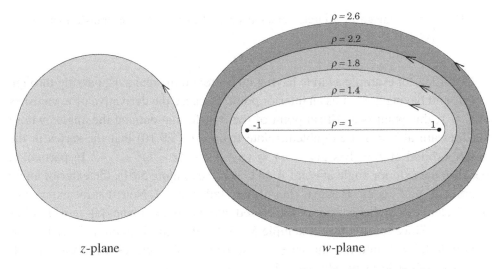

Figure 5.60 Transformation of circle onto ellipse for Example 5.9.3

The circles $r = \rho$ and $r = \rho^{-1}$ yield the same ellipse; if $\rho = 1$, the ellipse degenerates into the linear segment connecting $w = 1$ and $w = -1$. We note that because the ray $\theta = \varphi$ is orthogonal to the circle $r = \rho$, the above ellipses are orthogonal to the family of hyperbolae

$$\frac{u^2}{\cos^2 \varphi} - \frac{v^2}{\sin^2 \varphi} = 1, \qquad \varphi \text{ a constant,}$$

which are obtained from Eq. (5.9.9) by eliminating r.

Example 5.9.4 (Joukowski Profiles)   The transformation

$$w = \frac{1}{2}\left(z + \frac{1}{z}\right) \tag{5.9.10}$$

arises in certain aerodynamical applications. This is because it maps the exterior of circles onto the exterior of curves which have the general character of airfoils.

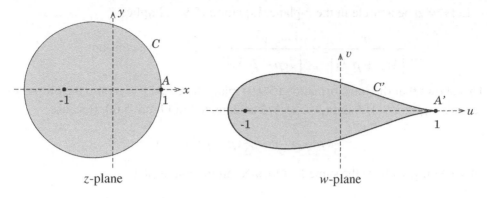

Figure 5.61 Image of a circle centered on real axis, under the transformation $w = \frac{1}{2}(z + 1/z)$

Consider, for example a circle having its center on the real axis, passing through $z = 1$, and having $z = -1$ as an interior point. Because the derivative of w vanishes at $z = 1$, this point is a critical point of the transformation, and the angles whose vertices are at $z = 1$ are doubled. (Note from Eq. (5.9.10) that the series in the neighborhood of $z = 1$ is $2(w - 1) = (z - 1)^2 - (z - 1)^3 + \cdots$.) In particular, because the exterior angle at point A on C is π (see Figure 5.61), the exterior angle at point A' on C' is 2π. Hence C' has a sharp tail at $w = 1$. Note that the exterior of the circle maps to the exterior of the closed curve in the w-plane; $|z| \to \infty$ implies $|w| \to \infty$. Since we saw from Example 5.9.3 that the transformation (5.9.10) maps the circle $|z| = 1$ onto the slit $|w| \le 1$, and because C encloses the circle $|z| = 1$, the curve C' encloses the slit $|w| \le 1$.

Suppose that the circle C is translated vertically so that it still passes through $z = 1$ and encloses $z = -1$, but its center is in the upper half plane. Using the same argument as above, the curve C' still has a sharp tail at A' (see Figure 5.62). But because C is not symmetric about the x-axis, we can see from Eq. (5.9.10) that C' is not symmetric about the u-axis. Furthermore, since C does not entirely enclose the circle $|z| = 1$, the curve C' does not entirely enclose the slit $|w| \le 1$. A typical shape of C' is shown in Figure 5.62.

By changing C appropriately, other shapes similar to C' can be obtained. We note that C' resembles the cross section of the wing of an airplane, usually referred to as an airfoil.

*5.9.2 The Modulus of a Quadrilateral

Let Γ be a positively oriented Jordan curve (i.e., a simple closed curve), with four distinct points a_1, a_2, a_3, and a_4 being given on Γ, arranged in the direction of

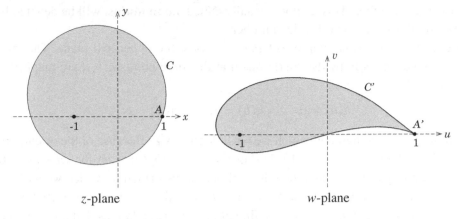

z-plane   w-plane

Figure 5.62 Image of a circle whose center is above real axis

increasing parameters. Let the interior of Γ be called Q. The system $(Q; a_1, a_2, a_3, a_4)$ is called a quadrilateral (see Figure 5.63).

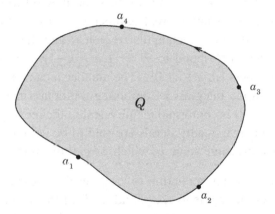

Figure 5.63 Quadrilateral

Two quadrilaterals $(Q; a_1, \ldots, a_4)$ and $(\tilde{Q}; \tilde{a}_1, \ldots, \tilde{a}_4)$ are called conformally equivalent if there exists a conformal map, f, from Q to $\tilde{Q}$ such that $f(a_i) = \tilde{a}_i$, $i = 1, \ldots, 4$.

If one considers trilaterals instead of quadrilaterals – that is, if one fixes three instead of four points – then one finds that all trilaterals are conformally equivalent. Indeed, it follows from the proof of the Riemann Mapping Theorem that in a conformal mapping any three points on the boundary can be chosen arbitrarily (this fact was used in Sections 5.6–5.8).

Not all quadrilaterals are conformally equivalent. It turns out that the equivalence class of conformally equivalent quadrilaterals can be described in terms of a single positive real number. This number, usually called the **modulus**, will be denoted by μ. We shall now characterize this number.

Let h be a conformal map of $(Q; a, b, c, d)$ onto the upper half plane. This map can be fixed uniquely by the conditions that the three points a, b, d are mapped to $0, 1, \infty$; that is,

$$h(a) = 0, \qquad h(b) = 1, \qquad h(d) = \infty.$$

Then, $h(c)$ is some number, which we shall denote by ξ. Because of the orientation of the boundary, $1 < \xi < \infty$. By letting $\tilde{z} = (az + b)/(cz + d)$ for $ad - bc \neq 0$, we can directly establish that there is a bilinear transformation, which we will call g, that maps the upper half plane onto itself, such that the points $z = \{0, 1, \xi, \infty\}$ are mapped to $\tilde{z} = \{1, \eta, -\eta, -1\}$. We find after some calculations that η is uniquely determined from ξ by the equation

$$\frac{\eta + 1}{\eta - 1} = \sqrt{\xi}, \qquad \eta > 1.$$

Recall the Example 5.6.9. We can follow the same method to show that for any given $\eta > 1$, there exists a unique real number $\mu > 0$ such that the upper half z-plane can be mapped onto a rectangular region R and the image of the points $z = \{0, 1, \xi, \infty\}$ which correspond to $\tilde{z} = \{1, \eta, -\eta, -1\}$, can be mapped to the rectangle with the corners $\{\mu, \mu + i, i, 0\}$. (The number μ can be expressed in terms of η by means of elliptic integrals.) Combining the conformal maps h and g, it follows that $(Q; a, b, c, d)$ is conformally equivalent to the rectangular quadrilateral $(R; \mu, \mu + i, i, 0)$. Thus two quadrilaterals are said to be conformally equivalent if and only if they have the same value μ, which we call the modulus.

Example 5.9.5 (Physical Interpretation of μ)   Let Q denote a sheet of metal of unit conductivity. Let the segments (a, b) and (c, d) of the boundary be kept at the potentials V and 0, respectively, and let the segments (b, c) and (d, a) be insulated. Establish a physical meaning for μ.

From electromagnetics, the current I flowing between a, b is given by (see also Example 5.4.3)

$$I = \int_a^b \frac{\partial \Phi}{\partial n} \, ds,$$

where $\partial/\partial n$ denotes differentiation in the direction of the exterior normal, and Φ is the potential. The function Φ is obtained from the solution of (see Figure 5.64)

$$\nabla^2 \Phi = 0 \text{ in } Q, \qquad \Phi = V \text{ on } (a, b), \qquad \Phi = 0 \text{ on } (c, d),$$

$$\frac{\partial \Phi}{\partial n} = 0 \text{ on } \quad (b, c) \text{ and } (d, a).$$

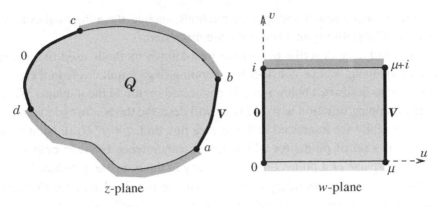

Figure 5.64 Electric current through the sheet Q for Example 5.9.5

In the w-plane we can verify the following solution for the complex potential: $\Omega(w) = \frac{V}{\mu}w$. From the definition of the potential (5.4.11), $\Omega = \Phi + i\Psi$, so

$$\Phi = \operatorname{Re}\Omega = \frac{V}{\mu}\operatorname{Re} w = \frac{V}{\mu}u.$$

At the top and bottom we have $\frac{\partial \Phi}{\partial v} = 0$; on $u = 0$, $\Phi = 0$; and for $u = \mu$, $\Phi = V$. Hence we have verified that the solution of this problem in the w-plane is given by $\Phi = V\mu^{-1}\operatorname{Re} w$. Furthermore, we know that (see Eq. (5.4.14)) the integral I is invariant under a conformal transformation. Computing this integral in the w-plane, we find

$$I = \mu^{-1}V \quad \text{or} \quad V = \mu I.$$

Therefore μ has the physical meaning of the resistance of the sheet Q between (a, b) and (c, d) when the remaining parts of the boundary are insulated.

*5.9.3 Computational Issues

Even though Riemann's Mapping Theorem guarantees that there exists an analytic function that maps a simply connected domain onto a circle, the proof is not constructive and does not give insight into the determination of the mapping function. We have seen that conformal mappings have wide physical application, and, in practice, the ability to map a complicated domain onto a circle, the upper half plane, or indeed another simple region is desirable. Towards this end, various computational methods have been proposed and this is a field of current research interest. It is outside the scope of this book to survey the various methods or even all of the research directions. Many of the well-known methods are discussed in the books of Henrici, and we also note the collection of papers (Trefethen 1986) where other reviews can be found and specific methods, such as the numerical evaluation of Schwarz–Christoffel transformations, are discussed.

Here we will only describe one of the well-known methods used in numerical conformal mapping. Let us consider the mapping from a unit circle in the z-plane to a suitable (as described below) simply connected region in the w-plane. We wish to find the mapping function $w = f(z)$ that will describe the conformal mapping. In practice, we really are interested in the *inverse* function, $z = f^{-1}(w)$. Numerically, we determine a set of points for which the correspondence between points on the circle in the z-plane and points on the boundary in the w-plane is deduced.

We assume that the boundary C in the w-plane is a Jordan curve that can be represented in terms of polar coordinates, $w = f(z) = \rho e^{i\theta}$, where $\rho = \rho(\theta)$, and we impose the conditions $f(0) = 0$ and $f'(0) > 0$ (Riemann's Mapping Theorem allows us this freedom) on the unit circle $z = e^{i\varphi}$. The mapping fixes, in principle, $\theta = \theta(\varphi)$ and $\rho = \rho(\theta(\varphi))$. The aim is to determine the **boundary correspondence points**; that is, how points in the z domain, $\varphi = \{\varphi_1, \varphi_2, \ldots, \varphi_N\}$, transform to points in the w domain which is parametrized by $\theta = \{\theta_1, \theta_2, \ldots, \theta_N\}$ (see Figure 5.65).

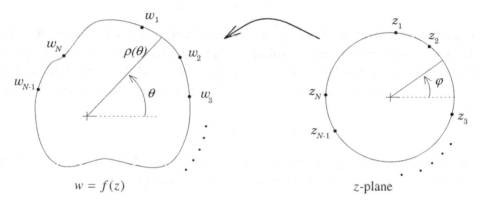

Figure 5.65 Boundary correspondence points

The method we describe involves the numerical solution of a nonlinear integral equation. This equation is a modification of a well-known formula (derived see Problem 12 of Section 2.6.4, and Problem 15 of Section 4.3.3) that relates the boundary values (on $|z| = 1$) of the real and imaginary parts of a function analytic inside the circle. Specifically, consider $F(z) = u(x, y) + iv(x, y)$, which is analytic inside the circle $z = re^{i\varphi}$, for $r < 1$. Then on the circle $r = 1$ the following equation relating u and v holds:

$$v(\varphi) = v(r = 0) + \frac{1}{2\pi} \int_0^{2\pi} u(t) \cot\left(\frac{\varphi - t}{2}\right) dt, \qquad (5.9.11)$$

where the integral is taken as a Cauchy principal value (we reiterate that both t and φ correspond to points on the unit circle). The integral equation we shall consider is derived from Eq. (5.9.11) by considering $F(z) = \log(f(z)/z)$, recalling that $f(0) = 0$, $f'(0) > 0$. Then, using the polar coordinate representation $f(z) = \rho e^{i\theta}$, we see that $F(z) = \log \rho + i(\theta - \varphi)$; hence in Eq. (5.9.11) we take $u = \log \rho(\theta)$. We require that $f'(0)$ be real, thus $v(r = 0) = 0$. On the circle, $v = \theta - \varphi$; this yields

$$\theta(\varphi) = \varphi + \frac{1}{2\pi} \int_0^{2\pi} \log \rho\,(\theta(t)) \cot\left(\frac{\varphi - t}{2}\right) dt. \tag{5.9.12}$$

Equation (5.9.12) is called **Theodorsen's integral equation**. The goal is to solve Eq. (5.9.12) for $\theta(\phi)$. Unfortunately, it is nonlinear and cannot be solved in closed form, though a unique solution can be proven to exist. Consequently an approximation (i.e., numerical) procedure is used. The methods are effective when $\rho(\theta)$ is smooth and $|\rho'(\theta)/\rho(\theta)|$ is sufficiently small.

Equation (5.9.12) is solved by functional iteration:

$$\theta^{(n+1)}(\varphi) = \varphi + \frac{1}{2\pi} \int_0^{2\pi} \log \rho\left(\theta^{(n)}(t)\right) \cot\left(\frac{\varphi - t}{2}\right) dt, \tag{5.9.13}$$

where the function $\theta^{(0)}$ is a starting "guess."

Numerically speaking, Eq. (5.9.13) is transformed to a matrix equation; the integral is replaced by a sum, and $\log \rho(\theta)$ is approximated by a finite Fourier series – that is, a trigonometric polynomial – because $\rho(\theta)$ is periodic. Then, corresponding to $2N$ equally spaced points (roots of unity) on the unit circle in the z-plane (t: $\{t_1, \ldots, t_{2N}\}$ and φ: $\{\varphi_1, \ldots, \varphi_{2N}\}$), one solves, by iteration, the matrix equation associated with Eq. (5.9.13) to find the set $\theta^{(n+1)}(\varphi_j)$, $j = 1, 2, \ldots, 2N$, which corresponds to an initial guess $\theta^{(0)}(\varphi_j) = \varphi_j$, $j = 1, 2, \ldots, 2N$, which are equally spaced points on the unit circle. As n is increased enough, the iteration converges to a solution that we call $\hat{\theta}$: $\theta^{(n)}(\varphi_j) \to \hat{\theta}(\varphi_j)$. These points are the boundary correspondence points. Even though the governing matrix is $2N \times 2N$ and ordinarily the "cost" of calculation is $O(N^2)$ operations, it turns out that special properties of the functions involved are such that fast Fourier algorithms are applicable, and the number of operations can be reduced to $O(N \log N)$.

Further details on this and related methods can be found in Henrici (1977), and articles by Gaier (1983), Fornberg (1980), and Wegmann (1988).

Appendix

Answers to Selected Odd-numbered Exercises

Chapter 1

Section 1.1

1(a). $e^{2\pi i k}$, $\quad k = 0, \pm1, \pm2, \ldots;$

1(b). $e^{(3\pi i/2)+2\pi i k}$, $\quad k = 0, \pm1, \pm2, \ldots;$

1(c). $\sqrt{2}e^{i(\pi/4)+2k\pi i}$, $\quad k = 0, \pm1, \pm2, \ldots;$

1(d). $e^{i(\pi/3)+2k\pi i}$, $\quad k = 0, \pm1, \pm2, \ldots;$

1(e). $e^{i(-\pi/3)+2k\pi i}$, $\quad k = 0, \pm1, \pm2, \ldots.$

3(a). $z_\kappa = 4^{1/3}e^{2\pi i \kappa/3}$, $\quad \kappa = 0, 1, 2;$

3(b). $z_\kappa = e^{i(\pi+2\kappa\pi)/4}$, $\quad \kappa = 0, 1, 2, 3;$

3(c). $z_\kappa = (c^{1/3}e^{2\pi i\kappa/3} - b)/a$, $\quad \kappa = 0, 1, 2;$

3(d). $z_1 = 2^{1/4}e^{i3\pi/8}$, $\quad z_2 = 2^{1/4}e^{i11\pi/8}$, $\quad z_3 = 2^{1/4}e^{5i\pi/8}$, $\quad z_4 = 2^{1/4}e^{13i\pi/8}.$

Section 1.2

1(a). Compact; 1(b). open, bounded;

1(c). Closed, unbounded;

1(d). Closed, unbounded;

1(e). Bounded, not open or closed;

5(a). $\displaystyle\sum_{j=0}^{\infty} \frac{(-1)^j z^{2j}}{(2j+1)!}$;   5(b). $\displaystyle\sum_{j=0}^{\infty} \frac{z^{2j}}{(2j+2)!}$;   5(c). $\displaystyle\sum_{j=0}^{\infty} \frac{z^{j+1}}{(j+2)!}$.

11. Circle passing through the north pole.

Section 1.3

1(a). 0;   1(b). $(1/z_o)^m$;   1(c). $i\sin(1)$; 1(d). 1;   1(e). Doesn't exist;

1(f). 1/9;   1(g). 0.

3(a). Differentiable for all z s.t. $|z| < \infty$;   3(b). For all z except $z = \frac{\pi}{2} + \kappa\pi$, $\kappa \in \mathbf{Z}$;

3(c). For all z except $z = \pm i$;   3(d). For all z except $z = 0$;

3(e). Not differentiable anywhere.

Section 1.4

1. $w = e^{-\alpha t}(A_1 \cos \mu t + A_2 \sin \mu t)$, $\mu = \sqrt{\omega_0^2 - \alpha^2}$, A_1, A_2 const.

Chapter 2

Section 2.1

1(a). C–R not satisfied;   1(b). C–R satisfied; $f(z) = i(z^3 + 2)$;

1(c). C–R not satisfied.

3(a). Analytic except at singular points $z = (\pi/2 + n\pi)$, $n \in \mathbf{Z}$;

3(b). Entire; singularity at ∞;   3(c). Analytic except at singular point $z = 1$;

3(d). Analytic nowhere;

3(e). Analytic except at $z = \exp(i(\pi/4 + n\pi/2))$, $n = 0, 1, 2, 3$

3(f). Entire; singularity at ∞.

Section 2.2

1(a). Branch points at $z = 1, \infty$. Possible branch cuts:

(i) $\{z : \text{Re } z \geq 1, \text{ Im } z = 0\}$ obtained by letting $z - 1 = re^{i\theta}, 0 \leq \theta < 2\pi$;

(ii)$\{z : \text{Re } z \leq 1, \text{ Im } z = 0\}$ obtained by letting $z - 1 = re^{i\theta}, -\pi \leq \theta < \pi$;

1(b). Branch points at $z = -1 + 2i, \infty$. Possible branch cut:

$\{z : \text{Re } z \geq -1, \text{ Im } z = 2\}$ obtained by letting $z + 1 - 2i = re^{i\theta}, 0 \leq \theta < 2\pi$;

1(c). Branch points at $z = 0, \infty$. Possible branch cut:

$\{z : \text{Re } z \geq 0, \text{ Im } z = 0\}$ obtained by letting $z = re^{i\theta}, 0 \leq \theta < 2\pi$;

1(d). Branch points at $z = 0, \infty$. Possible branch cut:

$\{z : \text{Re } z \geq 0, \text{ Im } z = 0\}$ obtained by letting $z = re^{i\theta}, 0 \leq \theta < 2\pi$.

3(a). $z = e^{2\pi in/5}, n = 0, 1, 2, 3, 4$;   3(b). $z = i(1 + (2n + 1)\pi), \; n = \text{integer}$;

3(c). $z = \pi/4 + n\pi, \; n = \text{integer}$;   3(d). $z = e^{i\pi/2}$.

7. $\begin{aligned} \phi &= \kappa \log \rho \\ \psi &= \kappa\theta \end{aligned}$ where $z - z_o = \rho e^{i\theta}$

Section 2.3

1(a). Branch points at $z = +i$.

Branch cut: $\{z : \text{Re} z = 0, \; -1 \leq \text{Im} z \leq 1\}$ obtained by letting

$z - i = r_1 e^{i\theta_1}$ and $z + i = r_2 e^{i\theta_2}$ for $-3\pi/2 \leq \theta_1, \theta_2 < \pi/2$.

1(b). Branch points at $z = -1, 2, \infty$.

Branch cut: $\{z : \text{Re} z \geq -1, \text{ Im} z = 0\}$ obtained by letting

$z + 1 = r_1 e^{i\theta_1}$ and $z - 2 = r_2 e^{i\theta_2}$ for $0 \leq \theta_1, \theta_2 < 2\pi$.

3. Branch points at $z = \pm i, \infty$.

Branch cut: $\{z : \text{Re} z = 0, \text{ Im} z \leq 1\}$ obtained by letting

$z + i = r_1 e^{i\theta_1}$ and $z - i = r_2 e^{i\theta_2}$ for $-\pi/2 \leq \theta_1, \theta_2 < 3\pi/2$.

Section 2.4

1(a). 0;   1(b). 0;   1(c). $2\pi i$.

3(a). 0;   3(b). πi;   3(c). $8i$;   3(d). $4i$.

11(b). $\Gamma = 0, F = 2\pi\kappa$.

Section 2.5

1(a). 0;   1(b). 0;   1(c). $2\pi i$;

1(d). 0;   1(e). 0;   1(f). 0.

3(a). 0;   3(b). $-2i$;   3(c). $-4i$;   3(d). $-2i$.

Section 2.6

1(a). 0; 1(b). 0; 1(c). 0; 1(d). $2\pi i$; 1(e). $-2\pi i$.

3. 0.

Chapter 3

Section 3.1 "UC" means uniformly convergent.

1(a). $\lim_{n\to\infty} n/z^2 = 0$: UC; 1(b). $\lim_{n\to\infty} 1/z^n = 0$ for $1 < \alpha \le |z| \le \beta$: UC;

1(c). $\lim_{n\to\infty} \sin(z/n) = 0$: UC; 1(d). $\lim_{n\to\infty} \frac{1}{1+(nz)^2} = 0$: UC.

3. $\lim_{n\to\infty} \int_0^1 nz^{n-1}dz = 1$; $\int_0^1 \lim_{n\to\infty} nz^{n-1}dz = 0$; not a counterexample because convergence is not uniform.

Section 3.2

1(a). $R = 1$; 1(b). $R = \infty$; 1(c). $R = 0$; 1(d). $R = \infty$; 1(e). $R = e$.

3(a). $R = \pi/2$;

3(b). $E_0 = 1$, $E_1 = 0$, $E_2 = -1$, $E_3 = 0$, $E_4 = 5$, $E_5 = 0$, $E_6 = -61$.

7. $\log(1 + z) = \sum_{k=0}^{\infty} \frac{(-1)^k z^{k+1}}{k + 1}$.

9. $\sum_{k=m-1}^{\infty} \frac{k(k - 1)\cdots(k - (m - 2))z^{k-(m-1)}}{(m - 1)!}$.

Section 3.3

1(a). $\sum_{n=0}^{\infty}(-1)^n z^{2n}$; 1(b). $\sum_{n=0}^{\infty} \frac{(-1)^n}{z^{2n+2}}$.

3(a). $\left(\frac{2}{5} - \frac{i}{5}\right) \sum_{n=0}^{\infty} \frac{(2i)^n - 1}{2^n} z^n$; 3(b). $\frac{-2 + i}{5} \sum_{n=0}^{\infty} \left(\left(\frac{z}{2}\right)^n - (-1)^n \left(\frac{i}{z}\right)^{n+1}\right)$;

3(c). $\frac{2 - i}{5} \sum_{n=0}^{\infty} \left(\left(\frac{2}{z}\right)^{n+1} + (-1)^n \left(\frac{i}{z}\right)^{n+1}\right)$.

5(a). $\frac{1}{(z - 1)^2} + (z - 1) + 1$; 5(b). $\sum_{0}^{\infty} \frac{(-1)^n}{(z - 1)^{2n+1}(2n + 1)!}$.

9. $\left(z - \frac{1}{2} - \frac{1}{8z} + \cdots\right)$.

Section 3.4

1(a). Yes; 1(b). Yes; 1(c). No; 1(d). Cauchy sequence for $\text{Re} z > 0$.

Section 3.5

1(a). $z = 0$ is a removable singularity; $z = \infty$ is an essential singularity;

1(b). $z = 0$ is a simple pole; $z = \infty$ is an essential singularity;

1(c). $z = \frac{\pi}{2} + n\pi$ are essential singularities for $n \in \mathbf{Z}$;

   $z = \infty$ is a non-isolated singularity; it is a cluster point;

1(d). Simple poles at $z = -\frac{1}{2} \pm i\frac{\sqrt{3}}{2}, \infty$.

1(e). Branch points at $z = 0, \infty$. Depending on which branch of $z^{\frac{1}{3}}$ is chosen, may have simple pole at $z = 1$.

   Letting $z = re^{i\theta}$ for $0 \le \theta < 2\pi$, $z = 1$ is a removable singularity;
   letting $z = re^{i\theta}$ for $2\pi \le \theta < 4\pi$, $z = 1$ is a simple pole;
   letting $z = re^{i\theta}$ for $4\pi \le \theta < 6\pi$, $z = 1$ is a simple pole;

1(f). $z = 0, \infty$ are branch points. If we let $z = re^{i\theta}$ for $2\pi \le \theta < 4\pi$, then $z = 1$ is an additional branch point;

1(g). $|z| = 1$ is a boundary jump discontinuity;

1(h). $|z| = 1$ is a natural boundary;

1(i). Simple poles at $z = i(\pi/2 + n\pi)$ for $n \in \mathbf{Z}$, and $z = \infty$ is a cluster point;

1(j). Simple poles at $z = i/(n\pi)$ for $n \in \mathbf{Z}$, and $z = 0$ is a cluster point.

3(a). Simple poles at $z = 2^{\frac{1}{4}}e^{i\pi(1+2n)/4}$, $n = 0, 1, 2, 3$, with strength

   $\frac{-i}{4\sqrt{2}}, i/4\sqrt{2}, -i/4\sqrt{2}, -i/4\sqrt{2}$, respectively;

3(b). Simple poles at $z = \pi/2 + n\pi$, $C_{-1} = -1$;

3(c). Poles of order two at $z = n\pi$; $C_{-2} = n\pi$, $n \ne 0, = \pm1, \pm2, \ldots$;
   pole order one at $z = 0$, $C_{-1} = 1$;

3(d). Pole of order two at $z = 0$; $C_{-2} = \frac{1}{2}$;

3(e). Simple pole at $z = \sqrt{2}$, $C_{-1} = 1/2$; simple pole at $z = -\sqrt{2}$, $C_{-1} = 1/2$.

Section 3.6

1(a). Converges for $|z| < 1$;   1(b). Converges for all z;

1(c). Diverges;   1(d). Converges for all z.

Section 3.7

1(a). $z = 0$ fixed singularity, no movable singularities; $w(z) = -z + cz^2$;

1(b). $z = 0$ fixed singularity, $z = e^c$ a movable singularity, where $w(z) = 1/(c - \log z)$.

1(c). No fixed singular points; movable singular point at z:

$$c - 2\int_{z_0}^{z} a(z)dz = 0,$$

$c = 1/\omega_0^2$ where $\omega(z_0) = \omega_0$, and

$$w(z) = 1/\left[(1/w_0^2) - 2\int_{z_0}^{z} a(z')dz'\right]^{1/2};$$

1(d). $z = 0$ fixed singularity, no movable singularities;

$$w(z) = [c_1 \sin(\log z) + c_2 \cos(\log z)].$$

Chapter 4

Section 4.1

1(a). $-\frac{3}{10}$;   1(b). 1;   1(c). 0;   1(d). $\frac{1}{2}\log\left(\frac{3}{2}\right)$;   1(e). $\frac{1}{2}$.

3(a). pole of order m;   3(b). branch point;

3(c). simple pole;   3(d). branch point;

3(e). branch point;   3(f). essential singularity;

3(g). simple pole;   3(h). analytic;

3(i). branch point;   3(j). branch point.

Section 4.2

1(a). $\dfrac{\pi}{2a}$;   1(b). $\dfrac{\pi}{4a^3}$;   1(c). $\dfrac{\pi}{2ab(b + a)}$;   1(d). $\dfrac{\pi}{3}$.

Section 4.4

1(a). $I = n$;   1(b). $I = 0$;

1(c). The Theorem can't be applied; singularity on contour;

1(d). $I = N - M$;   1(e). The Theorem can't be applied; essential singularity.

1(a). 3 times;   1(b). 3 times;   1(c). $2\pi N$.

Section 4.5

1(a). $\hat{F}(k) = \dfrac{2}{1 + k^2}$;   1(b). $\hat{F}(k) = \dfrac{\pi}{a}e^{-|k|a}$;   1(c). $\hat{F}(k) = \dfrac{\pi e^{-|k|a}}{2a^2}\left(\dfrac{1}{a} + |k|\right)$;

$$1(d). \hat{F}(k) = \begin{cases} \frac{\pi}{2ic}\left(e^{(-ib-c)(-k+a)} - e^{(-ib-c)(-k-a)}\right) & \text{for} \quad k < -a, \\ \frac{\pi}{2ic}\left(e^{(-ib-c)(-k+a)} - e^{(-ib+c)(-k-a)}\right) & \text{for} \quad -a < k < a, \\ \frac{\pi}{2ic}\left(e^{(-ib+c)(-k+a)} - e^{(-ib+c)(-k-a)}\right) & \text{for} \quad a < k. \end{cases}$$

$$5(a). \hat{F}(k) = \begin{cases} \pi, & |k| < \omega, \\ 0, & |k| > \omega. \end{cases}$$

7(a). $\hat{F}_s(k) = \dfrac{k}{\omega^2 + k^2}\sqrt{2}$;   7(b). $\hat{F}_s(k) = \dfrac{\pi}{\sqrt{2}}e^{-k}$;

7(c). $\hat{F}_s(k) = \begin{cases} \dfrac{1}{\sqrt{2}}\pi e^{-\omega}\sinh k & k < \omega, \\[2mm] \dfrac{1}{\sqrt{2}}\pi e^{-k}\sinh\omega & \omega < k. \end{cases}$

11(a). $f(x) = \cos\omega x$;   11(b). $f(x) = xe^{-\omega x}$;   11(c). $f(x) = \dfrac{x^{n-1}}{(n-1)!}e^{-\omega x}$;

11(d). $f(x) = \dfrac{1}{(n-1)!}\left[(n-1)x^{n-2}e^{-\omega x} - \omega x^{n-1}e^{-\omega x}\right]$;

11(e). $f(x) = \dfrac{e^{-\omega_1 x}}{\omega_2 - \omega_1} + \dfrac{e^{-\omega_2 x}}{\omega_1 - \omega_2}$;

11(f). $f(x) = \dfrac{x}{\omega^2} - \dfrac{\sin\omega x}{\omega^3}$;   11(g). $f(x) = \dfrac{e^{-\omega_1 x}(\sin\omega_2 x)}{\omega_2}$;

11(h). $f(x) = \dfrac{e^{\omega x}}{4\omega^3}(x\omega - 1) + \dfrac{e^{-\omega x}}{4\omega^3}(x\omega + 1)$.

Section 4.6

5. $u(x,t) = \dfrac{e^{-i\pi/4}}{2\sqrt{\pi t}}\displaystyle\int_{-\infty}^{\infty} e^{i(x-\xi)^2/4t}f(\xi)d\xi$.

9. $\Phi(x,y,t) = \displaystyle\int_{-\infty}^{\infty}\int_{-\infty}^{\infty} f(x',y')\dfrac{1}{4\pi t}\exp\left(\dfrac{-\left[(x-x')^2 + (y-y')^2\right]}{4t}\right)dx'dy'$.

The Green's function is obtained by taking $f(x,y)) = \delta(x,y)$.

Chapter 5

Section 5.2

3. $w = 2z - 2 + \frac{3i}{2}$ or $w = -2z + 2 + \frac{3i}{2}$.

Section 5.3

1. $x^2 - y^2 = c_1$, $xy = c_2$.

3(a). $w = (1+i)(z^2 + \bar{z}^2) + (2 - 2i)z\bar{z} + 8iz$;   3(b). $w = z^3$.

Only 3(b) can define a conformal mapping.

Section 5.6

1(a). $w = z^{5/4}$;   1(b). $w = \log z$.

5. $q^2 = u_0^2/\left(1 + \left(\dfrac{2\pi}{d}\right)^2 e^{\frac{4\pi x}{d}} + \dfrac{4\pi}{d}e^{\frac{2\pi x}{d}}\cos\dfrac{2\pi y}{d}\right)$, $w = z + e^{\frac{2\pi z}{d}}$,

$z = x + iy$, $w = u + iv$, q is the speed of the flow. An explicit formula for z in terms of w, i.e., $x = x(u,v)$ and $y = y(u,v)$, does not exist.

$$F(w) = \frac{hq}{\pi} f^{-1}(w) = \frac{hq}{\pi} z, \text{ where}$$

$$w = f(z) = \frac{h}{\pi} \left((z^2 - 1)^{1/2} + \cosh^{-1} z \right).$$

9. $w = f(z) = \left(\dfrac{\ell\pi}{\Gamma(\alpha_1)\Gamma(\alpha_2)\Gamma(\alpha_3) \sin \pi\alpha_3} \right) \displaystyle\int_0^z t^{\alpha_1 - 1} (1 - t)^{\alpha_2 - 1} dt.$

Section 5.7

1. $w = \dfrac{(8 - 3i)z + i}{(3 - i)z + 1}.$

3. Radius of the inner circle $\delta = \dfrac{\rho}{c + \sqrt{c^2 - \rho^2}} = \dfrac{\rho}{c + a}.$

5. $w = \dfrac{e^{\pi/z} - i}{e^{\pi/z} + i}.$

7. $\phi(x, y) = \dfrac{2}{\pi} \left(\dfrac{5\pi}{4} - \tan^{-1} \dfrac{x - y}{x(x - 1) + y(y - 1)} \right).$

Bibliography

Ablowitz, M.J., Bar Yaacov, D. and Fokas, A.S., *Stud. in Appl. Math.* **69**, 1983, p.135.

Ablowitz, M.J. and Clarkson, P.A., *Solitons, Nonlinear Evolution Equation, and Inverse Scattering*, London Mathematical Society Lecture Notes #149, Cambridge University Press, 1991.

Ablowitz, M.J. and Fokas, A.S., *Complex Variables, Introduction and Applications*, Cambridge University Press, 2003.

Ablowitz, M.J. and Segur, H., *Solitons and the Inverse Scattering Transform*, SIAM Studies in Applied Mathematics, 1981.

Buck, R.C., *Advanced Calculus*, McGraw-Hill, 1956.

Carrier, G.F., Krook, M. and Pearson, C.E., *Functions of a Complex Variable: Theory and Technique*, McGraw-Hill, 1966.

Chazy, J., *Acta Math.* **34**, 1911, p.317.

Corliss, G. and Chang, Y., *ACM Transactions of Math Software* **3**, 1982, p. 114.

Darboux, G., *Ann. Sci. Éc. Normal. Supér.* **7**, 1878, p.101.

Driscoll, T.A. and Trefethen, L.N., *Schwarz-Christoffel Mapping*, Cambridge University Press, 2002.

Fokas, A.S., *A Unified Approach to Boundary Value Problems*, SIAM, 2006.

Fokas, A.S., Its, A.R., Kapaev, A.A., and Novokshenov, V.Yu., *Painlevé Transcendents. The Riemann–Hilbert Approach.* Mathematical Surveys and Monographs, Vol. 128, Amer. Math. Soc., 2006.

Fornberg, B., *SIAM J. Sci. Stat. Computing* **1**, 1980, p.386.

Gaier, D., In *Computational Aspects of Complex Analysis*, edited by Werner, H. et al., Reidel, 1983, p.51.

Garabedian, P.R., *Amer. Math. Monthly* **98**, 1991, p.824.

Halphen, G.H., *C.R. Acad. of Science of Paris* **92**, 1881, p.1001.

Henrici, P., *Applied and Computational Complex Analysis*, Volumes I, II, III, Wiley–Interscience, 1977.

Hille, E., *Ordinary Differential Equations in the Complex Plane*, Wiley–Interscience, 1976.

Ince, E.L., *Ordinary Differential Equations*, Dover Publications, 1956.

Jeffreys, H. and Jeffreys, B., *Methods of Mathematical Physics*, Cambridge University Press, 1962.

Levinson, N. and Redheffer, R.M., *Complex Variables*, Holden-Day, 1970.

Lighthill, M.J., *An Introduction to Fourier Analysis and Generalized Functions*, Cambridge University Press, 1959.

Nehari, Z., *Conformal Mapping*, McGraw-Hill, 1952.

Rudin, W., *Theory of Real and Complex Functions*, McGraw-Hill, 1966.

Titchmarsh, E.C., *Introduction to the Theory of Fourier Integrals*, 2nd ed., Oxford University Press, 1948.

Trefethen, L.N., editor, *Numerical Conformal Mapping*, North-Holland, 1986.

Wegmann, R., *J. Comput. Appl. Math* **23**, 1988, p.323.

Whittaker, E.T. and Watson, G.N., *A Course of Modern Analysis*, 4th ed., Cambridge University Press, 1927.

The following is a short list of *supplementary* books which have basic and advanced material, applications, and numerous exercises.

Abramowitz, M. and Stegun, I., *Handbook of Mathematical Functions*, Dover Publications, 1965.

Ahlfors, L.V., *Complex Analysis: An Introduction to the Theory of Analytic Functions of One Complex Variable*, 3rd ed., McGraw-Hill, 1979.

Batchelor, G.K., *An Introduction to Fluid Dynamics*, Cambridge University Press, 1967.

Bender, C.M. and Orszag, S.A., *Advanced Mathematical Methods for Scientists and Engineers*, McGraw-Hill, 1978.

Bleistein, N. and Handelsman, R.A., *Asymptotic Expansions of Integrals*, Dover Publications, 1986.

Boas, R.P., *Invitation to Complex Analysis*, 1st ed., Random House, 1987.

Bowman, F., *Introduction to Elliptic Functions with Applications*, English Universities Press, 1953.

Churchill, R.V. and Brown, J.W., *Complex Variables and Applications*, 5th ed., McGraw-Hill, 1990.

Conway, J.B., *Functions of a Complex Variable*, Springer, 1973.

Copson, E.T., *An Introduction to the Theory of Functions of a Complex Variable*, Oxford University Press, 1955.

Copson, E.T., *Asymptotic Expansions*, Cambridge University Press, 1965.

Gakhov, F.D., *Boundary Value Problems*, Pergamon Press, 1966.

Hille, E., *Analytic Function Theory*, 2nd ed., Chelsea Publishing Co., 1973.

Markushevich, A.I., *Theory of Functions of a Complex Variable*, 2nd ed., Chelsea Publishing Co., 1977.

Saff, E.B. and Snider, A.D., *Fundamentals of Complex Analysis for Mathematics, Science, and Engineering*, Prentice-Hall, 1976.

Silverman, R.A., *Complex Analysis with Applications*, Prentice-Hall, 1974.

Spiegel, M.R., *Schaum's Outline of Theory and Problems of Complex Variables*, McGraw-Hill, 1964.

Thron, W.J., *Introduction to the Theory of Functions of a Complex Variable*, Wiley, 1953.

Index